W9-CUK-251

11TH INTERNATIONAL MEETING ON ELECTRO-OPTICS AND MICROELECTRONICS IN ISRAEL

הכנס הבין לאומי ה11- לאלקטרו-אופטיקה ומקרואלקטרוניקה בישראל

Annals of the Israel Physical Society Volume 14

Series Editors

Joan Adler Baruch Rosner Raoul Weil
Technion – Israel Institute of Technology, Haifa

ELECTRO-OPTICS AND MICROELECTRONICS

Proceedings of the 11th International Meeting on Electro-optics and Microelectronics in Israel held in Tel Aviv, November 9-11, 1999

Edited on behalf of the Israel Physical Society by
Raphael Lavi and Ehud Azoulay
Soreq NRC

published by

Institute of Physics Publishing, Bristol and Philadelphia
and
The Israel Physical Society, Jerusalem

Published jointly by:
Institute of Physics Publishing, Dirac House, Temple Back, Bristol, BSI 6BE, UK;
and
The Israel Physical Society, P. O. B. 16105 Jerusalem 91160, Israel
Editorial Office: c/o Physics Department, Technion, Haifa, 32000 Israel

ISSN: 0309-8710 $2.00 + $0.50

British Library Cataloguing in Publication Data
A catalogue record for this book is available from the British Library

Library of Congress Cataloging-in-Publication Data are available

ISBN 0-7503-0753-6

Printed by Ayalon Offset Ltd., Haifa, Israel.

TABLE OF CONTENTS

PREVIOUS VOLUMES IN THIS SERIES

Vol. 1 ATOMIC PHYSICS IN NUCLEAR EXPERIMENTS
Proceedings of the International Workshop on Topics in Atomic Physics Related to Nuclear Experimentation held at Haifa.
Editors: Baruch Rosner and Rafael Kalish, Technion - Israel Institute of Technology, Haifa.
638 pages. 1977, ISBN O-85274-355-6 ISSN 0309-8710

Vol. 2 STATISTICAL PHYSICS - "STATPHYS 13"
Proceedings of the 13th IUPAP Conference on Statistical Physics, held in Haifa.
Editors: Chanoch Weil (Executive Editor), Dario Cabib, Charles G. Kuper and Ilan Riess, Technion-Israel Institute of Technology, Haifa
1.087 pages. 2 parts. 1978, ISBN O-85274-356-4, ISSN 0309-8710.

Vol. 3 GROUP THEORETICAL METHODS IN PHYSICS
Proceeding of the VII International Colloquium on Group Theoretical Methods in Physics, held at Kiryat Anavim.
Editors: L. Horwitz, Y. Ne'eman, University of Tel Aviv.
417 pages, 1980. ISBN O-85274-424-2. ISSN 0309-8710

Vol. 4 MOLECULAR IONS, MOLECULAR STRUCTURE AND INTERACTION WITH MATTER
Proceedings of the 38th Bat Sheva Seminar held at the Weizmann Institute of Science, Rehovoth, at Ein Bokek, and the Technion-Israel Institute of Technology, Haifa.
Editor: Baruch Rosner, Technion-Israel Institute of Technology, Haifa.
290 pages. 1981. ISBN O-085274-441-2 ISSN 0309-8710

Vol. 5 PERCOLATION STRUCTURES AND PROCESSES
Invited Review Papers on the Theoretical and Experimental Aspects of Percolation Structures and Processes.
Editors: G. Deutscher, Tel Aviv University, R. Zallen, Xerox Corp. Rochester, N.Y., and J. Adler, Technion - Israel Institute of Technology, Haifa.
502 pages. 1982 ISBN O-85274-477-3 ISSN 0309-8710

Vol. 6 VACUUM ULTRAVIOLET RADIATION PHYSICS - VUV VII
Proceedings of the 7th International Conference on Vacuum Ultraviolet Physics, held at Jerusalem.
Editors. A. Weinreb and A. Ron, Hebrew University, Jerusalem.
622 pages. 1984. ISBN-85274-760-8. ISSN 0309-8710.

Vol. 7 ELECTROMAGNETIC PROPERTIES OF HIGH SPIN NUCLEAR LEVELS
Proceedings of the Workshop held at the Weizmann Institute of Science, Rehovoth and at Ein Bokek.
Editors: Gvirol Goldring and Michael Hass, Weizmann Institute of Science, Rehovoth.
347 pages. 1984. ISBN O-85274-775-6. ISSN 0309-8710

Vol. 8 FRAGMENTATION FORM AND FLOW IN FRACTURED MEDIA
Proceedings of the F^3 Conference, held at Neve Ilan
Editors: R. Engelman and C.Yaeger, Sorek Nuclear Research Center.
628 pages, 1986. ISBN O-85274-578-8. ISSN 0309-8710.

Vol. 9 DEVELOPMENTS IN GENERAL RELATIVITY, ASTROPHYSICS AND QUANTUM THEORY
A Jubilee Volume in Honour of Nathan Rosen
Editors: F. I. Cooperstock, Victoria, L. P. Horowitz, and J. Rosen, Tel Aviv University.
377 pages, 1990. ISBN O-7503-0053-1. ISSN 0309-8710

Vol. 10 CATACLYSMIC VARIABLES AND RELATED PHYSICS
Proceedings of the 2nd Technion Haifa Conference
Editors: O. Regev and G. Shaviv, Technion - Israel
Institute of Technology.
337 pages, 1993. ISBN O-7503-0282-8. ISSN 0309-8710

Vol. 11 ASYMMETRICAL PLANETARY NEBULAE
Proceedings of the University of Haifa at Oranim Conference
Editors: A. Harpaz and N. Soker, - University
of Haifa at Oranim
306 pages, 1995. ISBN O-7503-0330-1. ISSN 0309-8710

Vol. 12 THE DILEMMA OF EINSTEIN, PODOLSKY AND ROSEN
60 YEARS LATER
An International Symposium in Honour of Nathan Rosen
held in Haifa.
Editors: A. Mann and M. Revzen, Technion – Israel
Institute of Technology
314 pages, 1996. ISBN - 0-7503-0394-8. ISSN 0309-8710

Vol. 13 INTERNAL STRUCTURE OF BLACK HOLES AND SPACETIME
SINGULARITIES
Proceedings of the workshop held at Technion, June 29-July 3, 1997
Editors: L.M. Burko and A. Ori, Technion – Israel
Institute of Technology
534+ix pages, 1997. ISBN - 0-7503-0548-7. ISSN 0309-8710

FOREWORD

The Israel Physical Society provides a forum for interaction amongst Israeli physicists. There are two types of membership: individual and corporate. Individual physicists and scientists in fields related to physics are eligible for individual membership while educational and research establishments, companies, foundations, with an interest in the promotion of physics in Israel, are eligible for corporate membership.

The Society organizes an annual national general conference. In addition it co-sponsors many international conferences of a more specialized nature, held in Israel. The Society publishes the "Annals" which primarily serve as a vehicle for rapid publication of the proceedings of international conferences held under its sponsorship. Occasionally, an up-to-date review of a topical field in which many of our members are active, is published.

We thank Ms Liz Youdim, the editorial assistant of the Annals of the Israel Physical Society, for her professional and caring assistance in the production of this volume.

Joint Series Editors

CORPORATE MEMBERS OF THE ISRAEL PHYSICAL SOCIETY

Bar Ilan University
Ben Gurion University of the Negev
The Hebrew University of Jerusalem
Israel Atomic Energy Commision
Technion-Israel Institute of Technology
Tel Aviv University
The Weizmann Institute of Science

ON THE COVER: A waveguide deposited on a periodically poled $KTiOPO_4$ (KTP) sample is shown schematically on the cover. Periodically poled crystals have opened the possibility of implementing a large variety of frequency shifting devices for continuous laser sources and are a good example of the combination of electro-optic and microelectronic techniques. The periodically poled crystals are noncritically quasi-phase-matched. Therefore they are not affected by walk-off. Their effective nonlinear coefficient, d_{eff}, is generally larger than in the case of birefringent phase matched crystals. Waveguide structures deposited on these crystals allow a significant enhancement of the conversion efficiency.

PREFACE

This Proceeding of the Israeli Physical Society contains articles from the oral presentations at the 11th International Meeting on Electro-Optics and Microelectronics in Israel, which was held in Tel-Aviv 9-11 November 1999. The conference was sponsored by the Ministry of Science, Culture and Sport, the Israel Atomic Energy Commission, ElOp Electro-Optics Industries LTD, Israel Aircraft Industries LTD, Semi Conductor Devices, and RAFAEL. The conference was attended by more than 300 scientists, with more than 100 papers being presented. A large number of scientists, leaders in their areas, came from overseas.

In the eleven conferences held since the first meeting, there has been an evolution in the scope of the topics considered. The content of this volume presents the various subjects on which the meeting focused. Among these are - solid state lasers and their applications (1mm lasers, mid and far IR lasers, as well as ultrafast light sources), non-linear optics (quasi phase matching, frequency shifting and non-linear materials), thin films (coatings and micro-lithography), diffractive optics, fiber optics and sensors, machine vision and image processing, micro-mechanics and micro-optics. In the spirit of this era one of the sessions was devoted to start-ups that are connected with electro-optics and microelectronics. Two short courses were given prior to the conference, one on optical coatings by P. Baummeister, the other on quasi phase matching by M.M. Fejer.

We thank the other members of the Organizing and Program Committees, M. Oron, M. Barak, R. Finkler, A. Israel, N.S. Kopeika, K. Rabinovitch, S.R. Rotman, J. Schacham, J. Van Zwaren, S. Blit, N. Eisenberg, M. Guelman, G. Gordon, D. Haronian, M. Heiblum, A. Katzir, K. Rabinovitch, D. Ritter, G. Sarusi, R Shapira, O. Yadid-Pecht, A. Zigler, M. Zimmermann and Z. Ziony.

Ehud Azoulay, Raphy Lavi

Chair Program Committee, Chief Scientific Editor,

Soreq NRC, June 2000

PART 1

1 μM LASERS

Chairpersons: *K.L. Schelper, USA; S.M. Jackel, Israel*

Annals of the Israel Physical Society, v. 14

ELECTRO-OPTICS and MICROELECTRONICS
Eds: Raphael LAVI and Ehud AZULAY

□AXIAL AMPLIFIED SPONTANEOUS EMISSION MEASUREMENTS IN Nd:YAG OSCILLATOR-AMPLIFIER LASER CONFIGURATIONS

A.A. Ishaaya, G. Ravnitzky and I. Shoshan

Laser Products Operation, El-Op Electrooptics Industries
P.O.Box 1165, Rehovot 76111 Israel

Abstract

For establishing the design parameters of Nd:YAG laser systems, axial Amplified Spontaneous Emission (ASE) calculations and measurements were done with various laser configurations. In each configuration the small signal gain was measured, and the typical "look angle" was calculated, enabling the estimation of the axial ASE. The actual axial ASE was measured with a photodiode and also directly with an energy meter. The photodiode measurements were sensitive resulting in detection of weak ASE, well below the point where practical energy loss could be observed. The thresholds obtained with the energy meter were in good agreement with the calculated

1. Introduction

In high gain Q-switched (QS) laser configurations Amplified Spontaneous Emission (ASE) may pose serious limitations on the extracted energy. At high gain levels, spontaneous emission originating at the laser medium undergoes amplification causing thereafter considerable pre-depletion of the upper level via stimulated emission prior to the opening of the QS. ASE can be generated from the pump cavity by multiple reflections from the rod surfaces, by various transverse modes[1] or by amplification of the pump light. On the other hand, *axial* ASE can be generated by reflections from external elements or by passing through a long multi-element gain path. In this study we present axial ASE calculations and measurements with various laser configurations, for the purpose of establishing the design parameters of Nd:YAG laser systems.

2. ASE Model

For estimating the ASE practical thresholds, a model presented by Linford et al.[2] was used. This model relates the ASE flux I_{ASE} to the saturation flux in the material I_S, the active angular field Ω and the small signal gain for the longest possible path G:

$$\frac{\mathbf{I_{ASE}}}{\mathbf{I_S}} = \frac{\Omega}{\mathbf{4}} \frac{\mathbf{G}}{\sqrt{\mathbf{ln(G)}}} \qquad \mathbf{G > 1} \qquad (1)$$

The active angular field is defined as the largest solid angle in which flux emitted from a volume element at the beginning of the ASE path can escape from the system at the end of the path (the "look angle").

When the practical threshold $I_{ASE}/I_S \approx 1$ is reached, the ASE flux is considered to be large enough to cause appreciable energy loss in QS operation. For each of the examined configurations (Fig. 1) the active angular field was calculated. The single pass small signal gain as a function of the pump energy was determined by inserting each rod into a PL-PL oscillator and measuring the lasing threshold as a function of the output reflectivity. The results are presented in Fig 2. Inserting the gain and the look angle in Eq. (1) enabled the calculation of the practical ASE threshold for each configuration.

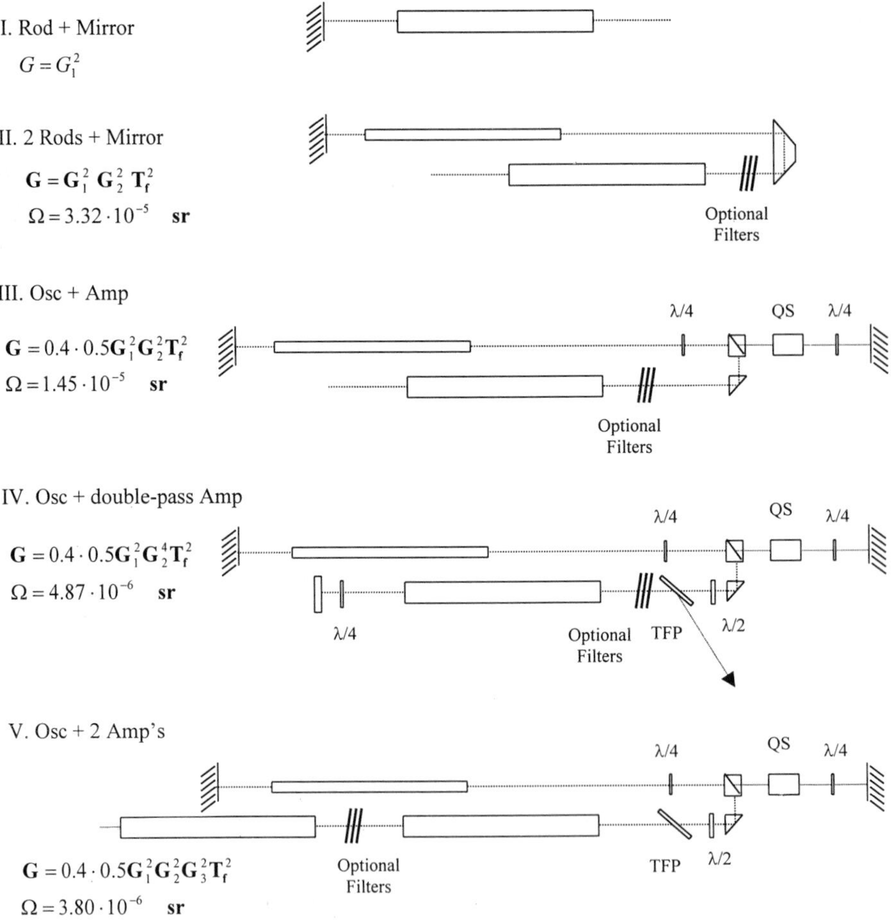

Fig. 1: Examined configurations

Fig 2: Measured small signal gain for each rod

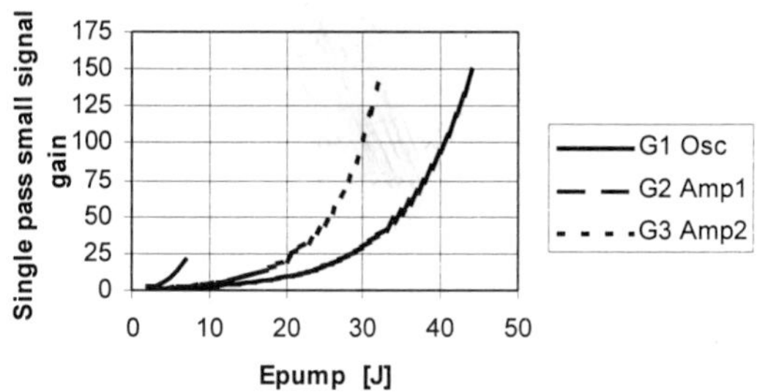

3. Results

Actual ASE measurements were done: (a) using a photodiode along with an interference filter on the optical axis for observing the fluorescence pulse from the rod. In the presence of ASE a deviation from the linear dependence of the detected pulse area on the pump energy was observed. (b) Absolute energy measurement with energy meter. This could be done only in configurations IV and V where high gain was employed (6 passes). (c) Observing the QS output energy as a function of the pump energy. The point at which the energy saturates can be considered as the practical ASE threshold.

ASE thresholds obtained from the photodiode measurements in configurations I, II and III, were significantly lower than the calculated thresholds, correlating with values of $I_{ASE}/I_S \approx 0.005$. In configuration I the ASE pump energy thresholds for a mirror at a distance of 22cm and 5.5cm from amplifier 1 were 35J and 28J respectively. The results for configuration II are presented in Fig. 3a. In configurations I-III, using the absolute measurement method and examining QS output (configuration III), *no* ASE was detected over all range of pump energies. It can be concluded that the photodiode measurements detect weak ASE, well below the practical threshold. Absolute ASE energy measurements for an oscillator and a double pass amplifier (config. IV) are presented in figure 3b. The parasitic output was measured by blocking the QS branch of the oscillator. This parasitic output was subtracted from the total output to obtain the energy in QS operation. Configuration IV is rather sensitive to *parasitic oscillations* because of the HR mirror in the amplifier channel. Hence it could not be established firmly that parasitic oscillation was not present, so the results should be regarded as a lower limit for onset of ASE. Figure 3c shows the results for configuration V. The calculated threshold with $I_{ASE}/I_S \approx 1$ is in good agreement with the measured result. Simulating the effect of saturable absorbers in the amplifier channel, Neutral Density filters were inserted in the amplifier chain (configuration V), resulting in higher practical ASE thresholds than expected (Fig. 3d). Use of intracavity etalons for ASE suppression resulted in higher ASE thresholds but lower oscillator output due to higher intracavity losses introduced by the etalon.

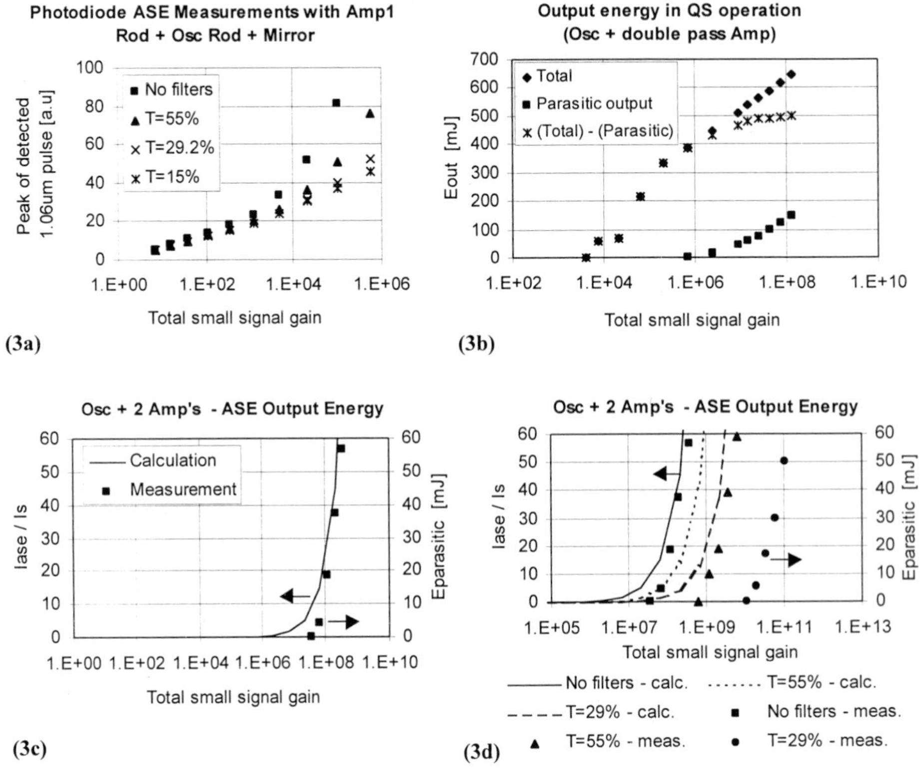

Fig. 3: ASE measurement results. (a) photodiode measurement in confg.II ; (b) parasitic output measurement in confg. IV ; (c) and (d) energy meter measurements and calculations in confg. V .

4. Conclusions

ASE thresholds obtained from the photodiode measurements are significantly lower than the practical thresholds for significant energy loss. The calculated thresholds based on the Linford et al.[2] model are in good agreement with those obtained with the absolute energy measurements. This suggests that this model can serve as an adequate design tool for optimizing configurations similar to those examined. Suppression of ASE can be achieved using suitable saturable absorbers in the amplifier channel or using intracavity etalons.

References:

[1] "Lasers", A.E. Siegman, University Science Books, 1986

[2] Linford et al., Appl. Optics Vol. 13, No. 2, p. 37, Feb. 1974

GROUND AND EXCITED STATE DIRECT PUMPING – A PREFERRED WAY FOR PUMPING LASING MATERIALS

R. Lavi, S. Jackel, M. Katz, Y. Tzuk, E. Lebiush, M. Winik, and I. Paiss

Non-Linear Optics Group, Electro-Optics Division
Soreq NRC, Yavne 81800, Israel
Tel: 972-8-9434513, Fax: 972-8-9434401, e-mail: raphy@ndc.soreq.gov.il

Abstract

An efficient pumping scheme involving direct excitation from either the ground (869nm) or thermally excited states (885nm) to the upper lasing level of the Nd^{3+} ion is demonstrated experimentally. Results obtained using those pumping schemes were compared to "traditional" ~808nm pump band excitation. Comparing direct pumping from thermally excited states to "traditional" pumping, slope efficiency increased 17% and threshold dropped 8%. These results are in agreement with theory.

1. Introduction

Pumping of Nd^{3+} based lasing materials at 808nm is today regarded as the "traditional" pump scheme. This pumping scheme takes advantage of the broad 790-820nm Nd^{3+} $^4F_{5/2}$ absorption band. However, it suffers from two main drawbacks which limit the overall light-to-light conversion efficiency: the Stokes shift, due to the difference between the pump and laser transition energies, and the quantum efficiency, due to incomplete energy transfer between the pump and upper laser levels. In Nd:YAG, these processes result in energy losses of 24% and 5%, respectively.

Potentially, pumping from the ground level directly to the upper lasing level should be the most efficient "four level laser" pumping scheme, because it reduces to a minimum the Stokes factor losses, and eliminates the quantum efficiency loss.(1) Direct pumping of solid-sate lasing materials from the ground state to the upper lasing level was demonstrated during early stages of laser development.(2)

A further Stokes shift reduction may be obtained through direct pumping from thermally excited Stark components of the ground state. This approach alongside "traditional" and direct pumping from the ground level is illustrated in figure 1.

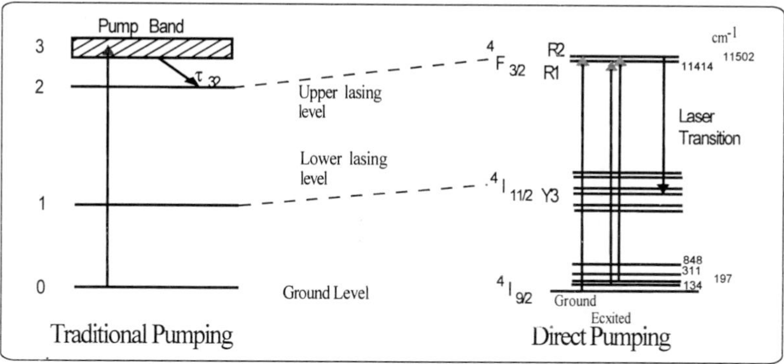

Fig. 1: Nd:YAG energy level diagram with schematic description of "traditional" and direct pumping from either the ground level or thermally excited Stark components of the ground state.

In this work we experimentally demonstrate the novel scheme of direct pumping from thermally excited Stark levels in the ground state to the upper lasing level, in a Nd:YAG oscillator. We characterized the pumping process with regards to bandwidths and absorption coefficients from various thermally excited ground state levels and to the R1/R2 sub-levels of the $^4F_{3/2}$ upper lasing level. In addition, we compared, under the same experimental conditions, efficiencies and thresholds obtained for "traditional" pumping and direct pumping from both ground and thermally excited ground levels.

2. Experimental results

Figure 2 shows the Nd:YAG ground and excited Stark level transitions to the upper lasing level. The absorption spectrum was measured using a tunable, narrow linewidth (0.04nm FWHM) CW Ti:S source. In addition to the ground to R1 and R2 transitions, excitation from the first, second and third thermally excited Stark levels were seen at room temperature.

The overlapping transitions around 885nm from the first thermally excited Stark level to R1, and the second thermally excited Stark level to R2, are especially significant for two reasons: their 1.8cm^{-1} absorption coefficients are adequate in many situations, and the 2.7nm bandwidth of the double peaked feature is within the bandwidth capabilities of laser diodes. This make 885nm a realistic candidate for diode pumping.

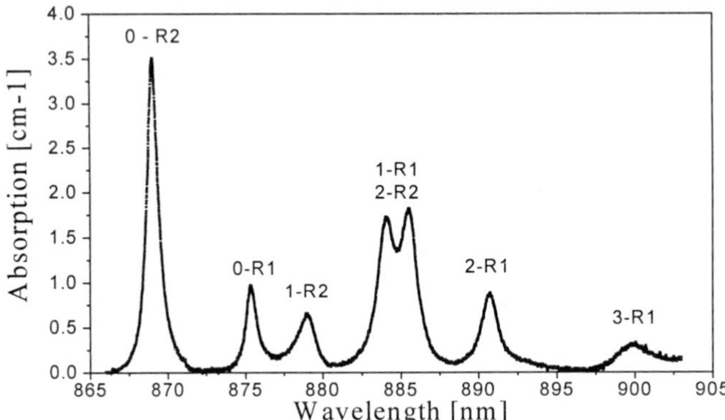

Fig. 2: The absorption spectrum of Nd:YAG (doping of 1.1 at.%) at room temperature describing excitation from ground (0) and thermally excited (1-3) states to the upper lasing level, both to the R1 and R2 sub levels.

The laser cavity, 8cm in length, was formed with two mirrors – a dichroic (HT between 800 and 900nm, and HR @ 1064nm) concave (25cm radius of curvature) back mirror, and a flat output coupler. The Ti:S beam was focused (f=20cm) to end-pump a 50mm long by 2mm diameter Nd:YAG rod (1.1 at.%) that was placed in the middle of the cavity. The YAG rod was AR coated for 1.06μm on both faces. The calculated fundamental laser mode diameter was 400μm ($1/e^2$), at the surface facing the pump beam and 380μm ($1/e^2$) at the other rod end. This dimension was ~1.5 larger than the 270-280μm ($1/e^2$) pump diameter measured throughout the absorption region, at all pump wavelengths. Thus, the overlap integrals were the same in all cases, and the entire light incident upon the laser materials was absorbed within the 1μm lasing volume.

Fig. 3 shows results obtained with a 95% output coupler for all pump wavelengths. The thermally boosted "direct" pumping slope efficiency was 1.17 times higher, and the threshold was, 0.92 times lower, than traditional pumping at 808nm. Compared to ground state pumping, the excited state pumping slope efficiency was 1.05 times higher, and the thresholds were similar. The same trend was observed using three other output couplers (85%, 90% and 98%). These results are in agreement with theory.

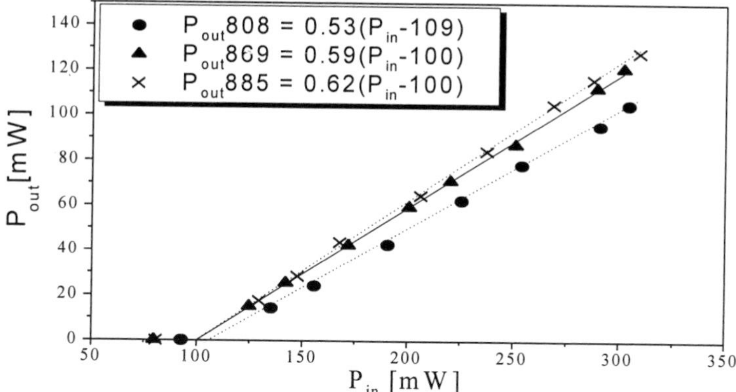

Fig. 3: Nd:YAG CW performance for direct ground level (triangle), thermally excited states (cross), and "traditional" band (circle) pumping, with a 95% output coupler. P_{in} is the pump power just interior to the rod's pumped face.

In summary, we have demonstrated pumping from thermally excited Stark ground state levels directly to the upper lasing level of Nd:YAG at a pump wavelength around 885nm. The process proved to be more efficient than both ground state direct pumping and band pumping. The combined transitions from the first and second thermally excited Stark ground state levels to R1 and R2 respectively, have characteristics that makes thermally boosted pumping a suitable candidate for use with diode lasers: reasonable absorption (1.8cm^{-1}), and reasonable bandwidth (2.7nm (FWHM)).

Acknowledgements: We thank the ministry of industry for supporting this project under the LESHED magnet consortium.

References:

[1] J.T. Verdeyen, *Laser Electronics,* 2nd .ed. (Prentice Hall, New Jersey, 1989) p. 232.

[2] M. Ross, "YAG laser operation by semiconductor laser pumping," Proc. IEEE, **56**, 196 - 197 (1968).

The High Power Diode Lasers and Diode Pumped Solid State Lasers Consortium enters its fourth year

Andrei Ben-Amar Baranga

Arava Laser Laboratory, Rotem Industrial Park, Mishor Yamin, D. N. Arava , Israel
e-mail: baranga@netvision.net.il

Abstract

The "LESHED" High Power Diode Lasers and Diode Pumped Solid State Lasers Consortium is targeted to establish the infrastructure for the production of advanced high-power diode lasers in Israel and to enable laser manufacturers to develop the generic technologies that will lead to locally produced diode pumped laser systems. Achievements of high-power diode laser stacks, lasing materials, pumping schemes and configurations are highlighted. (*http://magnet.consortia.org.il/DLDPL)*

Initiated in 1996, the "LESHED" consortium under the "Magnet" Program of the Israeli Ministry of Industry and Trade, aims to establish the technological and industrial infrastructure and to develop generic technologies for manufacturing of systems based on semiconductor lasers and on diode pumped solid state lasers (DPSS). With applications in material processing, medicine, telecommunications, image recording and more, the targeted generic technologies include high power diode lasers, arrays and stacks, DPSS lasers and systems. Although the many advantages of diode pumping solid state lasers, the market's increase depends on reducing the diode segment price of the system. The consortium, a cooperation between leading Israeli industries and laboratories in electro-optics, established a five years R&D program. This program includes basic research, diodes manufacturing, DPSS lasing materials with wide absorption band and high quantum efficiency, new and efficient pumping schemes and configurations. The first two years of action, have been successfully dedicated to learn the pertinent technologies needed to close technological gaps. The third year was characterized by the transition to original developments and state of art manufacturing. Following are examples of LESHED achievements.

Figure 1. SCD 540 W QCW, 8 bars stacks.

At the Laser'99 International Trade Fair in Munich, High Power Diode Lasers, made in Israel, have been exhibited for the first time. This was a result of intensive R&D over the past three years at SCD and at the Technion. The laser design, epitaxial structures based on AlGaAs, and reproducible end-reflectors have been stabilized to produce 60 Watts peak power QCW bars with >1.15 W/A slope efficiency at room temperature. A substantial elimination of residual "smile" before mounting, and the developed mounting techniques of bars onto submounts, produce a reasonable uniform thermal impedance across the bar, reaching spectral width with commercial potential. SCD proceed with the development of two dimensional bar stacks. The first stack prototypes comprising of 8 bars and based on the conductive back-cooling concept, have been tested for reliability and spectral width, with acceptable results. Individual bar testing for electrooptical parameters, in addition to visual inspection, is done previous to the stack assembly. The stacking technique at SCD assure low and uniform thermal impedance across the stack, high temperature heat sinking and a robust bar-submodule with good heat dissipation characteristics. Figure 1 shows 8-bars stack prototypes. The stacks are designed for QCW, 2% duty cycle, heat sinking temperature of ~50 °C. 540 Watts of peak power were produced at 80 A current, with a slope efficiency of 8.1 W/A, at 25 Hz, 0.2 ms pulse duration, and a heat sink temperature of ~25 °C. SCD acquired a MOCVD reactor and the completion of growth capabilities is due by the end of this year. Future plans include completion of the development of QCW bars towards 100 watts of peak power, high brightness, increased reliability, increased duty cycle, high active layer temperature operation; development of CW bars with 20 and 40 watts output; and microoptics coupling.

"Arava" laser laboratory of the Ben-Gurion University of the Negev, in cooperation with Soreq-NRC, developed a diode pumped Yb:YAG laser. The Yb:YAG lases at the $^2F_{5/2}$, $^2F_{7/2}$ transition of Yb^{3+} ion, with absorption bands around 940 nm and 970 nm. The crystal's lasing and pumping properties make it a leading candidate for DPSS high power lasers. Yb^{3+} ion absorption bands are wider than Nd^{3+}; Yb^{3+} exhibits a smaller quantum defect, a longer lifetime emitting level, no excited state absorption and no up-conversion; doping of more than 20% at. is possible. Diode lasers for pumping Yb:YAG at 940 nm or 970 nm are available, cheaper and reliable.

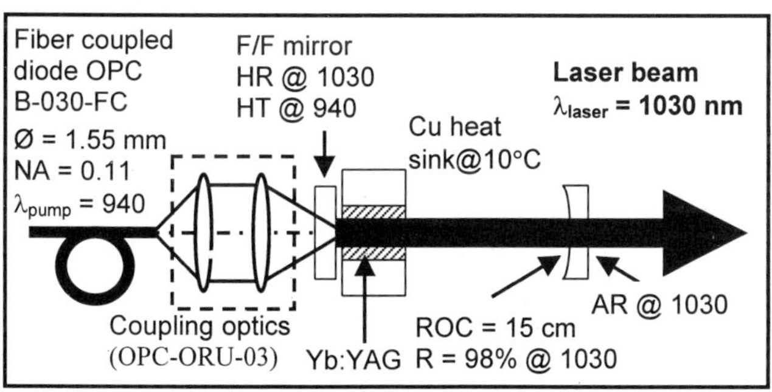

Figure 2. Schematics of Arava's Yb:YAG longitudinally pumped laser.

However, the Yb:YAG is a "quasi" three-level laser, with the ground state splinted into Stark components up to 612 cm^{-1}. These levels, thermally populated at room temperature, lower the population inversion and raise the lasing threshold. Fig. 2 shows Arava's Yb:YAG laser. A 30 W, 940 nm fiber-coupled diode laser end-pumps a 2 mm diameter, 6 mm length, AR coated, 8% doping Yb:YAG rod, through coupling optics. The rod is cooled by a water-cooled Cu heat sink at 10-20 °C. With a semi-spherical resonator, R=98% output coupler, the laser produces a maximum output power of 2.6 W, a slope efficiency of 17% and a beam quality factor $M^2<1.5$. The R&D continues at Arava, Soreq and El-Op, toward high power, high efficient 1 μm lasers.

A DPSS laser emitting blue light (473 nm) has been developed by Laser Industries in cooperation with Soreq. Blue light laser emission has been achieved by external frequency doubling of a 946 nm Nd:YAG. The 946 nm radiation was produced by a composite YAG-Nd:YAG-YAG laser rod (3+3+3 mm), 2 mm diameter, water cooled at 19 °C, diode end-pumped by a high-brightness 14 W CW diode laser array (OPC-D012-808-HB). The high–brightness pumping is essential for this laser because of its low stimulated emission cross section (about 10 times smaller than the 1064 nm line). The laser cavity was a 6 cm long flat-flat resonator, with an R=90% at 946 nm output coupler and an acusto-optic Q-switch. The doubling crystal was an external 10 mm long PPKTP crystal manufactured by Soreq, AR coated for both fundamental (946 nm) and second harmonic, and operated at 29 °C. The laser beam was focused to a ~0.1 mm diameter into the doubling crystal. As depicted in Figure 3, the maximum conversion efficiency was ~40%. Operated at 10 kHz, a maximum average power of 350 mW at 473 nm was reached at an intensity level of about 30 MW/cm^2 of the fundamental.

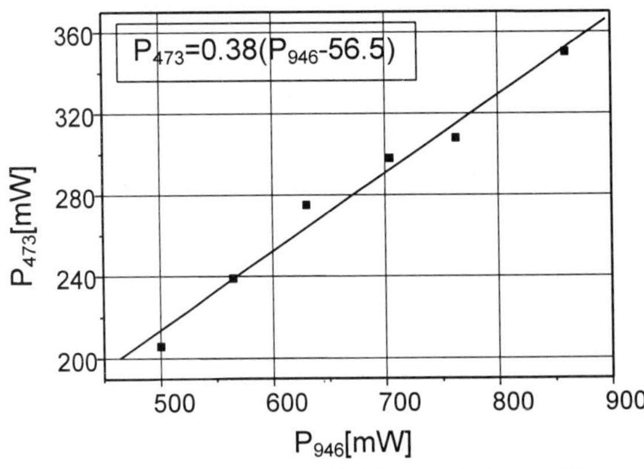

Figure 3. Laser Industries DPSS laser emitting at 473 nm.

Direct pumping of solid-sate lasing materials from the ground state and from thermally excited Stark components of the ground state, to the upper lasing level, was studied by Soreq-NRC. Direct pumping reduces to minimum the Stokes factor losses, and the quantum defect loss. Pumping from a thermally excited Stark component restores thermal equilibrium at a lower temperature, causing optical cooling. The benefits of direct pumping of Nd:YAG and Nd:YVO_4, over "traditional" 808 nm pumping were experimentally demonstrated. Figure 4 shows the results obtained with an R=95% output coupler for four pumping channels. The "direct" pumping slope efficiency was 1.17 times higher, and the threshold was 0.9 times lower, than traditional pumping at 808 nm. Compared to ground state pumping, the excited state pumping slope efficiency was 1.05 times higher, and the threshold was similar for both pumping schemes.

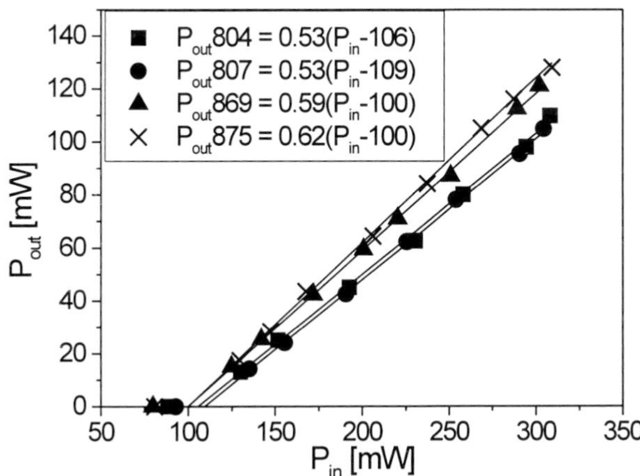

Figure 4. Results of Soreq direct pumping of Nd:YAG compared to "traditional" pumping.

El-Op developed a diode pumped Er:Yb:Glass eye-safe laser at 1.54 nm, with potential applications in medicine telecommunications, and military. The Yb ions absorbing at 940 nm, are the sensitizer for the Er lasing ions. Energy returned to Yb ions and excited state absorption lower the laser efficiency. A theoretical model was developed to calculate the pump efficiency. Operated in Q-Switch mode, 10 pps, ~25 ns pulse-width, the laser produces 15 mJ/pulse at 960 mJ/pumping-pulse (40 mJ/pulse in free running).

The LESHED consortium enters its forth year. The examples above emphasize the achieved technologies and the potential to develop original generic technologies. Work continues towards increased diodes' duty cycle and reliability, diode stacks and CW devices, MOCVD growths, improved pumping efficiency through novel schemes and configurations, new solid state materials, increased output power. Besides improving technologies, LESHED will address also in its forth year new generic topics, like fiber lasers and sub-nanosecond (pico and femtosecond) lasers.

ACKNOWLEDGMENT

The LESHED partners would like to thank the Israeli Ministry of Industry and Trade, the Office of the Chief Scientist, for its generous support of this program.

PART 2

MID AND FAR-IR LASERS

Chairperson: *I. Shoshan, Israel*

Annals of the Israel Physical Society, v. 14

ELECTRO-OPTICS and MICROELECTRONICS
Eds: Raphael LAVI and Ehud AZULAY

2μm DIODE PUMPED LASER

Sharone Goldring, Michael Winik, Raphy Lavi, Eyal Lebiush and Yitshak Tzuk

Non-linear Optics Group, Electro-OPtics Division
Soreq NRC

Stanley Rotman

Ben-Gurion University of the Negev; Department of Electrical and Computer Engineering

Abstract

We present a 2μm Tm:YAG laser operating at room temperature. A diffusion bonded crystal, consisting of Tm:YAG and undoped YAG end-caps, is end pumped by two fiber coupled laser diodes. We obtained a 2W CW laser output with $M^2 \approx 1$.

Introduction

In the last years there is a growing interest in 2μm lasers in a variety of fields such as medicine, atmospheric research and military applications [1,2].

Tm:YAG is a 2.0μm quasi 3-level laser material. At room temperature the lower laser level is thermally populated and its population is few percent of the ground level population [3]. Tm is suitable for pumping with commercially available laser diodes at a wavelength of ~800nm and has a unique phenomena called cross-relaxation or "two for one". One pumping photon excites 2 Tm ions to the upper laser level thus diode pumping is very efficient. The fluorescence lifetime (τ) of Tm is ~10ms however it has low cross-section (σ) so that its $\sigma\tau$ is about a third of that of Nd:YAG [3,4].

Tm:YAG laser

Fig. 1 describes our laser configuration. The laser rod is pumped from both ends by two fiber coupled laser diode at wavelength of 785nm. The light coming out of the fibers is focused into the crystal, passing through dichroic mirrors at 45°. Those mirrors are highly transparent to 785nm and highly reflective for 2μm. Two plain mirrors (one of which is the output coupler for 2.0μm) together with the 45^0 mirrors form a U shaped resonator. The crystal is diffusion bonded so that the central region (10mm) is 6% Tm doped while the end caps (5mm each) are undoped YAG (fig.2). This configuration allows efficient heat extraction and reduces thermal stress. The crystal is held by its end caps, the central region that heats up is water cooled to room temperature (25°C).

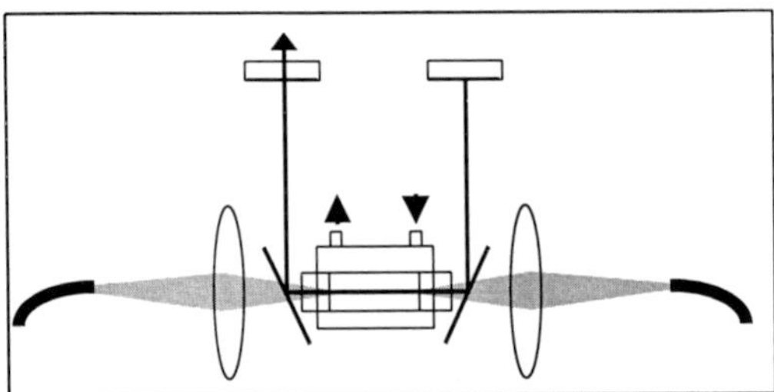

Figure 1. Pumping scheme and laser cavity

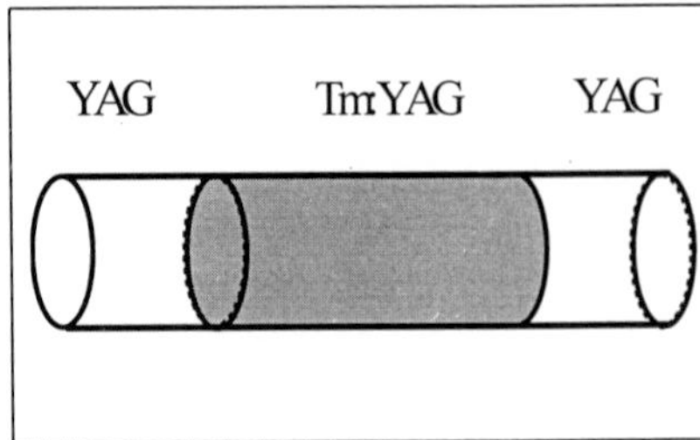

Figure 2. Tm:YAG and undoped YAG diffusion bonded crystal

Thermal Analysis

Since Quasi-3 level lasers are temperature sensitive, special care has to be taken in analyzing the temperature buildup in the crystal. To model the temperature distribution in the crystal, we assume that the pumping beam has a cone-like shape with a uniform intensity at any cross-section aria normal to its propagation axis. The intensity differs between cross-sections due to the deference in the cross-section arias and the absorption in the crystal. Another assumption made, is that the heat flow is mainly radial, so that every crystal

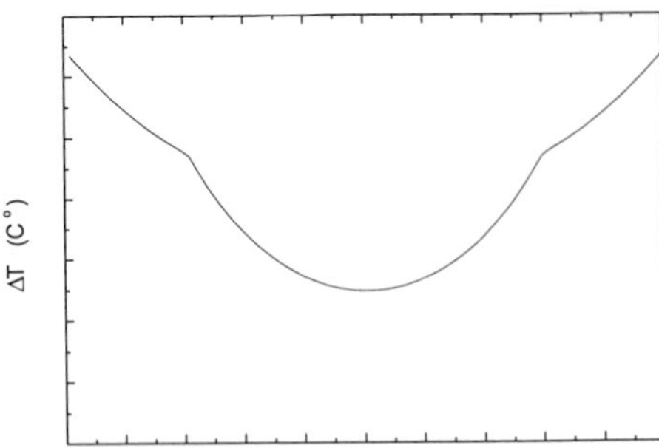

Figure 3. Calculated temperature difference between the center and circumference of a 10mm Tm:YAG laser rod.

slice normal to the laser axis, can be analyzed separately. Fig. 3 describes the calculated temperature difference between the crystal center and the crystal's circumference. It models a 10mm Tm crystal pumped from both ends with a pumping beam focused 2mm from each crystal end face and predicts temperature difference larger than 30^0C in the crystals first and last millimeters. When cooled to room temperature, an addition of $\sim 20^0$ has to be taken to account.

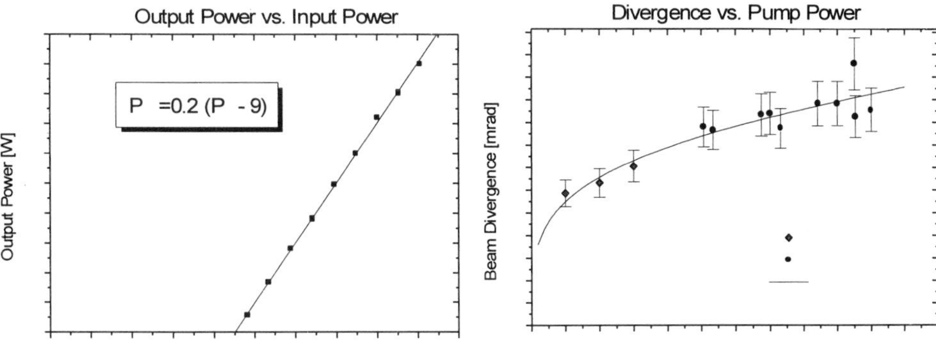

Figure 4. Tm:YAG laser pumped by laser diode at wavelength of 785 nm.

Figure 5. Measurements of divergence as a function of input power compared to calculations.

Experimental Results

Fig. 4 describes the laser output power versus the power of the pumping beam. We received a slope efficiency of 20% with a threshold at 9W. In fig.5 the measured beam divergence versus average input power is plotted. The higher the input power, the greater the thermal lensing of the crystal rods. Thus, the beam divergence increases. To see the lensing effects at powers below lasing threshold, we used pulsed pumping to give the desired average input power. The solid line represents the theoretical divergence expected from a Gaussian TEM_{00} beam. From comparing the calculations and the measurements, it is evident that the beam essentially remains at the TEM_{00} mode regardless of the pumping intensity. To calculate the M^2 we measured the beam waist on the flat output coupler. Multiplying by the divergence and dividing by $1.27\times\lambda$ we calculated $M^2 = 1.21\pm0.24$ for 10W input power and $M^2=1.12\pm0.22$ for 19W input power.

Summary

Utilizing diffusion bonded Tm:YAG crystals we developed a diode pumped laser operating at room temperature and received a 2W CW output power with slop efficiency of 20% and $M^2 \approx 1$.

References:

[1] E. C. Honea, R. J. Beach, S. B. Sutton, J. A. Speth, S. C. Mitchel, J. A. Skidmore, M. A. Emanuel and S. A. Payne, in IEEE Journal of Quantum Electronics. No. 9 Vol. 33 September 1997, pp.1592-1600

[2] C. Bollig, W. A Clarkson, R. A. Hayward and D.C Hanna, in Optics Communications Vol.154, 1998, pp. 35-38

[3] L. L. Chase, S.A. Payne, L.K. Smith, W. L. Kway and W. F. Krupke , in OSA Proceedings on Advanced Solid-State Lasers, Vol. 10, 1991 pp. 161-165

[4] S. A. Payne, L. L. Chase, L. K. Smith, W. L. Kway and W. F. Krupke in IEEE Journal of Quantum Electronics. No. 11 Vol. 28 September 1992, pp.2619-2630

MICROWAVE EXCITED CO_2-LASER

Avi Shahadi, Yoav Sintov[(a)], Eli Jerby[♠], and Shaul Yatsiv[(b)]

Tel Aviv University, Ramat Aviv 69978, Israel

(a) *El-Op, Electro-Optical Industries LTD, Israel*

(b) *Racah institute of physics, the Hebrew university, Jerusalem.*

Abstract

A slab CO_2 laser excited by a 2 kW, 2.45 GHz magnetron is presented. As shown by Yatsiv and others the slab configuration is suitable for microwave excitation of slow flow and sealed CO_2-lasers. These laser schemes are characterized by high average and peak powers. The experimental laser presented here operates with a slow gas-flow at a pressure of ~100 mbar. It generates ~ 400 W peak power with an overall efficiency of 6 %, in a duty cycle of 1 %. An average power of 33 W is attained. The overall efficiency, of 8.7 % in a duty cycle of 5 %, corresponds to ~ 22% microwave-to-laser power efficiency. Design considerations including microwave matching and heat removal are presented. This experiment leads to the development of a compact high-power sealed CO_2-laser at 300 W average and 3 kW peak power.

1. Introduction

Microwave excited CO_2-lasers are known in various schemes. Their advantages stem from the availability of highly efficient magnetrons, which enable high-peak pumping powers. Thermal instabilities and α-γ transitions are diminished in these schemes.

The development of fast-flow microwave-pumped CO_2-lasers in waveguide schemes, [1,2] were followed by improvements of heat dissipation rates in stripline lasers [3]. Reduced discharge gaps

♠ *jerby@eng.tau.ac.il, fax: 972-3-6423508, Tel: 972-3-6408048*

with the diffusion-cooling feature yield compact devices operating in a quasi-CW mode and low gas-pressures [4-6]. The high repetition-rate of the microwave pulses and the discharge stabilization by a ballast dielectric-strip enable high peak and average laser powers without a need for gas flow.

In this paper we present a slab CO_2-laser coupled to a rectangular resonator fed by a 2 kW, 2.45 GHz magnetron.

2. Experimental setup

The experimental device is illustrated in Fig. 1. The microwave source is a 2 kW magnetron (Hitachi 2M130) at 2.45 GHz. A maximal peak-power of 12 kW is available at short pulses from this magnetron. The microwave radiation is delivered through a circulator (Philips PDR-26) to a 60 dB coupler (Muegge MW-6971-0070). This allows the monitoring of the transmitted and reflected powers in this setup. An E-H tuner enables the basic matching of the radiation to a rectangular resonator. A double ridge waveguide is attached sideways to the rectangular resonator (Fig. 1), and the radiation is coupled through a slit. A standard Wilmad rectangular Pyrex-tube (WRT220, ID 2x20x475 mm) is placed between the ridges and attached to them by application of a silicon heat-conducting paste. The tube serves as the volume confining the gas, and as the microwave ballast dielectric-strip as well. The Pyrex tube is attached to two mirror holders by silicon O-rings. The gas flows through the mirror holders, hence, no changes are made in the standard Pyrex tube. A hemispheric optical-resonator employs a concave rear mirror (R = 5 m), and an output coupler with a reflectivity of 89 %. The fine tuning of the discharge electric-field is achieved by a variable short at one end of the rectangular-waveguide resonator, and by screws placed at the entrance to the resonator and along the slit connecting the two waveguides. The laser-head is cooled by circulating water through the ridges. A microwave power-meter (HP435A) measures the average delivered microwave-power. Two thermometers measure the cooling-water temperatures at the inlet and the outlet of the laser head (the temperature difference indicates the power absorbed by the plasma). An optical detector (Ophir F300A-SH) measures the average laser-power.

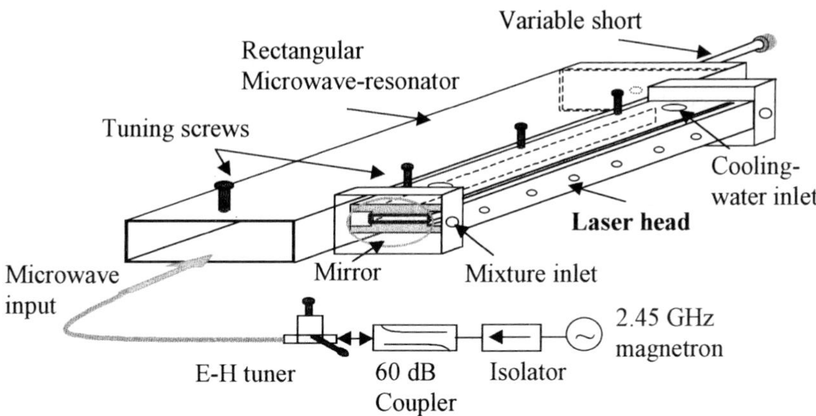

Fig. 1 The experimental setup.

3. Experimental results and discussion

Fig. 2 shows measurements of the average laser-power dependence on the average microwave-power at a pressure of 90 mbar. A mixture consisting of 18 % N_2, 6.5 % CO_2, 1.5 % Xe, and 74 % He is used, without any optimization. The pulse repetition-frequency does not exceed 5 kHz.

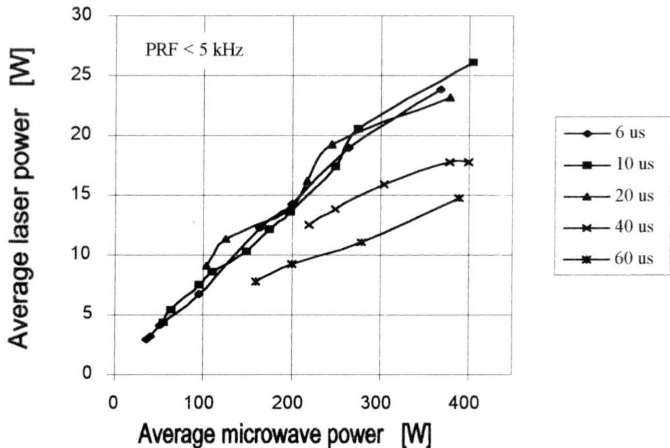

Fig. 2 The average output laser-power vs. the average microwave power for various pulse widths.

For a duty cycle below 10 %, a maximum peak laser-power of ~ 400 W is measured.

An optimized slab-laser design using this configuration for microwave excitation may result in a compact sealed high peak-power CO_2 laser.

References:

[1] B. Freisinger, H. Frowein, M. Pauls, G. Pott, J. H. Schäfer, J. Uhlenbusch, "Excitation of CO_2 lasers by microwave discharges," *SPIE, CO2 lasers and Applications II,* Vol. 1276, pp. 29-40, 1990.

[2] M. März and W. Oestreicher, "A versatile microwave plasma source and its application for a CO_2 laser," *Review of Scientific Instruments,* Vol. 65, pp. 2980-2983, 1994.

[3] S. Yatsiv, A. Gabay, Y. Sintov, "Diffusively cooled thin sheath high repetition rate TEA and TEMA lasers," *SPIE, Gas Flow and Chemical Lasers,* Vol. 1810, p. 157, 1992.

[4] Y. Sintov, A. Gabay, S. Yatsiv, "Self-activated, forced convective cooling in a pulse slab CO_2 laser," *Journal of Physics D, (Applied Physics),* Vol. 30, pp. 2530-2535, 1997.

[5] J. Nishime and K. Yoshizawa, "Development of CO_2 laser excited by 2.45 GHz microwave source," *SPIE, High-Power Gas Lasers,* Vol. 1225, pp. 340-348, 1990.

[6] M. März and W. Oestreicher, "Microwave excitation of a diffusion-cooled CO_2 laser," *Journal of Physics D, (Applied Physics),* Vol. 27, pp. 470-474, 1994.

PART 3

SOLAR PUMPED LASERS

Chairperson: *E. Hasman, Israel*

Annals of the Israel Physical Society, v. 14

ELECTRO-OPTICS and MICROELECTRONICS
Eds: Raphael LAVI and Ehud AZULAY

38 Watt Nd:YAG LASER PUMPED BY A 6.85 m^2 TARGET-ALIGNED SOLAR CONCENTRATOR

Mordechai Lando, Jacob Kagan, Boris Linyekin

Rotem Industries, Rotem Industrial Park, DN Arava 86800, Israel

Vadim Dobrusin

Arava Laser Laboratory and the Physics department

Ben –Gurion University, Beer-sheva, Israel

Abstract

We have constructed a *3.4m* diameter solar concentrator for solid state laser pumping. The primary mirror was segmented and mounted on a two-axis positioner, directed south, *32°* above the horizon. This Astigmatic Corrected Target Aligned design flattened the irradiation density variations along the day. The primary mirror focused the solar light through a deflecting plane mirror to a 8.9x9.1 cm^2 aperture 2-dimensional compound parabolic concentrator, which focused the light into a 1.0cm diameter Nd:YAG rod. The flat-flat resonator had a 98% reflectivity output coupler. The laser manifested 38 Watt power and up to 5 hours of continuous operation.

1. Introduction

Various applications were proposed for solar pumped lasers at a variety of space technologies as well as on earth applications[1]. High solar concentrations are needed to pump solid state lasers. The desired concentration may be obtained by combining a primary concentrator with a non-imaging optics concentrator[1,2]. For primary concentrator, we have designed and constructed an innovative Astigmatic Corrected Target Aligned (ACTA) concentrator[3], and here we report some preliminary laser pumping experiment using this design. The ACTA concentrator is a modification of the tower configuration, in which the first reflecting surface is also the primary concentrating mirror. This compacted design is thus highly efficient in terms of first reflecting surface area. A unique feature of the ACTA design, as compared to conventional tower configuration, is that the high efficiency is kept longer during the day. Thus, we have conducted a whole day experiment, in which the laser operated continuously for full five hours, standing a first durability study for any solar pumped laser.

2. The Astigmatic Corrected Target Aligned solar concentrator

The structure of our ACTA solar concentrator is shown schematically in Fig. 1. A segmented primary mirror is mounted on a two-axis ORBIT AL-4035-1SL positioner which tracks the solar orbit, following solar coordinates calculation by MICA1.5 astronomical software and OMEK OPTICS angular bisection and coordinate rotation code. The primary mirror focuses the incoming solar light and directs it towards a folding mirror, which deflects it downwards to the collecting laser head positioned on the focal plane. The fixed axis of the primary mirror mount is tilted southwards 32° above the horizon, to the direction of the folding mirror center. The novel target aligned configuration flattens the irradiation density variations along the day[3], comparing to the conventional configuration with a normal to the ground fixed axis. The 3.4 m diameter primary mirror is composed of 61 hexagonal segments, 360 mm between sides, each mounted on a separate two-axis base. The segments are spherically curved at 17 m radius of curvature, while their vertexes are placed on an 8.5 m radius spherical cap.

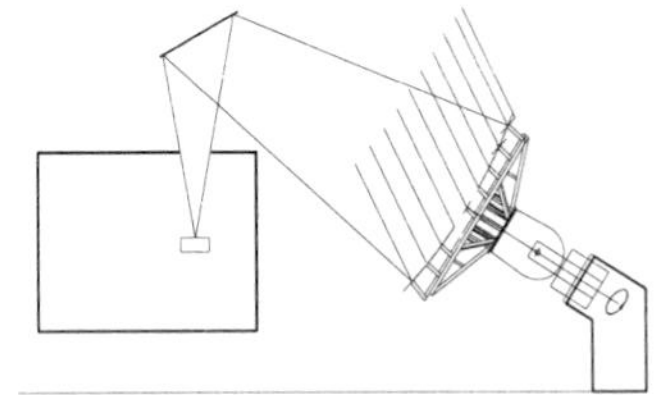

Fig. 1: Schematic view of the Astigmatic Corrected Target Aligned concentrator

3. The solar pumped laser

To reduce the thermal load to the laser rod, we removed part of the IR radiation which does not contribute to the laser pumping, and eliminated UV radiation bellow 500 nm which may cause rod solarization. The undesired spectrum portion was deflected by a 45° dielectric ELOP beamspliter towards a special thermal absorber, enabling future parallel use for different applications[2]. The pumping portion of the solar spectrum entered the 8.9x9.1 cm^2 aperture of a 2-dimensional compound parabolic concentrator (2-D CPC), which was manufactured by Jenkins *etal*[4] in the University of Chicago for solar pumped laser experiments in the National Renewable Energy Laboratory (Golden, Colorado). An anti-reflection coated 1.1% Nd:YAG laser rod, 10 mm in diameter and 130 mm long, was mounted inside a quartz flow-tube, which was located along the 2-D CPC axis.

The rod, the CPC, and the thermal absorber were water cooled, and the absorbed solar power in each part was calorimetrically measured. Water temperatures and flow rates were taken with National Instruments DAQ AT-MIO-64E card and the removed heat was calculated by a LabVIEW 5.01 software. The heat removed from the laser rod and CPC was used as a measure of the absorbed solar power.

As the laser resonator, we used high reflection flat back mirror, and a 98% reflectivity output coupler. The laser output power was measured with a calibrated Ophir L300A-LP-SH power meter head, connected to a computerized Ophir Laserstar monitor. For laser beam analysis we used a CCD camera connected to SPIRICON software.

4. Results

We have measured the thermal lensing as a function of the absorbed solar power, using a collimated He-Ne laser beam. The focal length was 102 cm at an absorbed solar power of 1000 watt, and the focal length obeyed the expected inverse law[5] dependence on absorbed solar power. To get a high quality beam, we have designed a convex-concave laser resonator, based on the focal length measurement. Nevertheless, the curved mirrors were not yet available and this preliminary experiment was done with the above flat mirrors. We have optimized the rear and front mirror distances from the laser rod face to 7 cm and 27 cm, respectively, and got output power of 38 watt at 936 Watt/m^2 insolation. The dependence of the output power on the absorbed solar power is depicted in the slope efficiency curve of Fig. 2. Higher power will be obtained upon completion of a two stage secondary concentrator of higher concentration.

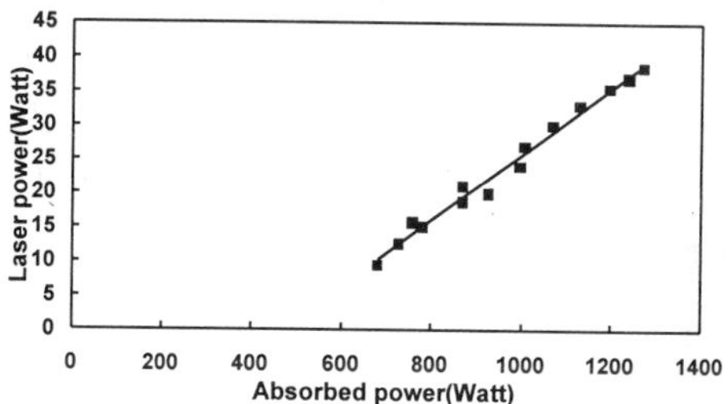

Fig. 2 : Slope efficiency curve at 936 Watt/m^2 solar insolation using 98% output coupler

A typical beam intensity image at 50 cm distance from the output coupler is shown in Fig. 3. The image was measured at absorbed solar power of 510 Watt and laser output power of 7 Watt, and is evidently spherically asymmetric. This asymmetry reflects the power absorption distribution as calculated by Jenkins[6] and as verified by us in a fluorescence measurement. A knife-edge beam quality measurements were made separately for the horizontal and vertical directions. The M^2 beam quality factor obtained was 77 and 144 for the horizontal and vertical directions, respectively. We expect beam quality improvement upon installation of the designed convex-concave resonator.

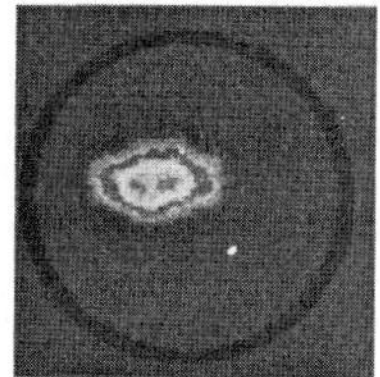

Fig. 3: A beam intensity image 50 cm from the output coupler. The circle diameter is 18.1 mm.

To demonstrate the solar pumped laser durability and stability, we operated it during May 19, 1999. The five-hour laser power and solar insolation measurements are presented in Fig. 4. The laser power followed fast changes in insolation, and slow variation due to the effect of changing solar beam deflection angle, which is indeed smaller for the ACTA design, but is still unavoidable.

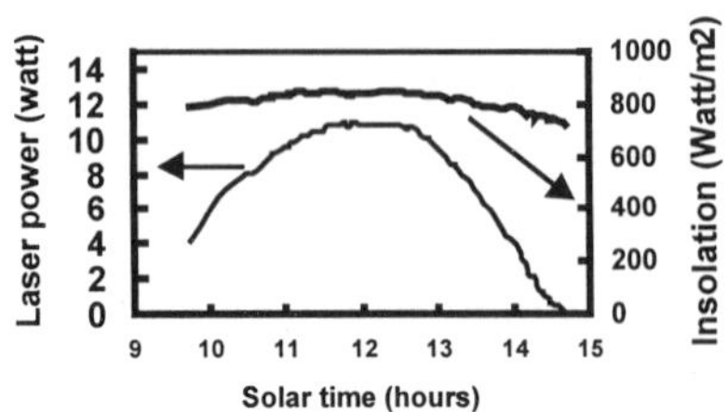

Fig.4: Output power and solar insolation as function of time during a whole day

Acknowledgements: The Israeli Ministry of Trade and Industry supported the research through the MAGNET program, CONSOLAR consortium for concentrated solar energy. We are indebted to Mr. A. Lewandowski for letting us use the CPC and laser rod, and Dr. R. Pitts for the CPC coating instructions. Dr. Y. Shimony provided us valuable advice as well as technical assistance. V.D. thanks Prof. S. Rosenwaks for his supervision.

References:

[1] M. Weksler and J. Shwartz, "*Solar pumped solid state* lasers", IEEE J. QE 24, 1222, 1988.

[2] V. Krupkin, Y. Kagan, A. Yogev, "*Non imaging optics and solar pumped laser pumping at the Weizmann Institute*", Proc. SPIE. 216,50-60,1993.

[3] M. Lando, J. Kagan, B. Linyekin, L. Sverdalov, G. Pecheny, Y. Achiam, "*An ACTA solar concentrator for solid state laser pumping*", ISES Solar World Congress, Jerusalem, July 1999.

[4] D. Jenkins, M. Lando, J. O'Gallagher, R. Winston, A. Lewandowski, C. Bingham and R. Pitts , "A *solar pumped Nd:YAG laser with a record efficiency of 4.7 watt/m^2 of primary mirror area*", Bull. of IPS (1996), P. 101.

[5] W. Koechner, Solid State Laser Engineering, 4th edition, Springer, Berlin, 1996, chap.7.

[6] D. Jenkins, "*A 57 W solar pumped Nd:YAG laser*", Ph.D. thesis, 1996.

SPECTROSCOPIC STUDIES OF DIMER GAS MOLECULES AS CANDIDATES FOR SOLAR PUMPED LASER

Idit Pe'er, Amnon Yogev*, Irina Vishnevetsky & Nir Naftali.
Weizmann Institute of Science, Rehovot, 76100 Israel.
Phone: 972-8-9343776/8, Fax: 972-8-9344117, E-Mail: jhidit@wis.weizmann.ac.il
**Stephen Meadow chair professor.*

Abstract.
The absorption spectrum of many dimer molecules posses a broad structural spectrum overlapping with the solar spectrum and can be good candidates for direct solar pumping. In the gas phase, the emission spectrum of the active medium offers tunability and high beam quality without significant thermal lensing or thermal induced birefringence. The analysis of the electronic transitions in dimer molecules as an optional system is presented. The focus is on Te_2, which was found to be the most promising material for direct solar pumping. Absorption and laser induced fluorescence were used to confirm the calculated electronic transitions. The experiments were conducted under different temperature and pressure conditions and different pump power. Experiments with different partial pressures of inert gas were carried out in order to study the energy transfer mechanism.

1. Introduction.

The possibility of building lasers based on gas medium, with solar excitation was first discussed in *1976* [1]. Since than, there were many experiments of optically pumped dimer vapor gases, pumped by laser line, but none of them was pumped using a non-monochromatic light source. In contrast to monochromatic pumping of the gas, when solar light pumps the gas, there are no selective mechanisms to create population inversion and each one of the different excited vibrational-rotational states of an electronic level can radiate. In order to have a selective population inversion, a rotational-vibrational relaxation mechanism should be imposed on the molecule. This can happen in different ways. The relaxation can occur due to collisions with the other molecules of the dimer gas - this process might lead to relaxation to different electronic levels and compete with the desired radiative laser transition. The relaxation can occur with the help of an additional inert buffer gas. When an Inert gas is added to the system, it collides with the active medium and the relaxation process begins. The rate of these relaxation processes depends on the energy gap, the rate of collisions, which directly depends on the partial pressure of the buffer gas, the mass of the colliding particles and the collisional cross section. The pressure of the Inert gas can be controlled in a way that the rotational-vibrational relaxation processes will occur faster than the radiative lifetime of those levels and most of the energy will be accumulated in a few vibrational levels at the electronic excited state. At higher pressure, relaxation of electronic states will also take place and laser emission will be reduced. Due to the different effects of the relaxation mechanisms, together with the necessity to have sufficient concentration of dimer molecules and absorption of pump radiation, the realization of a satisfactory dimer laser requires a careful optimization of parameters such as total pressure, temperature, dimer partial pressure and buffer gas partial pressure with respect to the available pump power.

The absorption and emission spectra of four materials Se_2, S_2, Bi_2 and Te_2 were studied as candidates for solar pumping. Te_2 was found to be the most suitable medium for solar pumped lasers. It absorbs light at the range of *340nm* to *520nm,* obtaining good overlap with the solar spectrum, and it's emission spectrum is at the range of *450nm* to *650nm*. The different transitions of Te_2 were analyzed using laser-induced fluorescence. Te_2 was pumped at the range of *451nm* to *570nm*. A large number of fluorescent bands were detected. The various transitions were analyzed and compared with theoretical calculations. Experiments were conducted at various pressures and the experimental results confirmed the estimated fluorescence lifetime and the calculated collisional time. The spectra were taken under various temperature conditions and a temperature dependence was observed only for wavelengths longer than *533nm*.

2. Experimental setup

The experimental system is described in Figure 1. A molecule cell was placed inside an oven that can heat the reservoir of the cell and the cell itself to different temperatures. The cell was illuminated by a doubled/tripled Nd:YAG Continuum laser, model Surelite I and a GWU OPO system, tunable at the visible range. A monochromator is placed perpendicular\parallel to the pump source, to measure emission\ absorption. The excitation is achieved by illuminating the Te_2 gas with a monochromatic light from an OPO system. The pump power out of the OPO ranged from *0.3mJ* to *5mJ* per *10nsec* pulse, at a repetition rate of *10Hz*. The OPO spectral width ranged from $18cm^{-1}$ to $40cm^{-1}$. In order to measure absorption, a halogen quartz lamp was placed perpendicular to the OPO beam, behind the cell. The absorption and emission spectra are observed, using a modified ¾ meter SPEX monochromator model 1702.

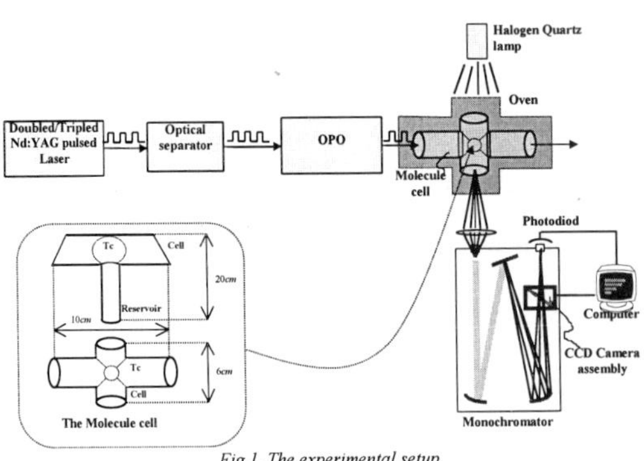

Fig.1. The experimental setup.

3. Results and Discussion

3.1. Materials selection.

Four materials were pre-analyzed, Se_2, S_2, Bi_2 and Te_2. Absorption spectra were measured at different temperature conditions. The results are summarized in Table I.

Figure 2 shows the absorption spectrum of Te_2 at different pressures. As one can see, the absorption spectrum of Te_2 has a good overlap with the solar spectrum. It's emission spectrum was found to be the strongest spectrum of the four spectra measured. In order to understand absorption and emission dynamics of Te_2, and in order to be able to estimate its performance as solar pumped laser, more detailed studies of Te_2 were performed.

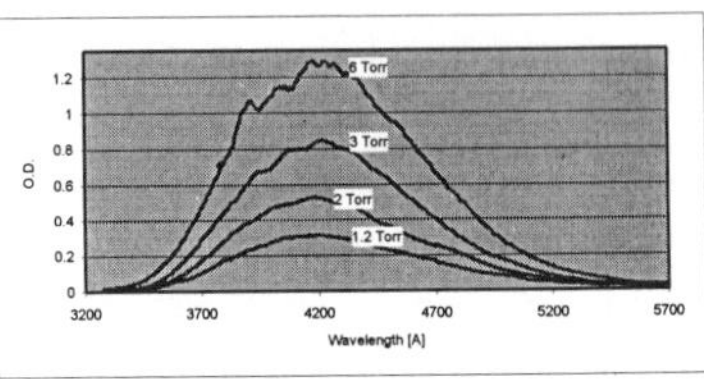

Fig.2 Absorption spectra of Te_2 at different pressures.

Table I. Measurements results of the absorption and emission of several materials.

Material	Pressure [*Torr*]	Temperature [°C]	Peak of Abs. [Å]	Abs. range [Å]	Pump Wavelength [Å]	Emission Range [Å]
Te_2	5	750	4320	3500-5200	5320	4700-6600
Bi_2[a)]	3	1050	3330	3000-3800		
			5200	4400-6600	5320	5100-5800[b)]
Se_2	0.6	480	3900	3500-4800	3550	3400-6000
S_2	2	400	4050	3860-4700	3550,5320	Not observed

a) Due to the high thermal excitation of the material, many emission lines appear due to thermal excitation and were observed without laser pumping. Absorption was also observed at wavelengths longer than 1micron.
b) Very weak.

3.2. Emission Measurements.

Te_2 fluorescence was measured under various temperature and pressure conditions and with different pump power, at the wavelength range of *451nm* to *570nm* with a step of about *0.5nm*. Various fluorescence bands were observed at different pump wavelengths. Each band was denoted with a letter from ***A*** to ***H***. Examples are given in Figures 3,4. In all cases, the emission spectra were shifted to longer wavelengths as the pump wavelength was shifted to longer wavelengths. This shift implies that fluoresces lines are emitted from discrete excited levels to various vibrational levels at the electronic ground state.

At pump wavelengths longer than *533nm* the intensity of the emission lines was decreased as the pump wavelength was shifted to longer wavelengths. That is due to the fact that in such a long pump wavelength, the absorption can occur only from high vibrational levels, which are populated at the operating temperature. The decrease in intensity is due to the low population at the higher vibration levels. When pumping at wavelengths in the range of *498.5nm* to *505.5nm,* groups *B* & *C* remain unchanged, but group *A* splits into two groups (see Fig.4.) and the envelope of these two groups is changing with the change of the pump wavelength. The different structures are consistent with the energy diagram that was obtained by calculations. Group *E* can be assigned to transitions from the $A0^+_u$ system to the $X1_g$ system, levels $v=3$ to *32*, and group *D*, to transitions to the $X0^+_g$ system, levels $v=0$ to *29*. (Identification of the various transitions is made on the basis of comparing the laser lines and the emission lines with possible energy gaps at different electronic structures).

When the excitation is at the range of *480nm* to *498.5nm*, three completely new groups appear (see Fig.4). However, the emission structure is very sensitive to the pump wavelength, and minor modifications of the pump wavelength can change completely the structure in a way that there is an alternation between the "old" structure and the "new" structure. At that pump wavelength range, the $B0^+_u$ system becomes accessible also from lower vibrational states at the ground state. The alternation between the different structures could be related to excitations from different upper states that are mainly governed by the Frank-Condon factors.

At a pump wavelength shorter than *480nm*, the tendency of alternation of the fluorescence structure, with minor changes of the excitation wavelength, is more dominant. When pumping at such short wavelengths, all the transitions $X0^+_g$□$A0^+_u$, $X1_g$□$A0^+_u$, $X0^+_g$□$B0^+_u$, $X1_g$□$B0^+_u$ are allowed, with no temperature constrains, and considering the fact that many vibrational levels can be involved from each one of the exited states and the ground states, many spectral combinations are allowed.

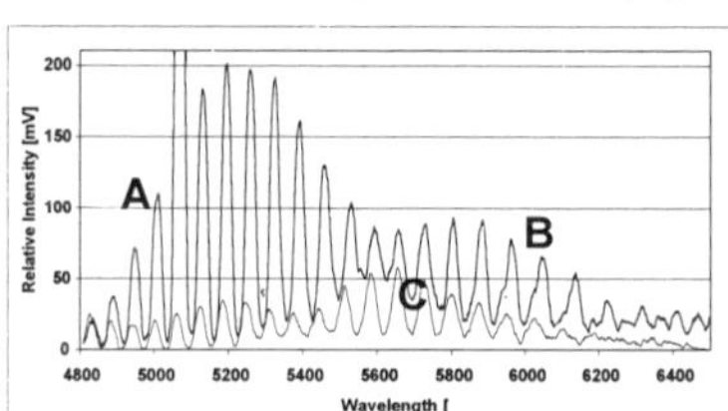

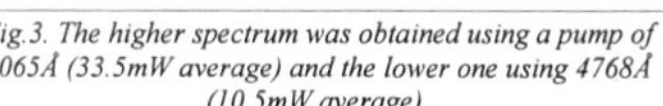

Fig.3. The higher spectrum was obtained using a pump of 5065Å (33.5mW average) and the lower one using 4768Å (10.5mW average).

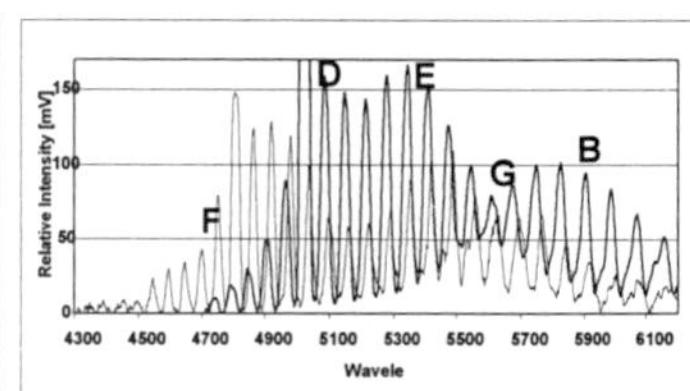

Fig.4. The darker spectrum was obtained using a pump of 5014Å (38mW average) and the lighter one using 4803Å (5.5mW average)

The minimum fluorescence intensity for a laser operation was calculated to be 4.86*w* in an equivalent four level laser system [2]. Calibration measurement showed that most of the measured peaks are much above threshold. If we succeed to accumulate all the excited molecules in a limited number of vibrational excited states, than the integrated intensity should be considered, and according to that calculation, is far above threshold.

3.3. Pressure & Temperature Dependence.

The various emission spectra were taken under different pressure conditions. The pressure was changed by changing the reservoir temperature. The overall structure of the spectra remained unchanged under the various pressure conditions. The emission lines intensity was measured for different peaks and there was a strong dependence of the emission lines intensity on pressure. The intensity increases with pressure due to the fact that at low pressures, there is a small amount of molecules, to absorb and emit light, inside the cell, so that the intensity at low pressures is weak, it reaches a maximum at 6*Torr* (reservoir temperature of 600℃ [3]) but than drops down again due to the fact that a) most of the light is being absorbed at the entrance to the cell; b) There is self-absorption of the emitted light, which increases with pressure.

In order to examine the collisional effect on the emission lines, the collisional lifetime was calculated for Te_2 at *1023°K* and *30Torr,* and it was found to be $5.3 \cdot 10^{-7} sec$. That time period between collisions can be compared with the lifetime of the fluorescence level. Fluorescence lifetime of few vibrational levels was measured to be at the order of few tenths of nanoseconds [4-6]. Since the collisional lifetime is longer than the fluorescence lifetime, the emission structure was not disturbed by the pressure changes, so that the molecule fluorescent before it decays to a lower vibrational level by collisional relaxation.

Following the experimental results, it is seen that as expected, many different excited states are formed with different excitation wavelengths. Adding a buffer gas at a sufficient pressure, can transfer, by collisional relaxation, most molecules to a lower vibrational state at the electronic excited state and thus assist the use of most of the excitation energy to obtain emission at a single wavelength from a solar pumped laser. In order to shorten the collisions lifetime to about $10 \cdot 10^{-9} sec$ and to accumulate all excited molecules in a limited energy region, one should work with *He* as a buffer gas at a pressure of *176Torr*. Using *He* as buffer gas will enable fine-tuning of the energy distribution of the excited Te_2 molecules. The probability for a collisional vibrational relaxation was calculated, using the Landau-Teller theory [7]. For Te_2-*He* collisions, at a temperature of *1023°K* and energy between two vibrational levels of $137cm^{-1}$ the calculation gives a probability value of *0.0093* per collision, which means that for high efficiency, high pressures would be needed. In order to examine the hypothesis, a preliminary experiment was performed, with *He* as a buffer gas, at a pressure of *70Torr @ 750°C*. At that pressure, a 52% decrease of the emission lines intensity was observed while only 10% of the integrated emission power of all the fluorescence lines occurred. That observation indicates that an energy transfer between energy levels has occurred.

The temperature dependence for different emission groups was also studied. At pump wavelengths longer than *533nm*, strong temperature dependence was observed. The fluorescence lines intensity increases with temperature. At this pump wavelengths range, the absorption is from high vibrational energy levels at the electronic ground state in which the population is strongly dependent on temperature

3.4. Laser head.

The main design considerations of a laser pump head for the dimer gas medium are: a) Short path of the laser beam inside the active media, due to self absorption losses; b) long path of the solar pump light inside the active media, due to small specific absorption; c) isothermal conditions of the cell, to prevent condensation on one of the cell optical windows; d) very small absorption losses in the preferred lasing wavelength.

The preliminary design concept of a laser head is described in Figure 5. The sunlight would be concentrated by a parabolic dish, into a 3D-CPC (Compound Parabolic Concentrator) with a wide exit angle. The light will enter a wave-guide that contains a Te_2 cell placed in the center, see Figure 5. The light will perform multiple passes through the Te_2 cell. The wave-guide has a conical end structure to ensure second transverse of the sunlight through the cell. That way, most of the pump light would be absorbed inside the cell. The cavity mirrors are placed outside the wave-guide.

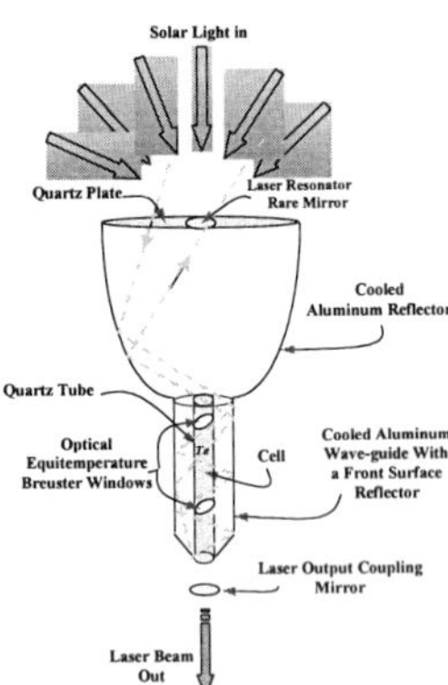

Fig.5. The solar pumped dimer gas laser configuration.

4. Acknowledgments.

This research was supported by a research grant from Dr. and Mrs. Robert Zaitlin, LA, California.

5. References.

1. B.F. Gordiets, L.I. Gudzenko, V.Ya. Panchenko JEPT Letters, vol. 26, pp. 152 , (1977).
2. Amnon Yariv, *"Optical Electronics"* forth edition, California Institute of Technology, Saunders College Publishing 1991.
3. A. Roth *"Vacuum technology"* third edition, Elsevier Science B.V. 1990.
4. I. P. Klincare and M. Ya. Tamanis *"Lifetime and Land'e-factor measurements of $A1^+_u$ and $A0^+_u$ states of $^{130}Te_2$ by laser induced fluorescence"* Chemical Physics Letters, **Vol. 180** number 1,2, pp. 63-67, May 1991.
5. R.S. Ferber, O. A. Shmit and M. Ya. Tamanis *"Lifetime and Land'e factors in the $A0^+_u$ and $B0^+_u$ states of $^{130}Te_2$"* Chemical Physics Letters, **Vol. 92**, No' 4, pp. 393-397, October 1982.
6. Ya. A. Harya, R. S. Ferber, N. E. Kuz'menko, O. A. Shmit and A. V. Stolyarov, *"Intensities of the Laser induced fluorescence of $^{130}Te_2$ and electronic transition strengths for the $A0^+_u$ - $X0^+_g$ and $B0^+_u$ - $X0^+_g$ systems"*
7. Karl F Herzfeld and Theodore A. Litovitz *"Absorption and dispersion of ultrasonic waves"* Academic Press, New York and London 1959, Ch. *VII "Theory of Vibrational and Rotational energy exchange"*, pp. 260-278.

PART 4
ULTRAFAST LIGHT SOURCES

Chairperson: *Z. Henis, Israel*

Annals of the Israel Physical Society, v. 14
ELECTRO-OPTICS and MICROELECTRONICS
Eds: Raphael LAVI and Ehud AZULAY

SUB-TWO-CYCLE PULSE GENERATION: TECHNIQUES, DYNAMICS AND APPLICATIONS

U. Morgner, F.X. Kärtner, W. Drexler[*], Y. Chen[*], S.H. Cho[*], H.A. Haus[*],

J.G. Fujimoto[*], E.P. Ippen[*]

Department of Electrical Engineering and Information Technology,

High-Frequency and Quantum Electronics Laboratory,
Karlsruhe University, Kaiserstr. 12, D-76128 Karlsruhe, Germany,

[*] *Department of Electrical Engineering and Computer Science and*

Research Laboratory of Electronics,
Massachusetts Institute of Technology, Cambridge 02139, USA

Abstract

Pulses shorter than two optical cycles, i.e. 5.2 fs, with bandwidths in excess of 400 nm have been generated from a Kerr-Lens mode-locked Ti:sapphire laser with a repetition rate of 90 MHz and an average power of 200 mW. An analysis shows that the dominant pulse shaping dynamic in sub-10 fs laser sources is dispersion managed soliton formation. Use of this ultrabroadband laser in high resolution optical coherence tomography leads to in vivo imaging of biological tissue with micron resolution.

Over the last years improved methods of dispersion compensation based on chirped mirrors have led to the generation of pulses as short as 4 and 4.5 fs from amplified or cavity-dumped low-repetition rate Ti:sapphire laser systems [1-4]. Recently, high repetition rate pulses directly from the laser oscillator with pulse width slightly shorter[5] and longer[6] than two optical cycles have been demonstrated. For the sub-two cycle pulses the use of low dispersion calcium-fluoride prisms in combination with double-chirped mirrors showing a controlled group delay dispersion (GDD) and high reflectivity over 400 nm is crucial. The corresponding bandwidth extends from 650 nm to 1050 nm. In this paper, we analyze the important dynamics underlying the pulse generation process in sub-10 fs lasers and report on an initial first application of this ultrabroadband laser source to optical coherence tomography (OCT), a new biomedical imaging technique.

The set-up of this laser, which is laid out in detail in ref. 5, can be decomposed into two negatively dispersive resonator arms and the positively dispersive laser crystal which also exhibits self-phase modulation (SPM) due to the focus of the beam inside the crystal. Thus the energy preserving dynamics in sub-10 fs lasers can be mimicked by the dispersion map shown in Fig. 1 together with SPM only acting in the positive dispersion section. The Kerr-Lens-Modelocking action can be modeled as a fast saturable absorber. At fixed pulse energy, the pulse length in these laser systems is well controlled by the amount of average negative dispersion in the laser cavity, assuming equal distribution of the negative dispersion in both resonator arms.

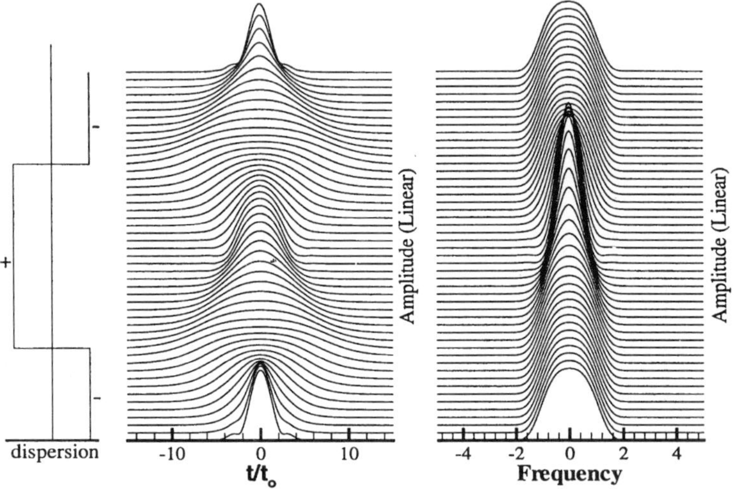

Fig. 1 Temporal evolution of the pulse through alternating positive dispersion with SPM and negative dispersion without SPM and net zero dispersion.

For pulses longer than 20 fs the steady state pulse formation process is governed by the well known concept of soliton modelocking [7], where the pulse width is directly determined by the amount of net negative dispersion in the cavity. The width of the soliton is proportional to the absolute value of the net dispersion. This concept is valid under the condition of weak pulse shaping, which means that the pulse shape does not change significantly during one roundtrip in the laser cavity. It is noteworthy, that in this case the steady state pulse shape is determined by energy preserving effects. The KLM action is only responsible for the pulse built-up and pulse stabilization against the filter losses imposed on the pulse by the finite gain and mirror bandwidth. For pulses shorter than 20 fs this assumption is no longer justified. Because of the large dispersion swing, which is the difference between the dispersion in the positive and the negative dispersion cell, the pulse width breathes for pulses approaching 5 fs by a factor of up to five, see Fig. 1. Despite this breathing, the general theory of dispersion managed solitons predicts again chirp-free steady state pulses in the center of the dispersion cells [7] having a shape only determined by the dispersion map and SPM. Furthermore, the shortest pulse width decreases as the dispersion swing decreases, i. e. with a shorter crystal.

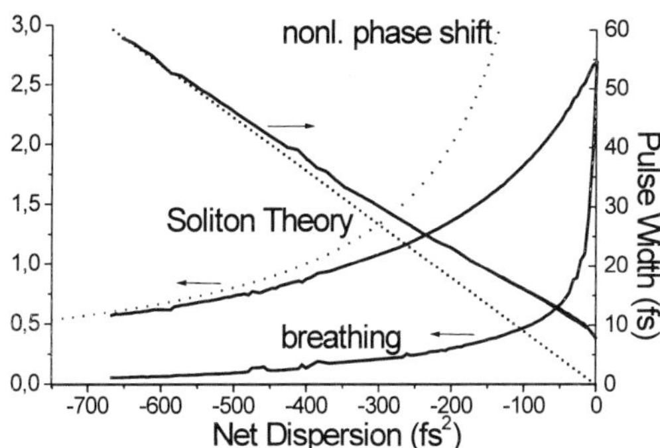

Fig. 2 Comparison solitary modelocking against dispersion managed modelocking. The breathing of the pulse limits the peak nonlinear phase shift and allows stable operation even at zero average dispersion.

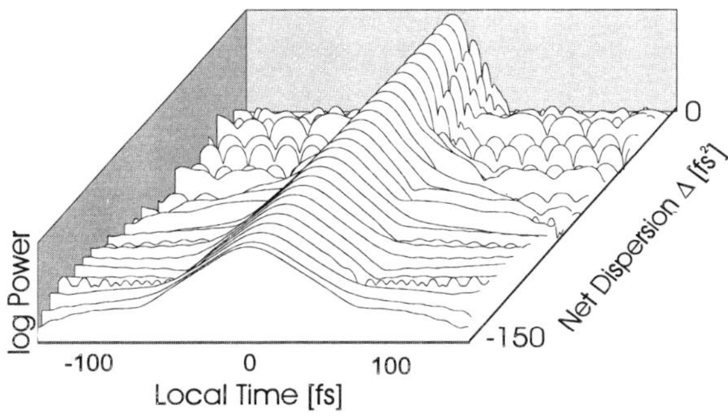

Fig. 3 Time domain pulse shapes on a logarithmic scale for different values of net cavity dispersion from the numerical simulation. The transition from the soliton ($sech^2$) to the dispersion managed soliton ($sinc^2$) is obvious.

This is true for operation even at zero average dispersion. Therefore, the well known concept of soliton modelocking, where the pulse is shaped by the usual soliton dynamics and only kept stable by the saturable absorber action, as discussed before, can be extended to sub-10-fs lasers by allowing for the formation of dispersion managed solitons. One may call this regime of operation dispersion managed modelocking[8].

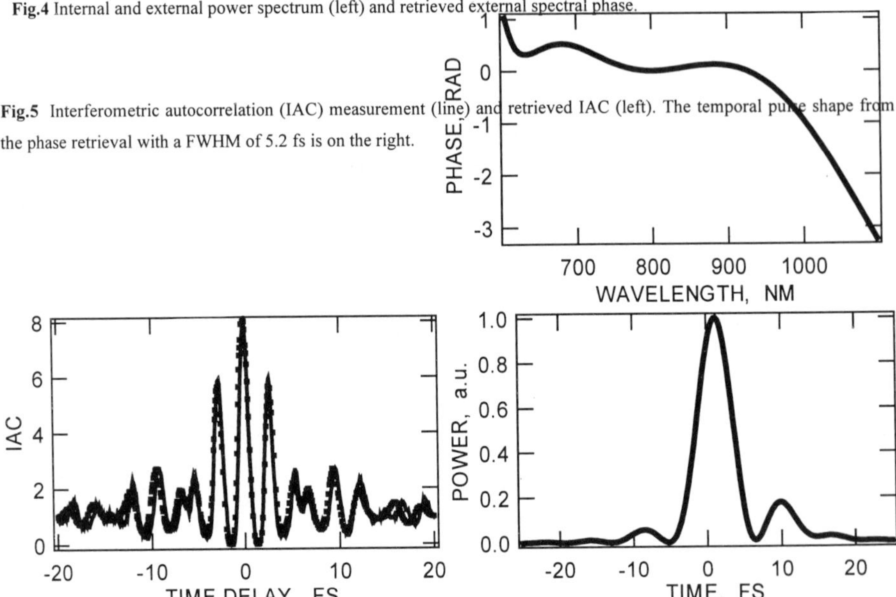

Fig.4 Internal and external power spectrum (left) and retrieved external spectral phase.

Fig.5 Interferometric autocorrelation (IAC) measurement (line) and retrieved IAC (left). The temporal pulse shape from the phase retrieval with a FWHM of 5.2 fs is on the right.

Simple one-dimensional numerical simulations of the pulse shaping process in these lasers show the transition from the solitary laser to the dispersion managed modelocked laser, Fig. 2: In the case of conventional soliton-modelocking the pulse width decreases with decreasing net dispersion and the peak nonlinear phase shift is inversely proportional to the net dispersion. Therefore, the nonlinear phase shift would diverge when zero net dispersion is approached (dotted lines in Fig. 2). The

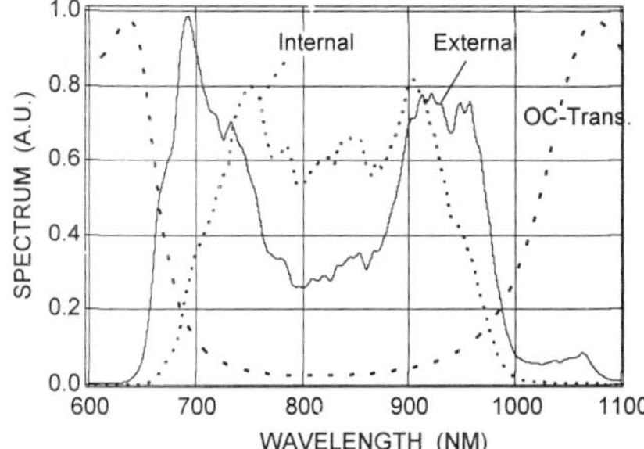

formation of a dispersion managed soliton results in a significantly reduced nonlinear phase shift directly related to the breathing of the pulse (solid lines). Even at net zero dispersion and slightly positive dispersion a stable steady state solution is possible.

Figure 3 shows the steady state pulse shapes obtained in the simulation for different values of the net dispersion. The pulse shape of the soliton is described by a $sech^2$ function, while the pulse shape of the dispersion managed soliton is between a Gaussian and a $sinc^2$ shape with sidelobes in the time domain, which is a consequence of the flat top spectra[8].

Fig. 4 (a) shows the experimentally obtained intra- and extracavity spectra[5]. We clearly find the flat top intracavity spectrum resulting from the pulse formation via dispersion managed soliton formation. The structure in the output spectrum, typically observed in this laser, is explained by the transmission of the output coupler, which is also shown in Fig. 4 (a). The increased output coupling in the wings of the intracavity spectrum strongly emphasizes the wings. The intracavity spectrum, if assumed to be Fourier-limited, i.e. zero spectral phase, results in a 6 fs pulse. The Fourier-limited extracavity spectrum would result in a 4.9 fs pulse. Fig. 5 (a) shows the simultaneously measured interferometric autocorrelation of the output pulse, which when fitted to a pulse with a $sinc^2$-shape gives a pulse width of 5.4 fs. Applying a recently demonstrated phase retrieval technique[9] to determine the electric field based on the measured spectrum, Fig. 4 (a), and interferometric autocorrelation, Fig. 5 (a), results in the temporal intensity envelope, Fig. 5 (b), and spectral phase shown in Fig. 4 (b). This phase is mainly due to the phase of the output coupler transmission. The full width at half maximum of the intensity envelope indicates a pulse width of 5.2 fs which is shorter than two cycles, 5.33 fs, of the carrier frequency derived from the center wavelength of 800 nm.

Such short pulses will have many applications in coherent control [10], detecting novel nonlinear optical effects based on the phase slip between envelope and carrier [11], and frequency comb generation for optical spectroscopy[12] . We have used the broad spectrum and spatial coherence of the light emitted from this source to improve imaging of biological tissue with optical coherence tomography to a resolution of only 1 micron. Optical coherence tomography[13] (OCT) has recently emerged as a powerful imaging technique for biological tissue. OCT can perform real time, in situ imaging of tissue cross sectional microstructure with micron resolution. The spatial resolution of OCT depends directly on the coherence length and thus on the optical bandwidth of the used light source.

Using the sub-two-cycle KLM Ti:sapphire laser as a low coherence source, in-vivo imaging of biological tissue can be demonstrated with a longitudinal resolution around 1 μm and a transversal resolution of 3 μm. To our knowledge, this is the highest resolution ever obtained with OCT. Figure 7 shows in vivo sub-cellular resolution tomograms of an anesthetized African tadpole. The ultrahigh resolution allows us to resolve the interior of living cells.

Fig. 4: *In vivo* ultrahigh-resolution OCT image of an African tadpole (Xenophus laevis) taken by use of the Ti:sapphire laser as broadband, high brightness light source. Multiple mesanchymal cells of various sizes and nuclear-to-cytoplasmic ratios, the mitosis of two cells as well as melanocytes are clearly shown.

In conclusion, we have generated what are currently the shortest pulses directly from a KLM Ti:sapphire laser at a 90 MHz repetition rate with a corresponding spectrum extending from 650 nm to

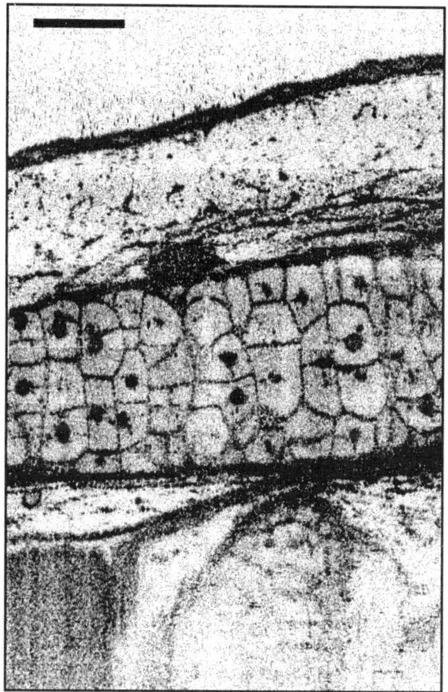

more than 1050 nm. The pulse width of 5.2 fs has been determined by a phase retrieval technique based on a fit of the interferometric autocorrelation and the spectrum. We identified the main pulse shaping mechanism as dispersion-managed soliton formation, which avoids overdriving the SPM when the pulse width evolves to the sub-10 fs range. The pulses have been used to significantly enhance the resolution of optical coherence tomography.

References

1. Z. Cheng, G. Tempea, T. Brabec, K. Ferencz, C. Spielmann, F. Krausz, in Ultrafast Phenomena XI, (Springer.Verlag, Berlin, 1998), pp. 8-10.
2. M.S. Pshenichnikov, A. Baltuska, R. Szipöcs, and D.A. Wiersma, in Ultrafast Phenomena XI (Springer-Verlag, Berlin, 1998), pp. 3-7.
3. A. Baltuška, Z. Wei, M.S. Pshenichnikov, and D.A. Wiersma, Opt. Lett. **22**, 102 (1997).

4. M. Nisoli, S. D. Silvestri, O. Svelto, R. Szipöcs, K. Ferencz, C. Spielmann, S. Sartania, and F. Krausz, Opt. Lett. **22**, 522 (1997).
5. U. Morgner, F.X. Kärtner, S.H. Cho, Y. Chen, H.A. Haus, J.G. Fujimoto, E.P. Ippen, V. Scheuer, G. Angelow, T. Tschudi, Opt. Lett **24**, 411 (1999)
6. D.H. Sutter, G. Steinmeyer, L. Gallmann, N. Matuschek, F. Morier-Genoud, U. Keller, V. Scheuer, G. Angelow, T. Tschudi, Opt. Lett **24**, 631 (1999)
7. T. Brabec, C. Spielmann, F. Krausz, Opt. Lett. **16**, 1961 (1991)
F. Krausz, M.E. Ferman, T. Brabec, P.F. Curley, M. Hofer, M.H. Ober, C. Spielmann, E. Wintner, A.J. Schmidt., IEEE Journ. of Quantum Electron. **QE 28**, 2097 (1992)
C. Spielmann, P.F. Curley, T. Brabec, F. Krausz, IEEE Journ. of Quantum Electron **QE 30**, 1100 (1994)
8. Y. Chen, H.A. Haus, Journ. Opt. Soc. Am. **B 16**, 24 (1999)
Y. Chen, H.A. Haus Opt. Lett. **23**, 1013 (1998)
9. A. Baltuška, A. Pugžlys, A.S. Pshenichnikov, D.A. Wiersma, B. Hoenders, H. Fewerda, Ultrafast Optics Conference, Th10, Ascona, Switzerland 1999
10. A. M. Weiner, in Progress in Quantum Electronics 1995, **19**, pp. 161-237.
11. F. Krausz, T. Brabec, C. Spielmann, Optics Photonics News **9**, 46 (1998)
12. T. Udem, J. Reichert, R. Holzwarth, T.W. Hänsch, Phys. Rev. Lett. **82,** 3568 (1999)
13. D. Huang, E.A. Swanson, C.P. Lin, J.S. Schuman, W.G. Stinson, W. Chang, M.R. Hee, T. Flotte, K. Gregory, C.A. Puliafito, J.G. Fujimoto, **Science 254**, 1178 (1991).

COMPACT QUASI-CW DIODE-PUMPED NEODYMIUM LASERS FOR HIGH ENERGY PICOSECOND PULSE GENERATION

A. Agnesi, S. Dell'Acqua, G.C. Reali and A.Tomaselli

INFM – Dipartimento di Elettronica, Università di Pavia
Via Ferrata 1 - 27100 Pavia (Italy)
http://ele.unipv.it/laser/laser.html

Abstract

A compact and reliable quasi-cw all-solid-state source of high power ps pulses is presented, that is significantly simpler than traditional master oscillator-power amplifier systems with same power output.

All-solid-state compact sources of stable, intense picosecond pulses were recently reported[1], employing nonlinear mirror (NLM) mode-locking[2] which had proved an effective and convenient technique in the picosecond regime. These devices are attractive for several scientific and practical applications such as nonlinear frequency conversion and micro-machining, being significantly simpler than traditional master oscillator-power amplifier systems; they may even be made tunable by using parametric generators and amplifiers[1].

In this work we report on a quasi-CW, 200Hz end-pumped diffraction-limited picosecond laser based on either Nd:YAG (@1064 nm) or Nd:$YAlO_3$ (@1079 nm) with a 5-times improved efficiency with respect to our previous side-pumped design[1], made possible by the development of a simple and cost-effective pump geometry. We provide for the first time an extensive characterization of the NLM operation and we interpret our results with a numerical model of the transient dynamics of the picosecond train evolution.

Fig. 1 shows the resonator layout.

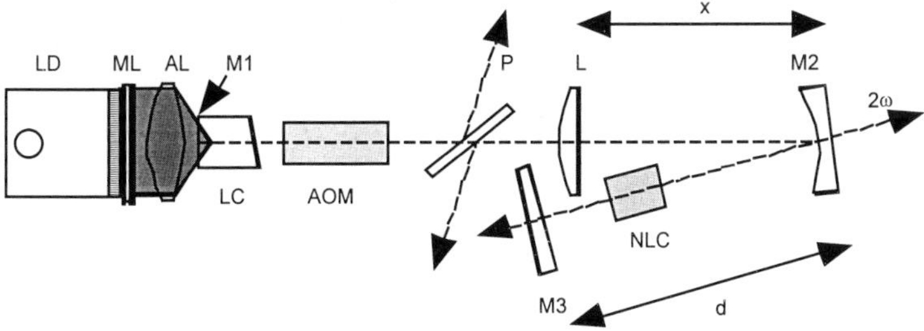

Fig. 1. Layout. LD: diode bar; ML: microlens; AL: acylindrical lens; LC: laser crystal; M1: input mirror; AOM: 60-MHz acousto-optic modulator; P: glass plate; L: f=250 mm lens; M2: r=150 mm mirror; NLC: nonlinear crystal; M3: dichroic OC.

The 0.001x10 mm^2 diode bar (Thomson TH-1301-S) generated 80 W peak power in a rectangular 200-μs optical pulse at 809 nm up to 200 Hz. The fast axis was collimated by a cylindrical 400-μm diameter GRIN lens, and the slow axis direction was focused by an acylindrical lens (focal length = 8.1 mm). About 80% of the emitted energy was properly conditioned and focussed into a 0.8x1.5 mm^2 spot size in the laser crystal.

The acousto-optic modulator initiated the mode-locking process generating ≈200-ps active-modelocking pulses which were then effectively compressed by the NLM based on a second harmonic (SH) nonlinear crystal NLC and a dichroic output coupler (OC) M3. The lens L and the mirror M2 were chosen to optimize the focussing in NLC and for mode-matching the pump spot size (the separation *x* was chosen to optimize the resonator stability parameter). The non-back-converted SH was extracted from M2, whose folding angle was kept small to limit astigmatism. The polarizer P could be rotated around the Brewster angle, thus making a second (variable) OC, allowing the study of the temporal pulse evolution inside the NLM laser.

Indeed, former numerical computations[3] validated by our specific model show that in the first pass through the SH crystal, pulses at the fundamental wavelength can undergo strong depletion and distortion, hence M3 may not always be regarded as the optimum OC element. In our experiment the single-pass SH energy conversion was always ≤75%, and in fair agreement with our numerical model we found the pulses exiting from the polarizer to be ≈20-30% shorter than those leaving the dichroic OC.

We chose either a 3-mm type II KTP or a 5-mm type I BBO (cut for SH generation at 532 nm), as the effective nonlinearity was nearly the same, and the group velocity mismatch in both cases was < 1.3 ps, not affecting significantly the typical build up dynamics of passively mode-locked Nd:YAG or Nd:$YAlO_3$ pulsed lasers. However, BBO had a smaller acceptance angle, hence a larger phase mismatch, which reduced the SH efficiency and optimized pulse durations from P and M3, in agreement with our model.

With the KTP crystal and the Brewster-angled plate P, a train of 63 pulses of 45 ps duration (FWHM) and total energy ≈370 μJ was generated by the Nd:YAG laser with 12-mJ absorbed pump energy. Rotating P, through the additional 8.5% output coupling provided by the Fresnel reflection we detected 29-ps pulses in a 31-pulses train, with a total energy of 133 μJ, while the train energy from M3 (R_ω=69%) reduced to 267 μJ.

The minimum pulse duration was observed with the Nd:YAG laser using the BBO crystal, together with a dichroic OC with reflectivity R_ω=69%: a train of 23 pulses of 25 ps FWHM and total energy 120 μJ was outcoupled through P, whereas the energy of the 31-pulses train exiting M3 was 267 μJ, with a 32-ps pulse width (Fig. 2). The amplitude fluctuations of the pulse trains were always less than 2%.
The Nd:$YAlO_3$ laser could be tested only with the BBO crystal, showing comparable results. The slightly improved cross-section ratio allowed the generation of shorter trains of 30-ps pulses, exhibiting a more pronounced Q-switching action. The reduced build-up time in this case was likely to prevent a better exploitation of the larger gain bandwidth available from Nd:$YAlO_3$.

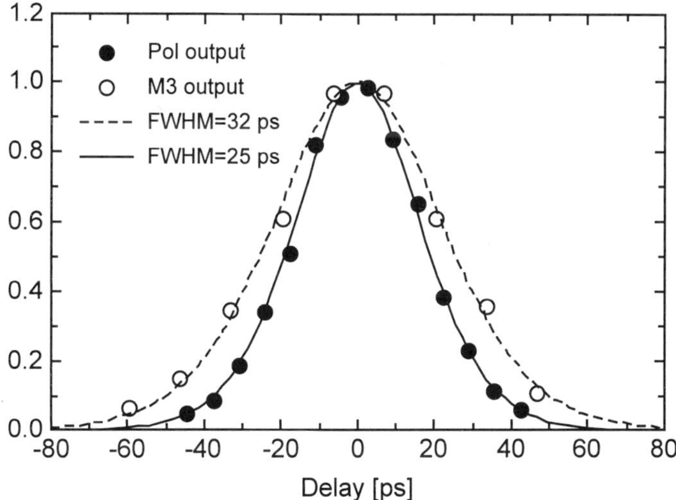

Fig. 2. Pulse autocorrelation (assuming $sech^2$) of the signal from M3 (dashed line) and P (continuous line).

The autocorrelation shape was always consistent with a $sech^2$ intensity profile, and within the SH conversion efficiency regimes we exploited, the pulse extracted from the dichroic OC never showed appreciable distortion. Therefore, when a picosecond pulse train is desired, with our NLM design the output can effectively be extracted from the dichroic mirror, at the price of a moderate increase in pulse length. In a preliminary test on cavity dumping using a Pockels, we generated a single intense well-shaped sub-30-ps pulse of ≈60 μJ energy (≈2-MW peak power), expecially interesting in high power, single pulse applications.

References:

[1] A. Agnesi, E. Piccinini, G.C. Reali and C. Solcia, Opt. Lett. 22, 1415 (1997)

[2] K.A. Stankov, Appl. Phys. B45, 191 (1988)

[3] I. Buchvarov, G. Christov and S. Saltiel, Opt. Commun. 107, 281 (1994)

COMPACT DIODE-PUMPED FEMTOSECOND CR:FORSTERITE LASER SOURCE FOR NONLINEAR MICROSCOPY

A. Agnesi, E. Piccinini, G.C. Reali and A.Tomaselli

INFM – Dipartimento di Elettronica, Università di Pavia
Via Ferrata 1 - 27100 Pavia (Italy)
http://ele.unipv.it/laser/laser.html

Abstract

We present an optimization study of a fs Cr:Forsterite laser for application in nonlinear microscopy.

Cr:Forsterite is an interesting vibronic laser material for its broad tunability range, 1130-1370 nm [1,2]: it can be conveniently pumped using Nd-lasers (at ≈1-μm), and allows generation of ultrashort pulses down to 25 fs [3]. However, Cr:Forsterite crystals generally suffer serious drawbacks, such as poor thermal characteristics, excited-state absorption (ESA), low figure-of-merit (FOM≈20-40) and a strong fluorescence thermal quenching. The FOM is defined as the ratio between the absorption coefficient at the pump wavelength and the absorption coefficient at the laser wavelength, which introduces parasitic intracavity losses. In this work for the first time we apply double-pass (DP) pumping to reduce effectively the pump threshold in a cw Cr:Forsterite laser. A crystal with length l pumped in a single-pass (SP) exhibits approximately the same absorption and gain as a crystal of length $l/2$ which is pumped collecting the transmitted pump after the first pass and re-focussing it inside the crystal: it turns out that the effective FOM doubles because the intracavity loss due to crystal re-absorption is only half.

The Cr:Forsterite crystal characteristics (Brewster-cut, n=1.635, α_p=0.8 cm^{-1}, FOM=35), and the pump and laser spot radii (w_p = 23 μm, w_l = 26 μm) were input according to the parameters actually available or measured in our setup. The laser cavity (Fig. 1) was a 1.5-m long bow-tie resonator with 100-mm radius of curvature folding mirrors M1, M2, a flat HR mirror M3 and a 3.5% output coupler OC. We calculated threshold powers and slope efficiencies for both the single- and the double-pass scheme by varying the crystal length. With the given parameters, it turned out that a 5-mm Cr:Forsterite crystal optimized the operation in DP, yielding nearly the same slope efficiency as a 10-mm crystal in SP, but with a ≈30% lower threshold. The threshold reduction is even more interesting for low output coupling (<1%), as a low pump power ≈600 mW is predicted for a 5-mm crystal and all-HR cavity. Furthermore, using folding mirrors with shorter radius of curvature (i.e. 50-mm) to produce smaller mode waists, threshold pump powers as low as ≈300 mW (≈150 mW absorbed) can be achieved with a 1% OC. This will allow an effective DP pumping with a single microchip laser emitting 800-1000 mW, yielding an output power of 30-40 mW.

Two Brewster-cut Cr:Forsterite crystal, 5-mm and 10-mm long, were chosen for the experiment. The excitation source was a 5-W homemade diode-pumped Nd:YVO_4 laser, emitting a TEM_{00} linearly polarized beam. A Faraday isolator (OFR IO-2-YAG) prevented the residual pump re-injection in the vanadate oscillator. The Cr:Forsterite crystal was kept at a constant temperature of 10 °C with a thermoelectric cooler. Mirrors M1 and M2 transmitted ≈94% of the incident power at 1064 nm and exhibited ≈0.2% transmittance loss at 1240 nm. A 200-mm focal length, AR-coated lens L2 with the HR mirror M4 re-collimated the residual pump power to allow the second pass into the Cr:Forsterite crystal. The experimental results are summarized in Fig. 2.

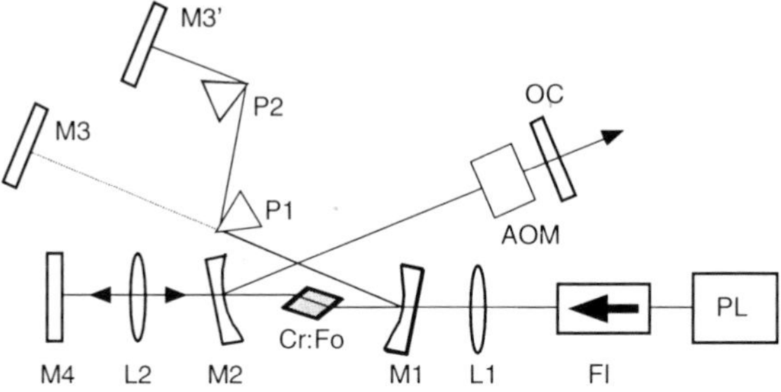

Fig. 1. Experiment layout: M3-M1-M2-OC is the cw Cr:Forsterite oscillator. M3'-P2-P1-M1-M2-OC is the KLM Cr:Forsterite oscillator. L2-M4 provide the 2nd pass through the laser crystal. PL: pump laser; FI: Faraday isolator; AOM: acousto-optic modulator.

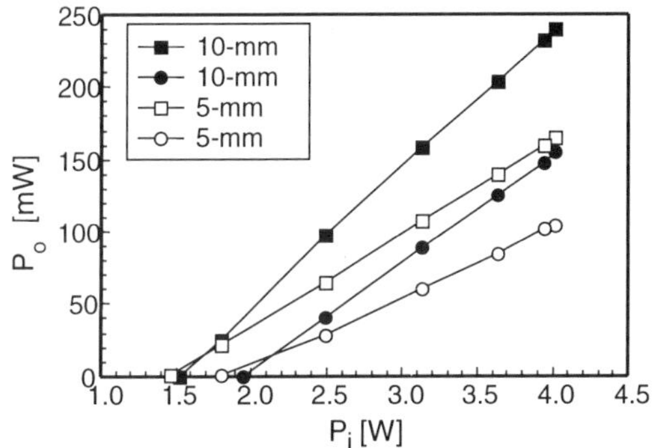

Fig. 2. DP vs. SP with 5-mm and 10-mm Cr:Forsterite crystals.

A maximum output power of 240 mW and 13% slope efficiency was obtained with 3.84 W of incident pump power on the 10-mm crystal in the DP configuration and the 3.5% OC. In DP, the 5-mm crystal generated 165 mW, 31% less than the longer crystal, with nearly the same slope efficiency as the 10-mm crystal in SP. However, in DP the 5-mm crystal did outperform the 10-mm with SP and demonstrated a 25% threshold reduction, as well. Inserting a prism for tuning the output wavelength, we observed smooth operation in the range 1200-1300 nm. We note that the slope efficiency can be approximately doubled by choosing a higher OC, although the laser threshold would increase significantly, as well.

Replacing our OC with a fourth HR flat mirror, we measured a minimum threshold as low a 630 mW (in DP) corresponding to only 330 mW absorbed power. To our knowledge, this is the lowest threshold measured at 10 °C in a Cr:Forsterite laser excited near 1 μm. The output power in this configuration was only few milliwatts. Inserting a couple of prisms (P1-P2) to compensate the group velocity dispersion (GVD) we obtained 180 fs pulses (sech^2 fit), transform limited, at 100 MHz repetition rate with an output power of 100 mW, as shown in figure 3. The mode locking is not self starting; in order to start its operation we inserted in the cavity an acousto-optic modulator, that is turned off as soon as the KLM starts.

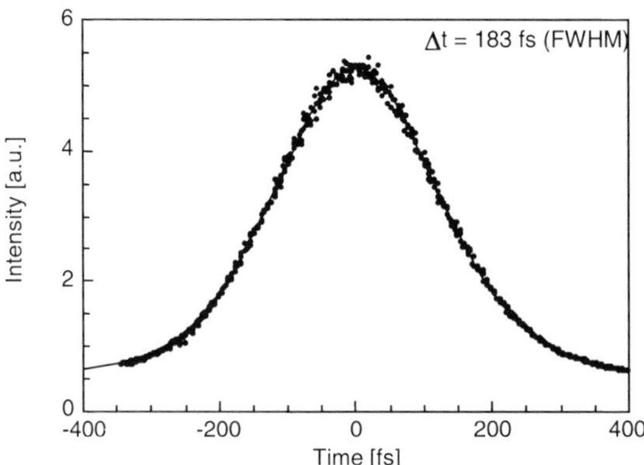

Fig. 3. Auto-correlation pulse. Best fit correspond to $sech^2$ pulse shape.

In conclusion, we have demonstrated a significant threshold reduction in all-solid-state Cr:Forsterite lasers pumped in a DP scheme, which is very interesting for the realization of compact and efficient diode-pumped tunable and ultrafast sources in the near-IR region (1200-1300 nm). The reduction of intracavity losses eliminating the AOM and the reduction of group velocity dispersion obtained using shorter Cr:Forsterite crystals, also allows the generation of shorter pulses.

References:

[1] V. Petricevic, S.K. Gayen, and R.R. Alfano, Appl. Phys. Lett. 53, 2590 (1988)

[2] V. Petricevic, S.K. Gayen, and R.R. Alfano, Opt. Lett. 14, 612 (1989)

[3] V. Yanovsky, Y. Pang, F. Wise, and B.I. Minkov, Opt. Lett. 18, 1541 (1993)

TIME TUNABLE FEMTOSECOND LASERS THROUGH USE OF VARIABLE PULSE COMPRESSORS

M. Werdiger, S. Jackel, B. Arad, S. Eliezer, Z. Henis.

Soreq NRC, Yavne, 81800, Israel.

Abstract

Short pulse high power lasers are in common use in various laboratories. A method is described here which enables one to vary the pulse width of the laser, in a continuous manner over an appreciable range. The variable range spans from the original width of several tens of femtoseconds up to several picoseconds.

1. Introduction

In many ultra-short pulse, high power laser experiments, pulse durations of femtoseconds to picoseconds are necessary. (In most of the laboratories there is a different laser for ps and fs regimes). This wide temporal regime can be obtained by starting with a femtosecond oscillator and amplifying the output with a Chirped Pulse Amplification (CPA) system [1,2]. In a CPA laser, the pulse is first stretched in time, amplified, and then compressed. Stretching and compression is achieved using dispersive gratings. Two gratings are often used in the pulse compressor. Changing the distance between the gratings or the entrance angle to the first grating, continuously changes the time width of the pulse from a minimum value to a large value.

When the stretched pulse, having wavelength in the range λ_1 to λ_2 ($\lambda_2 > \lambda_1$), incidents on the compressor grating (see fig. 1), the dispersion of the grating causes the optical path length of λ_2 to be larger than the path length of λ_1, i. e. λ_1 takes over λ_2. The dispersion of the second grating (parallel to the first one) stops the path length variation. As during the stretching the interplay of λ_2 and λ_1 takes place in the reverse order, the present variation causes the stretched pulse to shorten up to its possible minimum (after which it is stretched again).

A simple model of the behavior of the gratings has been developed and a computer program written. The duration of the compressed pulse was calculated as a function of the distance between the gratings, the grating constant, the entrance angle, the wavelength and bandwidth of the pulse, and the duration of the stretched pulse. The dimensions of the components and their locations in the compressor were calculated. Optical properties of the output beam were also determined. A sensitivity analysis was performed to determine the tolerance to an alignment error. In section 2 we present the calculation of the pulse duration, in section 3 we show some results, and end with a discussion and conclusion.

2. The duration of the compressed pulse.

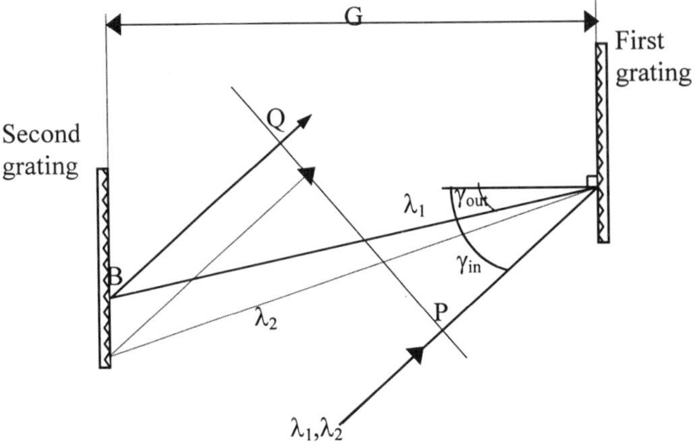

FIG. 1 Geometrical arrangement of the compressor.

Let γ_{in} and γ_{out} be the angles of the incident and the diffracted rays relative to the first grating as shown in fig. 2. The relation between these angles for first -order diffraction is

$$\sin(\gamma_{out}) = \frac{\lambda}{d} - \sin(\gamma_{in}) \tag{1}$$

where λ is the wavelength of the ray and d is the grating constant. The ray path length PABQ for a quasi-monochromatic wave is given by

$$p = b[1+\cos(\gamma_{in}-\gamma_{out})] = G/\cos(\gamma_{out})[1+\cos(\gamma_{in}-\gamma_{out})] = c\tau \tag{2}$$

were b is the slant distance AB between the gratings, G is the perpendicular distance between the gratings, c is the velocity of light and τ is the group delay. (Treacy [3] showed that that $\tau = \frac{\partial\phi(\omega)}{\partial\omega}$ where ϕ is the additional phase caused by the compressor). Differentiating p/c (equ. 1) with respect to λ and using equ. 2 yields:

$$\tau = \tau_0 + \frac{\partial \tau}{\partial \lambda}\Delta\lambda + \ldots \approx \tau_0 + \frac{G\lambda\Delta\lambda}{cd^2[1-(\lambda/d - \sin(\gamma_{in}))^2]^{1.5}} \tag{3}$$

where $\tau_0 = \tau(\lambda_1)$ (from equ. 1), $\Delta\lambda=\lambda_2-\lambda_1$ is the bandwidth of the pulse ($\lambda_2 > \lambda_1$). From the last equation it can be concluded that t, the duration of the pulse after the compressor, is

$$t = t_0 - (\tau - \tau_0) \tag{4}$$

where t_0 is the duration of the pulse before the compressor (assuming a square pulse).

For compression using two gratings and a mirror (the pulse travels twice through the compressor) we get:

$$t = t_0 - \frac{2G\lambda\Delta\lambda}{cd^2[1-(\lambda/d - \sin(\gamma_{in}))^2]^{1.5}} \tag{5}$$

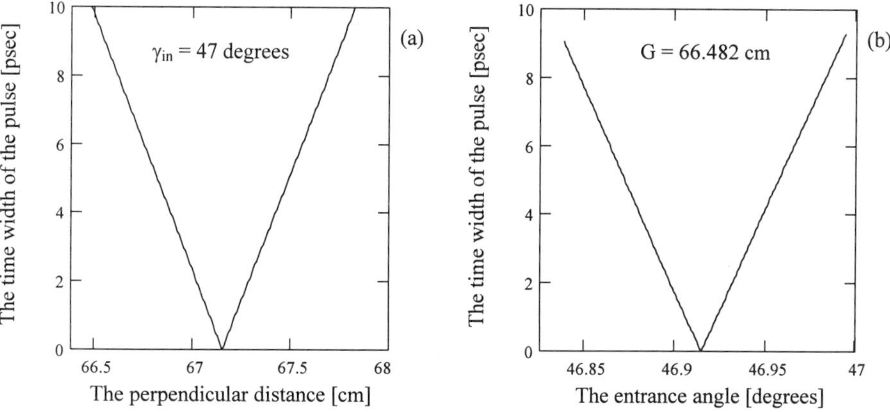

FIG. 2 The dependence of the time width of the compressed pulse for a system with λ_0=0.7953μm, $\Delta\lambda$=9.4nm, 1/d=2000gr/mm (a) on the perpendicular distance between the gratings. (b) on the entrance angle.

In fig. 2a and 2b the dependence of the time width of the compressed pulse on the perpendicular distance between the gratings and on the entrance angle is shown. These are the only parameters that can be changed for a specific system. The results are for a system with 1/d=2000gr/mm, τ_0=1nsec, γ_{in}=47degrees, G=66.482cm, λ_0=0.7953μm, $\Delta\lambda$=9.4nm (appropriate to the duration of a transform limited pulse of 100fs). In our model the pulse can be compressed to zero. The duration is linear with G as can be seen in fig. 2a. For a small change with the duration of the pulse, it is also linear with γ_{in} as seen in fig. 2b. Both figures show that the duration can change from zero up to 10ps.

3. Discussion and conclusions

The optical properties of the output beam (for an input beam with a divergence angle of 1mrad and diameter of 8.33cm) were also analyzed to see if changing G changes the quality of the beam. Some results are shown in table 1. Changing the entrance angle produces the same optical properties of the output beam.

Table 1 The duration and optical properties of the output beam for different values of G

γ_{in} [degrees]	**G** [cm]	**Duration of the pulse** [fsec]	**ellipticity**	**Divergence angle** [mrad]	**ΔG** [μm]
47	67.147	100	1.025	1.058	0
	67.146	109			10
	67.140	199			70
	66.482	10^4			6650

To compare between the methods, calculation of grating changes for a compression equal to the duration of a transform limited pulse (from zero to 100fs) was done and seen in table 2. Changing the entrance angle is so sensitive that it is not practical.

Table 2 Sensitivity analysis for max. compression

Method	**Changing to 100fs**
Changing the spacing	65 μm
Changing the entrance angle	15.71 μrad

We conclude that small changes in the grating separation or of the beam input angle, can change the time width of the pulses continuously from femtoseconds to picoseconds without changing the beam quality. Changing the grating separation is preferred because it requires a lower precision than changing the beam input angle.

Reference

[1] M. D. Perry and G. Mourou, Science **264**, 917 (1994).

[2] J. Squier et. al., Appl. Opt. **37**, 1638 (1998).

[3] E. B. Treacy, IEEE J. Quant. Electr. **5,** 454 (1969).

PART 5

NON-LINEAR OPTICS - QUASI PHASE MATCHING

Chairpersons: *A. Arie, Israel; M. Rosenbluh, Israel*

EFFICIENT RESONANT QUASI-PHASE-MATCHED FREQUENCY DOUBLING WITH PHASE COMPENSATION BY A WEDGED CRYSTAL

I. Juwiler, A. Arie, A. Skliar, and G. Rosenman

Department of Electrical Engineering – Physical Electronics,
Tel-Aviv University, Tel-Aviv 69978 Israel
Email: iritha@post.tau.ac.il, Fax: 972-3-6423508

Abstract

In multiple-pass nonlinear frequency conversion devices, the interacting waves may accumulate different phases owing to dispersive elements in the system. Phase compensation is therefore necessary for efficient frequency conversion. We experimentally demonstrate phase compensation in a compact standing-wave frequency-doubling cavity by using a wedged periodically-poled KTP crystal. The highest conversion efficiency and second harmonic power obtained by pumping with a 1064 nm continuous-wave Nd:YAG laser were 69.4% and 268 mW, respectively.

Resonant frequency doubling[1,2,3] can be used to significantly improve the conversion efficiency of low and medium power laser sources. The simplest configuration for external pump enhancement is the standing-wave Fabry-Perot cavity. However, in this case, second harmonic light is generated in both directions of propagation. One can coherently add the forward and backward generated harmonic waves by using a cavity mirror that reflects both the fundamental and second harmonic waves, so that the frequency-doubled light exits only through the other mirror of the cavity. However, for coherent addition of the two waves, the relative phase between them should be a multiple of 2π. An improper relative phase may completely null the second harmonic output signal.

Recently, a new method was demonstrated for compensating the phase in double-pass quasi-phase-matched (QPM) second harmonic generation[4]. In a dispersive material, the refractive indices of the fundamental and second harmonic waves are different, and as a result, the relative phase between these waves accumulates a difference of π every coherence length, L_c. This phase difference is kept close to zero throughout the interaction region by reversing the sign of the nonlinear susceptibility every L_c. By selecting appropriate width for the last domain of the crystal, the relative phase between the pump and second harmonic waves at the crystal output can be adjusted. A practical method of selecting the width of the last domain is to polish the output facet of the crystal at a small angle with respect to the QPM domain boundary, and to translate the crystal in a direction perpendicular to the beam direction of propagation[4]. As a result, the length of the last domain will vary between 0 and L_c, and the total phase difference in a double-pass configuration will vary between 0 and 2π. In this case, the second harmonic power will vary between 0 and 4 times the single-pass power. In this paper we utilize this technique and experimentally demonstrate phase compensation in resonant frequency doubling using a periodically-poled KTP (PP-KTP) crystal. Part of this work was published in Ref. 5.

The KTP crystal was 10-mm-long, 2-mm-wide and 0.5-mm-thick. Periodic poling with a 9.0 μm period ($= 2\times L_c$) was performed by the technique of low-temperature electric-field poling[6]. The crystal was polished in both faces with a small wedge angle in order to enable phase correction by transverse movement. The input and the output facets of the crystal were antireflection coated for the pump and second harmonic wavelengths.

A microscope picture of the crystal angled facet is shown in Fig. 1. Changing the transverse position by $\Delta x = 0.378$ mm changed the last domain width by a single coherence length. Independent measurements of the last domain width using an optical microscope confirmed this result. From these measurements we deduce that the wedge angle was $\approx L_c/\Delta x \approx 11.9$ millirad.

We have measured the second-harmonic power as a function of the temperature in single-pass and double-pass configurations, see Fig. 2. A higher conversion efficiency was obtained in the case of double-pass configuration, whereas the temperature acceptance bandwidth was half of the single-pass bandwidth, in accordance with the theory[4].

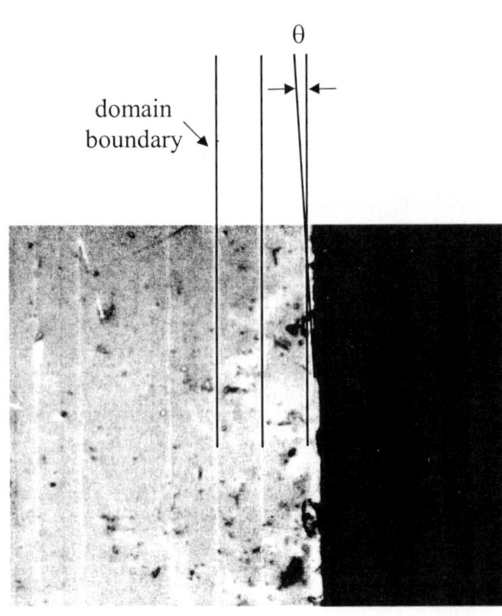

Fig. 1: Microscope picture of the PP-KTP crystal

The experimental setup for resonant frequency doubling is shown in Fig. 3. The pump source was a 1064 nm monolithic single-frequency Nd:YAG ring laser. The standing-wave cavity included an input coupler with transmission coefficients of 5.4% and 90% at 1064 nm and 532 nm, respectively, and a second mirror that reflected both the fundamental and second harmonic waves. The radius of curvature of each of the two mirrors was 2.5 cm, and the physical distance between them was ≈ 5.4 cm. The wedged crystal was positioned at the focal point between the two mirrors. The phase difference between the reflected fundamental and second harmonic waves was controlled by changing the transverse position of the crystal using a mechanical micro-positioner. The linear loss in the cavity was ≈ 2.0%, as determined from measurements of the finesse.

In order to obtain high efficiency, the crystal transverse position was set so that constructive interference was obtained. In this position, the FM-sideband technique[7] was used in order to keep the laser frequency at the cavity resonance frequency. In a measurement which lasted nearly one hour, see Fig. 4, the standard deviation of the second harmonic power with a measurement bandwidth of 7 kHz was 1.2%.

The second-harmonic power and the conversion efficiency as a function of the pump power were measured[5]. The highest internal green power level and conversion efficiency of 268 mW

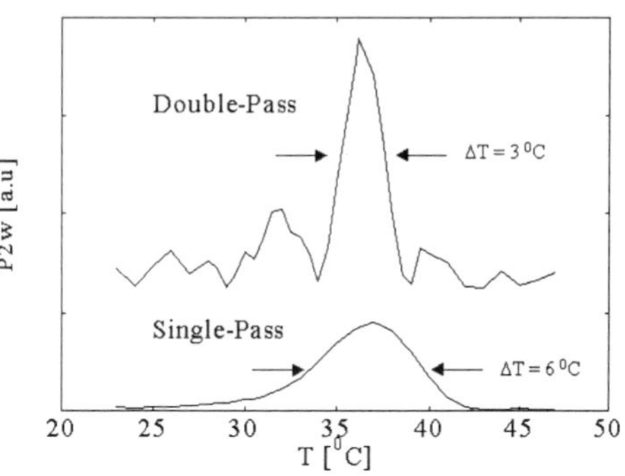

Fig. 2: Single-pass and double-pass second harmonic power as function of the temperature

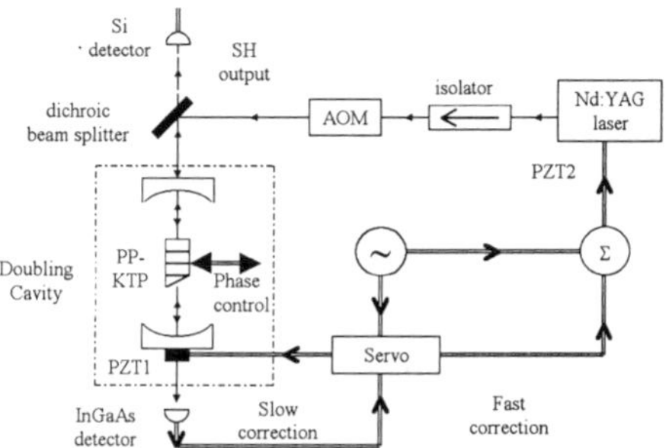

Fig. 3: Experimental setup for standing-wave resonant frequency doubling of Nd:YAG with a wedged PP-KTP crystal

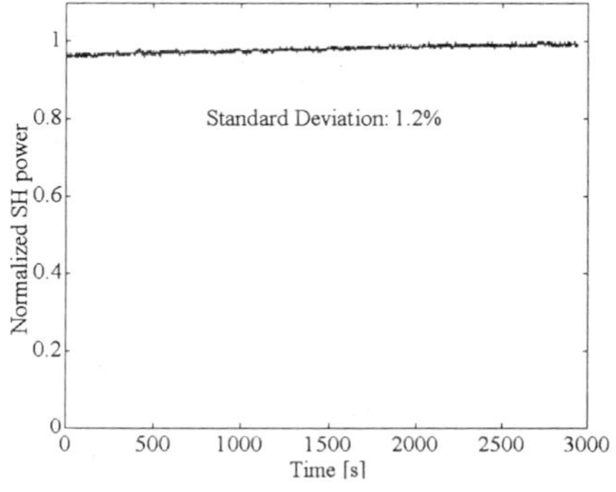

Fig. 4: Stability of the resonant doubler

and 69.4%, respectively, were obtained at the highest pump power of 386 mW. The measured results are in reasonable agreement with a theoretical calculation[8], with parameters that were derived independently: round-trip conversion efficiency of 1.6 $\%W^{-1}$ and loss of 2.0%.

In summary, we have demonstrated the ability to compensate the phase difference between forward and backward second-harmonic waves in a standing-wave cavity by using a wedged PP-KTP crystal. High frequency conversion efficiency was achieved in a compact configuration. Further improvement in efficiency and size can be obtained by using a semi-monolithic cavity[9], in which the input facet of the crystal acts as the input coupler.

This work was partly supported by the Israeli Ministry of Science.

References:

[1] A. Ashkin, G. D. Boyd, and J. M. Dziedzic, IEEE J. Quantum Electron. **2,** 109 (1966).

[2] W. J. Kozlovsky, C. D. Nabors, and R. L. Byer, IEEE J Quantum Electron. **24,** 913 (1988).

[3] A. Arie, G. Rosenman, A. Korenfeld, A. Skliar, M. Oron, M. Katz, and D. Eger, Opt. Lett. **23,** 28 (1998).

[4] G. Imeshev, M. Proctor, and M. M. Fejer, Opt. Lett. **23,** 165 (1998).

[5] I. Juwiler, A. Arie, A. Skliar, and G. Rosenman, Opt. Lett. **24,** 1236 (1999).

[6] G .Rosenman, A. Skliar ,D. Eger, M. Oron, and M. Katz, Appl. Phys. Lett. **73,** 3650 (1998).

[7] R. W. P. Drever, J. L .Hall, F. V. Kowalski, J. Hough, G. M. Ford, A. J. Munley and H. Ward, Appl. Phys. B **31,** 97 (1983).

[8] E .S. Polzik and H. J. Kimble, Opt. Lett. **16,** 1400 (1991).

[9] K. Schneider, S. Schiller, J. Mlynek, M. Bode, and I. Freitag, Opt. Lett. **21,** 1999 (1996).

PART 6

NON-LINEAR OPTICS IN POLYMERS AND ITS APPLICATIONS

Chairperson: *G. Berkovic, Israel*

Annals of the Israel Physical Society, v. 14

ELECTRO-OPTICS and MICROELECTRONICS
Eds: Raphael LAVI and Ehud AZULAY

MEASUREMENT OF THE SECOND ORDER OPTICAL NONLINEARITY OF 1-DIMENSIONAL AND 2-DIMENSIONAL ORGANIC MOLECULES

G. Meshulam[a], G. Berkovic[a] , Z. Kotler[a] A. Ben-Asuly[b], R. Mazor[b], L. Shapiro[b] and V. Khodorkovsky[b]

[a]Photonic Materials Group, Soreq NRC, Yavne 81800, Israel
[b]Department of Chemistry, Ben-Gurion University, Beer-Sheva 84105, Israel.

Abstract

The second order nonlinearity (β) was measured for several structurally similar organic π-conjugated molecules. A unique transition from "1-dimensional" to "2-dimensional" behavior is observed for molecules containing the carbazole donor group. These "2-D" molecules exhibit two β components which are detected by a combination of coherent and incoherent experimental techniques. Another aspect reported is the dependence of β on the frequency; the resonance enhancement of carbazole molecules with two β components is much larger than for the 1-D molecules.

1. Introduction

The second order nonlinear optical response is a sensitive indicator of electronic structure in polar organic molecules that consist of electron donor and electron acceptor groups, linked by a π-conjugation bridge. The second harmonic coefficient (β) can be measured with two techniques, electric field induced second harmonic (EFISH) and Hyper-Rayleigh scattering (HRS). In the EFISH technique, molecules are dissolved in a solvent, an electric field is applied to align the molecules, and coherent second harmonic generation (SHG) is measured along the same direction as the laser. In HRS, no electric field is applied, and scattered second harmonic light is detected. The setups have been described previously [1].

2. Results

Using the EFISH technique, we observed the influence of changing the acceptor to increase the intermolecular charge transfer. As demonstrated in Table 1, the change results in increasing the SHG efficiency of the molecule when better donor groups (CN) are added.

Table 1. EFISH $\mu\beta$ values (10^{-48} esu) measured in CH_2Cl_2 for molecules with different acceptors.

	Molecular formula	$\mu\beta_0$ (10^{-48} esu)
1a	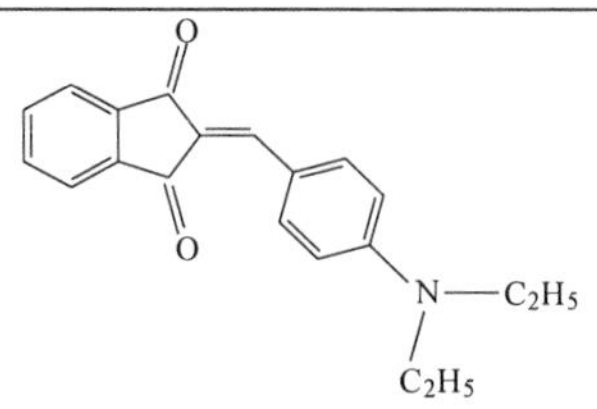 O O N—C_2H_5 C_2H_5	170
1b	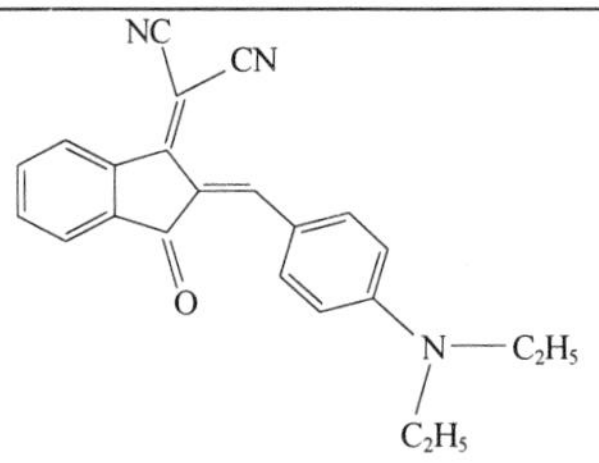 NC CN O N—C_2H_5 C_2H_5	220
1c	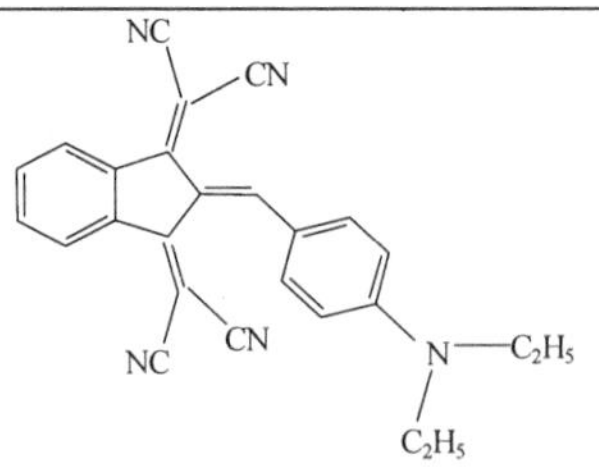 NC CN NC CN N—C_2H_5 C_2H_5	310

When we varied the donor group in these type of molecules, we found a completely different behavior. Increasing the length of R (Table 2) caused little difference in the β value, but a carbazole in the donor site resulted in a drastic reduction. This effect characterizes all the carbazole molecules that we studied [1]. In order to understand why is β so low for carbazole, we used a standard quantum chemical program to calculate the expected behavior of β. The calculations suggested [1] that in typical molecules like **1a-1f** only one significant β component is developed along the long axis of the molecule. On the other hand, molecule **2a** was calculated to show β components in two dimensions: a positive β tensor component β_{zzz} along the long molecular axis, and a negative component β_{zxx} perpendicular to it.

Table 2. EFISH μβ values (10^{-48} esu) measured in CH_2Cl_2 for molecules with different donors

Dialkylamino donor		**Carbazole donor**	
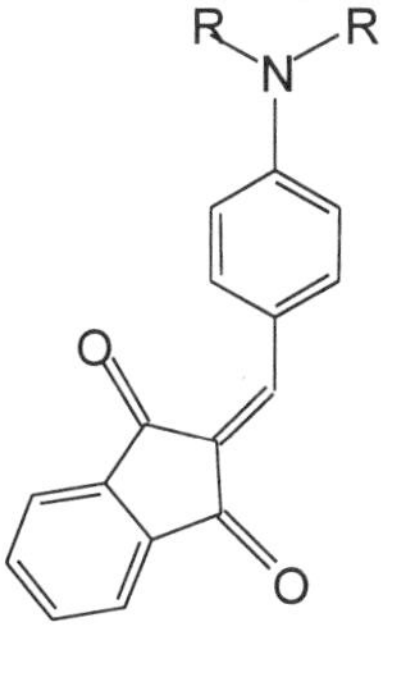	**1d**: R = CH_3 $\mu\beta_0$ = 130 **1a** : R = C_2H_5 $\mu\beta_0$ = 170 **1e** : R = C_4H_9 $\mu\beta_0$ = 150 **1f** : R = C_6H_4 CH_3 $\mu\beta_0$ = 150	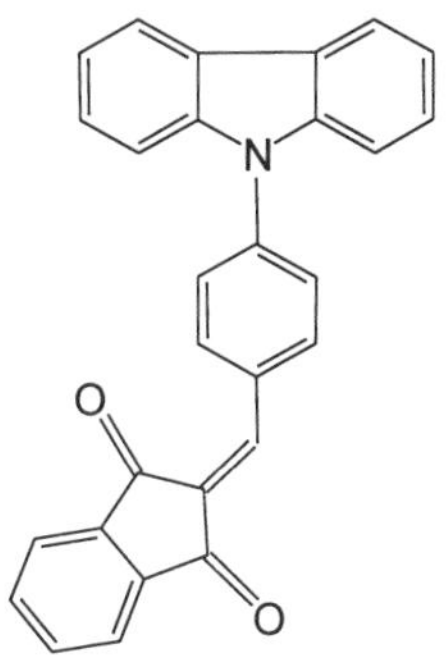	**2a**: $-8<\mu\beta_0<0$

Since EFISH measure the sum of β coefficients [2] the positive and negative terms will cancel out. On the other hand, HRS measures the sum of squares of β coefficients, so we still expect a strong signal if one term is negative and the second contribution is positive. And this is indeed observed. As presented in Table 1, the EFISH signal for the 2-D molecule **2a** is negligible compared to that of the corresponding 1-D molecule **1a**. However the HRS signal of these molecules are about the same [1].

The exact formulas [2] relating the two β components to the EFISH and HRS signals are shown in equations (1) and (2). This allows us to extract the values of the two components.

$$\beta^{EFISH} = \beta_{zzz} + \beta_{zxx} \quad (1)$$

$$<\beta^{HRS}>^2 = \frac{6}{35}\beta_{zzz}^{\;2} + \frac{16}{105}\beta_{zzz}\beta_{zxx} + \frac{38}{105}\beta_{zxx}^{\;2} \quad (2)$$

Analyzing the results obtained from the two experiments shows the following : the 1-D molecules have one dominant component, β_{zzz}, equal to about 50×10^{-30} esu for molecule **1a** (which has μ=3.7 Debye). For the 2-D molecule **2a**, there are two opposite components, a positive β_{zzz} and negative β_{zxx}

having approximately the same absolute value. These results agree very well with the quantum calculations.

3. Resonance enhancement

The dependence of β on the frequency has a significantly different behavior for the 1-D and the 2-D molecules. The results presented above were measured at low frequencies where 2ω of the laser is much lower than the absorption energy of the molecule. We have also measured *resonant* EFISH β values using a laser frequency whose harmonic frequency coincides with the energy of the lowest molecular excited state.

For typical organic molecules in solution, the absorption peaks are rather broad. For 1-D molecules, theory predicts [3] that the ratio between β measured on resonance to β measured far from resonance equals the ratio of the resonance frequency to the half width at half maximum of the peak (typically about 15 for these molecules). Indeed, for an analog of **1a,** we experimentally measured 100 for β_0 out of resonance and 1750 for β at resonance [3].

For the 2-D molecule **2a** (μ=2 Debye) we observed, as predicted, a much larger enhancement. Whereas the amplitude of β_0 is at most 5, we find a resonance β value of around 1000, an enhancement of 200. This result shows another aspect of the unique behavior of the 2-D molecules that distinguishes them from the 1-D molecules.

References:

[1] G. Meshulam, G. Berkovic , Z. Kotler A. Ben-Asuly, R. Mazor, L. Shapiro and V. Khodorkovsky, Proc. SPIE 3796, 279, 1999.

[2] C. Boutton, K. Clays, A. Persoons, T. Wada and H. Sasabe, Chem. Phys. Lett. 286, 101, 1998.

[3] G. Berkovic, G. Meshulam, Z. Kotler to be published.

PART 7

OPO AND NONLINEAR MATERIALS

Chairperson: *A. Ben-Amari, Israel*

MID-IR LONG PULSE GENERATION WITH INTRA-CAVITY PPLN OPO

Yosi Ehrlich, Idan Paiss and Raphy Lavi

Non-Linear Optics Group, Electro-Optics Div. Soreq NRC, Yavne 81800, Israel.
Phone: 972 8 9434651 Fax: 972 8 9434401 Email: Ehrlich@ndc.soreq.gov.il

Abstract

We propose a new approach for mid-IR wavelength generation, using a free-running pump laser and an intra-cavity OPO based on a periodically poled crystal. This configuration takes advantage of the intense fluence inside the cavity, and the large effective interaction length that is due to many round trips of the pump inside the OPO cavity. Using a PPLN OPO placed in a diode pumped Nd:YAG free running resonator, mid-IR pulses were generated with idler energies of ~1mJ/Pulse. The pulse duration was a few hundreds μsec. The spectral width was 6-7 times narrower than that of a similar extra-cavity OPO.

1. Introduction

Periodically poled crystals with a high non-linear coefficient had lowered significantly the threshold for parametric processes and increased their efficiency. Highly efficient, extra-cavity optical parametric oscilators (OPO) based on periodically poled Lithium Niobate (PPLN) crystals, operating in either Q-switched or continuos-wave mode, have already been demonstrated [1,2]. Presently available PPLN crystal are very thin (0.5 – 1mm), and have moderate damage threshold (about 100MW/cm^2 for 10nsec pulses). Therefore, the allowable pump energy in Q-switched mode is low, and the pulse energy of the generated mid-IR radiation is accordingly low.

In order to overcome the input energy limitation of the PPLN crystals one can use a long pumping pulse. A free-running laser pumping an extra-cavity OPO is problematic due to the relatively low peak power and the "spiked" temporal behavior. On the other hand, free-running intra-cavity pumping of a periodically poled nonlinear crystal is an attractive method for parametric conversion because of the following reasons: 1) The intense fluence inside the cavity allows low power pump lasers. 2) The improved efficiency owing to the many round trips of the pump inside the OPO cavity. 3) The parametric conversion acts as a power dependent attenuator inside the pump cavity, and as such, suppresses the pump's high intensity spikes to a desired continuous level just above the OPO threshold [3]. This configuration has already been demonstrated [4], but those early results suffered from "spiked" temporal behavior and poor conversion efficiency.

In this paper we present experimental results of wavelength conversion by a free-runing intra-cavity PPLN OPO. Those results show the advantage of the proposed method by means of high output pulse energy and high conversion efficiency, improved temporal behavior and narrow output linewidth.

2. Experiment

The experimental setup is shown schematically in fig. 1. The pump cavity was a conventional plano-concave hemispherical resonator. The lasing medium was a 2.5mm diameter Nd:YAG rod, side pumped by a 4-bar quasi-CW diodes array (SDL 3254-A4) in a close coupled configuration. The pulse duration of the diodes' current was up to 400μsec, at a rate of 40hz. Both ends of the rod were cut at the Brewster angle, which define the polarization state of the pump beam. The YAG rod was located near the curved mirror at the optimal position for TEM_{00} mode extraction. The PPLN crystal was placed near the flat output mirror, and the cavity length was adjusted to reduce the pump waist in order to fit the 0.5mm wide PPLN crystal. The PPLN crystal was heated to 170°C in order to avoid photo-refractive damage.

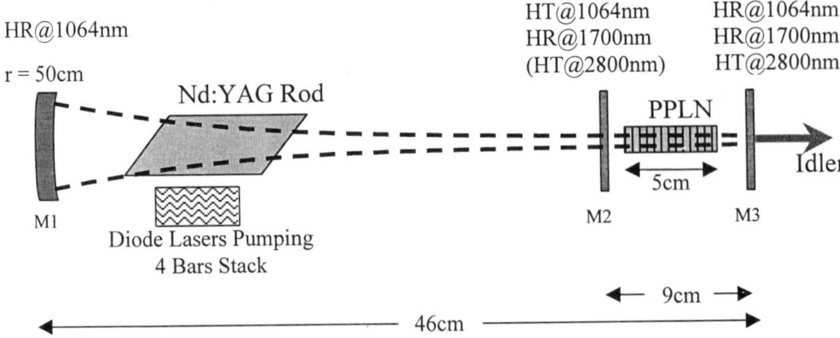

FIG. 1 Schematic layout of the intra-cavity OPO system.

The OPO cavity (M2 and M3 mirrors) resonated the signal wavelength at 1700nm – 1950nm. The idler wavelength at 2800nm – 2300nm escaped from both mirrors (T≈99% at M3, T≈90% at M2), but theenergy measurements refer to the idler output from M3 only.

3. Results

The diode to idler energy conversion efficiency is shown on fig. 2. For 200μsec pulse duration, the slope efficiency from diodes to idler at 2780nm was 2%. For 400μsec pulses, the slope efficiency was about 3.6% which is equivalent to 9.5% signal+idler efficiency (these values exclude the idler energy escaping out of mirror M2, therefore the total efficiency is higher).

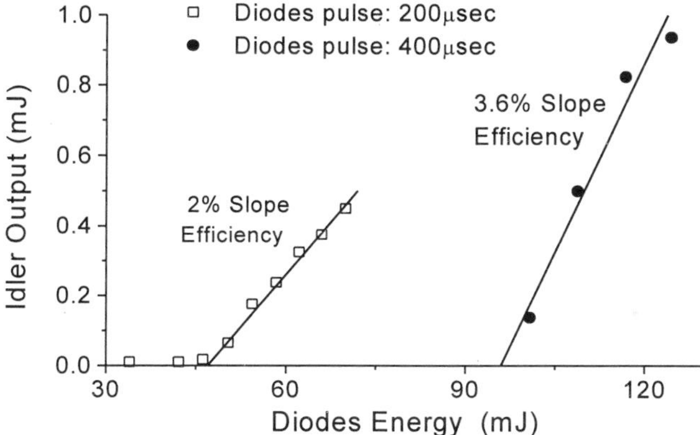

FIG. 2 Conversion efficiency of diodes light into idler output at 2783nm.

The temporal behavior of the pump, the signal and the idler has shown much better stability compared to the extra-cavity free-running configuration. Fig 3 presents the temporal behavior of the pump and the signal (measured by a fast InGaAs photodiode). During the first few μsecs of the pulse there are a few spikes of the pump, producing high signal spikes. Later on the pump intensity becomes stable and the signal intensity rises slowly approaching a steady state level.

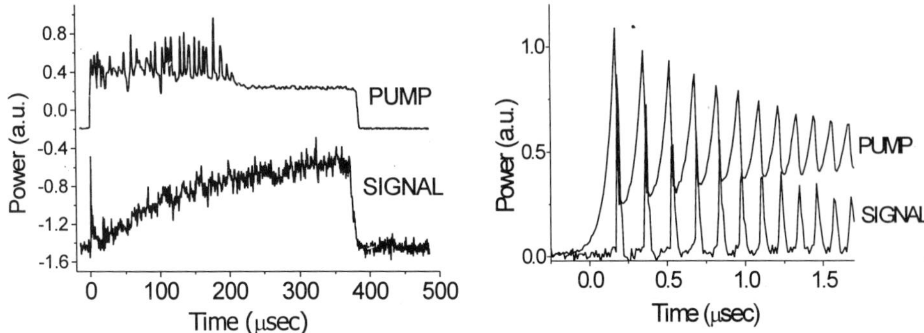

FIG. 3 Temporal behavior of pump at 1064nm and signal at 1724nm. Two measurements at two different time scales.

The spectral linewidth of the idler output was measured in both intra-cavity and extra-cavity configurations. The intra-cavity idler linewidth at 2780nm was 4.5 cm^{-1}, which is 6-7 times narrower than the extra-cavity OPO output at similar conditions.

4. Conclusion

Intra-cavity free-running pumped OPO, based on a periodically poled $LiNbO_3$ crystal, is a simple and efficient configuration for mid-IR wavelength generation. The described experiment shows high idler output pulse energy, great improvement of temporal behavior over a usual free-running laser, and narrower signal and idler linewidth than in the extra-cavity configuration.

References:

[1] L.E. Myers, R.C. Eckardt, M.M. Fejer, R.L. Byer, W.R. Bosenberg and J.W. Pierce, J. Opt. Soc. Am. B **12**, 2102 (1995).

[2] R.W. Bosenberg, A. Drobshoff, J.I. Alexander, L.E. Myers and R.L. Byer, Opt. Lett. **21**, 713 (1996).

[3] R. Lavi, A. Englander and R. Lallouz, Opt. Lett. **21**, 800, (1996).

[4] I. Paiss, A. englander and R. Lavi, OSA TOPS Vol. 19, Advanced Solid State Lasers, 253 (1998).

SIMPLE ANALYTICAL SOLUTIONS FOR SINGLY RESONANT OPO ENGINEERING

Er'el Granot, Shaul Pearl and Michael.M. Tilleman

Electrooptics Department, Soreq Nuclear Research Center, Yavne 81800, Israel.

Abstract

We present a simple analytical solution for singly resonant OPO. The present analysis permits calculating the depletion efficiency of the OPO even when the resonated signal (or idler) suffers from strong intracavity losses. To the best of our knowledge, this is the first model, which yields an analytical formula in the *depleted* signal case. The model can be a useful tool to design cavity mirror reflectivity for a given pump intensity and intracavity losses.

Although more than three decades have past since the first OPO invention there is no model, to the best of our knowledge that predicts the efficiency of an OPO analytically for any mirror reflectivity or losses. In the present work the case of a singly resonant OPO is theoretically analyzed. The analysis considers a single pass pump configuration, and assumes that there is no signal, idler, or pump absorption in the nonlinear crystal. A simple formalism, resulting in a relatively simple analytic expression for the SRO OPO efficiency as a function of the pump intensity, is developed.

We begin with the treatment of a continuos wave (CW) SRO OPO. One of the cavity mirrors is perfectly reflecting for the signal, while the output coupler has reflectance coefficient R at the signal wavelength. Both mirrors are completely transparent to both the pump and the idler wavelengths. Due to the CW nature of the pump beam, within each pass inside the resonator the pump intensity at one side of the crystal (say, the left one) is a constant, while the idler initiates from noise.

We follow the analysis put forth by Baumgartner and Byer [1]. In their notation $I_1(z), I_2(z)$ and $I_3(z)$ are the intensities of the signal, idler and the pump waves with the optical frequencies $\omega_1, \omega_2, \omega_3$

respectively. Denoting the fraction of the signal remaining inside the resonator after complete round trip by $\ell = R(1-l)$ where l is the scattering loss, the steady state implies: $I_1(0) = \ell I_1(L)$ which is valid because the cavity mirrors are completely transparent to the pump and the idler. Practically, the interesting and plausible case is where the phase mismatch is very small, i.e., $\Delta k = k_3 - k_1 - k_2 \cong 0$.

$I_1(L)$ can be obtained directly from Ref.1, and thus, the steady state relation, after some tedious albeit straightforward calculations, reads

$$\sqrt{I_3 / I_0} = \frac{1}{\sqrt{1+x}}\left[F\left(\cos^{-1}\sqrt{\left(\ell^{-1}-1\right)x},\gamma\right) + K(\gamma)\right]. \qquad (1)$$

where x is an internal conversion efficiency defined by:

$x \equiv I_1(0)\omega_3 / I_3(0)\omega_1$, $\gamma^2 = 1/(1+x)$ and $I_0 \equiv \left(\lambda_1\lambda_2\varepsilon_0 c n_1 n_2 n_3^2\right)/8\left(d_{eff}\pi L\right)^2$

where the λ's and the n's are the wavelengths and the corresponding refractive indices of the three interacting waves, d_{eff} is the nonlinear crystal coefficient, and finally ε_0 is the dielectric coefficient in vacuum. $K(\gamma)$ and $F(\phi,\gamma)$ are the complete elliptic integral and the standard elliptic integral respectively. Eq.(1) is a simple analytical expression, which relates the internal conversion efficiency x to the pump intensity I_3 by only two parameters, I_0 and ℓ.

The threshold intensity is achieved by evaluating Eq.(1) for $x \to 0$, which yields $I_3^{th} = I_0\left(\cosh^{-1}\sqrt{1/\ell}\right)^2$. Maximum internal efficiency (x_{max}) is achieved where dI_3/dx (from Eq.(1)) diverges. It occurs for $x = x_{max} \equiv \ell/(1-\ell)$, a value corresponding to the maximum conversion efficiency (outside the cavity):

$$\eta_{max} = x_{max}(1-R)/\ell = [1-R]/[1-R(1-l)].$$

By substituting x_{max} in Eq.(1), one can obtain the pump intensity that yields the maximum conversion efficiency (as a function of pump power)

$$I_3^{max} = I_0(1-\ell)K^2\left(\sqrt{1-\ell}\right) \qquad (2)$$

Therefore, $1 \le I_3^{max} / I_3^{th} \le \pi^2/4 \cong 2.46$. Eq.(2) is plotted in Fig.1.

In the regime $\ell \cong 1$ (i.e., $R \cong 1$ and no scattering loss) $I_3^{max} / I_3^{th} \cong (\pi/2)^2$ is almost a constant, and the only main difference with the non-depleted signal approximation is the value of the maximum conversion efficiency η_{max}, which takes also the loss into account.

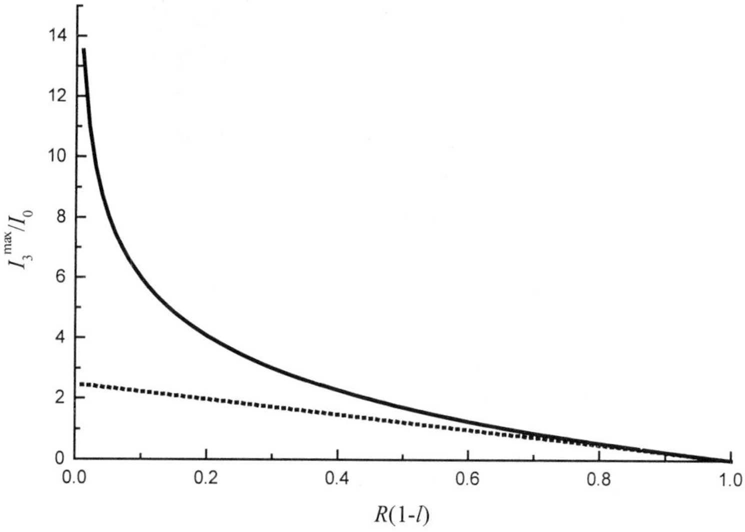

FIG.1: The pump intensity, which is required to obtain maximum efficiency for a given mirror reflectivity and loss, is represented by the solid curve, while the dotted gray one represents the non-depleted signal approximation.

The conversion efficiency in the Gaussian case is obtained by integrating the plane-wave efficiency $\eta(I_3)$ over the cross section of the Gaussian beam. After changing the integration variable one obtains, $\eta_{Gaussian} = \frac{1}{I_3^0}\int_{I_3^{th}}^{I_3^0} dI_3 \eta(I_3)$ where the Gaussian intensity profile is $I_3 = I_3^0 \exp\left[-(r/\rho)^2\right]$ (while r is the transversal radial coordinate and ρ corresponds to the beam's radius). In order to implement this relation, one should invert Eq.1 for $N \equiv I_3^0 / I_3^{th} < 10$ to $\eta(N) \cong 2.67\eta_{max}(\ln N)^{0.83}/N$. Integration of this expression is readily obtained $\eta_{Gaussian} = 1.459\eta_{max}(\ln N)^{1.83}/N$. This simple efficiency expression yields a maximum efficiency of $\eta_{Gaussian}^{max} \cong 0.7\eta_{max}$ at $N_{max} \cong 6.2$. That is, the efficiency of an SRO OPO, which is pumped by a Gaussian beam, cannot exceeds 70%, independent of how low the losses are.

Similarly, for a very long pulse, which has a Gaussian shape, the efficiency is reduced to $\eta_{pulse} = 0.9\eta_{max}(\ln N)^{2.33}/N$. This efficiency function reaches a maximal value of $\eta_{pulse}^{max} = 0.63\eta_{max}$ at

$N \cong 10.3$, independently of the pulsewidth. It is important to note that the limit $\tau \to \infty$ is not identical with the cw case. This is evident from the fact that the average intensity of a Gaussian beam is for every τ lower than its peak intensity, while they are identical in the cw case (for further discussions see Ref.2).

One of the major drawbacks of Optical Parametric Oscillators (OPO) is their wide spectral width. In order to show that these findings are applicable even for a multimode OPO, we adopt the coupled differential equations scheme took forth by Cassedy and Jain [3]. We also use the assumption that for $t \to \infty$ (when the OPO almost saturates) the signal and idler modes can be expressed by $S_n = S_0 \exp(-\Gamma_n t)$ and $I_n = I_0 \exp(-\Gamma_n t)$ respectively, where n is the mode number. After calculating the decay coefficients Γ_n, it is easy to show that $|S_0|^2 \sim t^{1/2}$ and the spectral width $\Delta\omega \sim t^{-1/2}$ while the total signal intensity $|S|^2 = \int dn |S_0|^2 \exp(-2\Gamma_n t)$ is time independent and is exactly the same expression as if only a single mode is present.

When putting an interferometer inside the cavity the depletion coefficient should be modified to include the losses caused by the interferometer (for specific wavelengths). This simple treatment allows calculating the dynamics of the mode in the interferometer method of spectral narrowing.

We would like to thank Dr. Y. Tzuk for helpful discussions.

References:

[1] R. Baumgartner and R. Byer, IEEE J. Quan. Elect. **QE-15**, 432-444 (1979).

[2] E. Granot, S. Pearl, and M.M. Tilleman, J. Opt. Soc. Am. B, in press.

[3] E.S. Cassedy, and M. Jain, IEEE J., QE-15, 1290-1301 (1979).

REMOTE GAS DETECTION USING TUNABLE MID-IR OPO

Idan Paiss, Shlomo Fastig, Yosi Ehrlich, and Raphy Lavi

Electro-Optic Div. Soreq NRC, Yavne 81800, Israel
Tel:08-9434747, Fax:08-9434401, idan@ndc.soreq.gov.il

Abstract

In this work we demonstrate a relatively simple and reliable two stages OPO system that can generate ~10mJ pulses in the 1.5-4μm wavelength range. The system was used for remote detection of CO gas with the OPO tuned on and off one of the absorption lines in the 2-0 vibrational band, near 2.3 μm.

1. Introduction

Tunable laser sources based on narrow-linewidth Optical Parametric Oscillator (OPO) devices can be used in Lidar systems to detect various gas species over broad spectral ranges.[1] However, the use of such devices is limited by difficulties in generating energetic narrow linewidth pulses. By separating the OPO into two parts: narrow linewidth Master Oscillator (MO) and high gain parametric amplifier, one can achieve both good spectral features and high energy output. In this work we describe a two-stages OPO system that operated in the 2-2.5μm IR range. Such a source enables measurements of CO and Methane overtone bands. The spectral range can be expanded up to 4μm by simple modifications.

2. Experimental setup

The transmitter was based on a master oscillator power oscillator (MOPO) configuration. It was pumped by a Q-switched single longitudinal mode (SLM) Nd:YAG laser (Continuum model Powerlite 8000). The first stage MO was a narrow-linewidth KTP OPO with an intra-cavity etalon, pumped by ~60mJ of the second harmonic at 532nm. The signal in the visible range was used for monitoring the wavelength and the spectral shape of the pulse. The line-shape consisted of 3-5 closely spaced longitudinal modes within an envelope of about 0.15 cm^-

[1] FWHM. The idler from the MO was used as a seeder to the second stage power oscillator, a ring $LiNbO_3$ OPO , pumped by ~150mJ of the fundamental Nd:YAG wavelength. The narrow linewidth output energy in the mid-IR was 5-10mJ. A schematic layout of the transmitter is shown in Fig.1.

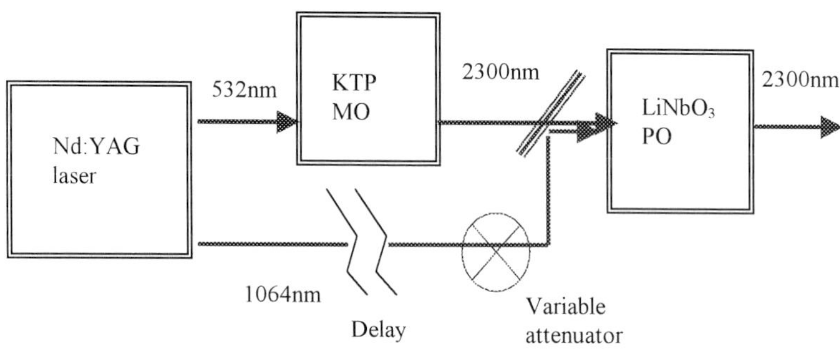

Figure 1: Schematic layout of the OPO transmitter

The output beam was sampled by two pyroelectric detectors, one of them behind a CO gas cell. The OPO was tuned to the 2-0 absorption band of CO at 2.3 μm, and was matched to one of the strongest rotational lines in the R branch (R(4)-R(9)), which was clear from water vapor disruption. The ratio between the pyroelectric detectors was used to measure the OPO line position relative to the absorption line of the gas. We used 10cm long gas cell with 500Torr of pure CO. The spectral width of the self-broadened lines was ~ 0.09 cm^{-1} FWHM. [2] The maximal absorption through the cell was ~ 70%, corresponding to an effective absorption cross section of about $5 \cdot 10^{-21}$ cm^2, about a factor 4 less than the theoretical line absorption peak calculated by the Hitran data base. [3]

The laser beam was transmitted from our lab, aiming at few remote ground targets at distances of 150-700 meters. The back-reflected energy was measured by a InGaAs fast detector sensitive up to 2.6μm. The receiver aperture was 40mm in diameter. The signals from the output sampling and the receiver were sampled by an A/D card and processed by a computer. The presence of the gas in the atmosphere was simulated by placing a small CO gas cell in the path of the beam, in front of a corner-cube retro-reflector. A layout of the Lidar measurement setup is shown schematically in Fig. 2.

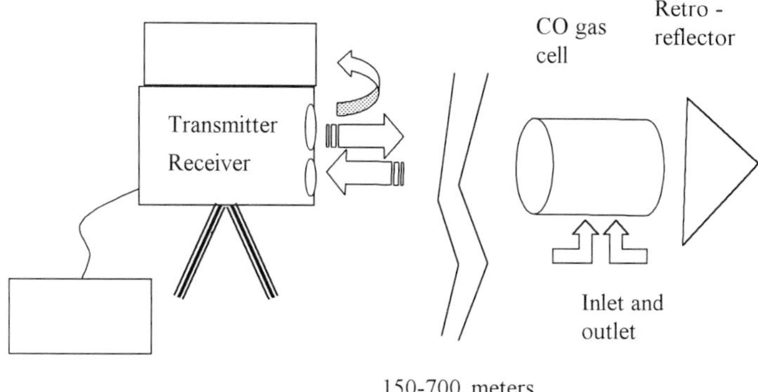

Figure 2: Schematic layout of the measurement setup

3. Experimental results

Fig.3 shows the absorption coefficient of the atmosphere and the simulation cell as a function of the absorption of the reference cell. The transmitted wavelength was tuned by the etalon in and out of the absorption line center. The gas concentration in the simulation cell was proportional to the slope of the curve, and could be calculated by an adequate look-up table. It is shown that the sensitivity of the measurement is in the order of 0.1 ppm*km.

4. Conclusions

The configuration of narrow linewidth KTP Master OPO, seeding a $LiNbO_3$ Power OPO, is useful for generating narrow linewidth, high energy pulses. We have demonstrated detection of light molecules in the atmosphere by measuring the absorption of a single rotation line. The same configuration can be used for different wavelengths in the 1.5- 4μm range, using crystals cut at appropriate angles and suitable dichroic coatings.

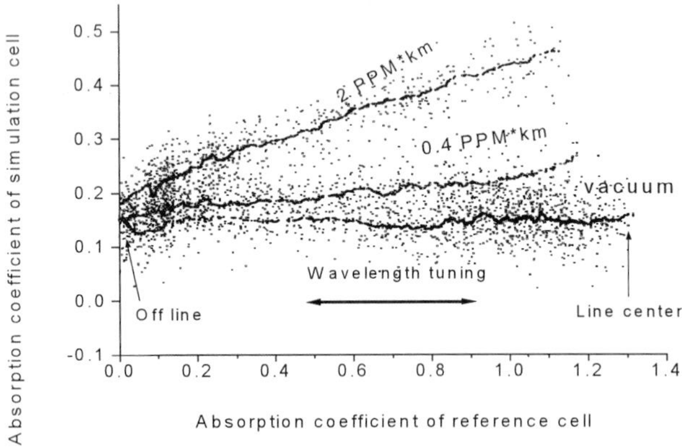

Figure 3: Atmospheric absorption for several values of CO integral densities as a function of absorption trough calibration cell

References

[1] E. Zanzottera, "Differential Absorption Lidar Techniques in the determination of trace pollutants and physical parameters of the atmosphere", Analytical Chem. 21(4), 279, 1990.

[2] R. H. Hunt, R. A. Toth and and E. K. Plyler, "High resolution determination of the width of self-broadened lines of Carbon Monoxide", J. of Chem. Phys. 49(9), 3909, 1968.

[3] L.S. Rothman et al, "The Hitran molecular database: Editions of 1991 and 1992", J. Quant. Spectrosc. Radiant. Transfer 48, No. 5/6, pp. 469-507, 1992.

DEVELOPMENT OF KTP CRYSTALS FOR LASER APPLICATIONS

M. Roth[1], N. Angert[2], L. Weizman[2], A. Chachehlidse[2], M. Shachman[2], D. Remennikov[2], M. Tseitlin[3] and A. Zharov[3]

[1]*School of Applied Science, The Hebrew University of Jerusalem, Jerusalem 91904, Israel*
[2]*"Raicol Crystals Ltd.", Industrial Zone, Yahud, Israel*
[3]*The Research Institute, College of Judea and Samaris, Ariel 44837, Israel*

Abstract

Correlation between conditions of TSSG of KTP crystals from self-fluxes and their optical uniformity has been studied with respect to the crystal morphology, variation of stoichiometry and formation of point defect. It has been shown that growth with pulling on x-oriented seeds yields single sector and single domain crystals exhibiting ideal transverse uniformity necessary for OPO and electrooptic elements. Growth from the $K_6P_4O_{13}$ self-flux at lower temperatures is found to reduce the concentration of potassium vacancies and their gradients and thus to improve the crystal optical uniformity along the x-direction as well as to suppress the detrimental gray-tracking phenomenon during SHG.

1. Introduction

The ferroelectric potassium titanyl phosphate (KTP) is one of the best nonlinear optical crystals for frequency conversion of 1064 nm Nd:YAG lasers, such as second harmonic generation (SHG) and optical parametric oscillations (OPO). In addition to large optical nonlinearity, its high optical damage threshold and excellent thermal stability has made it the material of choice in laser systems utilizing electrooptic amplitude modulation and Q-switching. KTP belongs to the family of isomorphic compounds with the general composition of $MTiOXO_4$, where X = {P or As} and M = {K, Rb, Tl or Cs (for X = As only)} exhibiting the *mm2* point group symmetry at room temperature[1]. However, many high power industrial and medical applications require further improvement of the bulk KTP crystal quality in terms of chemical homogeneity, defect structure and optical uniformity.

Both in hydrothermal and flux-grown KTP crystals a photochromic damage (gray-tracking) may occur during high peak power 1064 nm SHG[2], presumably due to the excitation of point defects comprising Ti^{3+} ions with adjacent $[K^+]$ vacancies by the frequency doubled green (532 nm) light. A better insight into the chemical composition of the crystals in general and their stoichiometry in particular is, therefore, needed to explore the avenues of reducing the point defect concentration and increasing the durability of KTP frequency doublers. In OPO, a nonlinear interaction of the pump, signal and idler beams is required[3]. This process is most efficient when the three beams overlap spatially and temporally and are in phase over a significant interaction length in the material. Large aperture applications require optical uniformity across the wave propagation plane in order to preserve the beam profile. Crystals with homogeneous distribution of the refractive indices within the entire volume are thus required for OPO as well as for electrooptic applications. In the present work, optical performance of nonlinear optical and electrooptic KTP elements has been studied in close relation to the structural and compositional variation of the crystal properties during its growth from self-fluxes. The role of growth methods in shaping the crystal habit is pointed out with respect to the requirements of optical uniformity in particular devices. Stoichiometry and distribution of point defects along the crystals has been considered in direct relation to the varying self-flux composition. Suggestions for improvement of the optical uniformity of KTP devices are made and the enhancement of their performance is demonstrated.

2. Crystal growth, morphology and stoichiometry

KTP single crystals were grown by the top-seeded solution growth (TSSG) method with and without pulling on crystallographically oriented seeds. A resistance heated furnace (ID 105 mm) equipped with a programmable Eurotherm controller was used. The temperature was controlled within 0.1 °C. Solutions of KTP in the $K_6P_4O_{13}$ (K6) solvent were prepared using the Aldrich 3N purity TiO_2 and KPO_3 obtained by thermal decomposition of KH_2PO_4 (Merck Suprapur). The charge was loaded into a 300 ml Pt crucible and subjected to 24 h soaking with flux homogenization aided by a Pt stirrer. Seed rotation rates varied from 60 to 100 rpm and pulling rates, when applied, from 0.07 to 0.12 mm/h. Ramped reduction of the solution temperature was carried out at a rate of 0.02 to 0.3 °C.

The solubility of KTP in different self-fluxes varies greatly, as well as the crystal morphology[4]. A typical habit of an immersion-seeded KTP crystal grown from the K6 solution is demonstrated in Figure1. Such crystals exhibit fourteen facets belonging to four families of crystallographic planes, namely: {100}, {110}, {011} and {201}. Accordingly, fourteen growth sectors develop simultaneously on the submerged seed. Cutting optical elements across the growth sectors results in their optical nonuniformity, in terms of the distribution of refractive indices, since each type of planes is characterized by its particular growth kinetics and different impurity incorporation mechanisms.

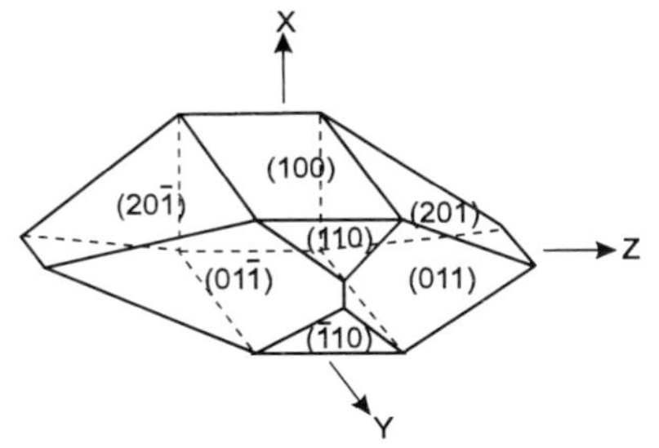

FIG. 1. Typical morphology of KTP crystals grown from the $K_6P_4O_{13}$ flux.

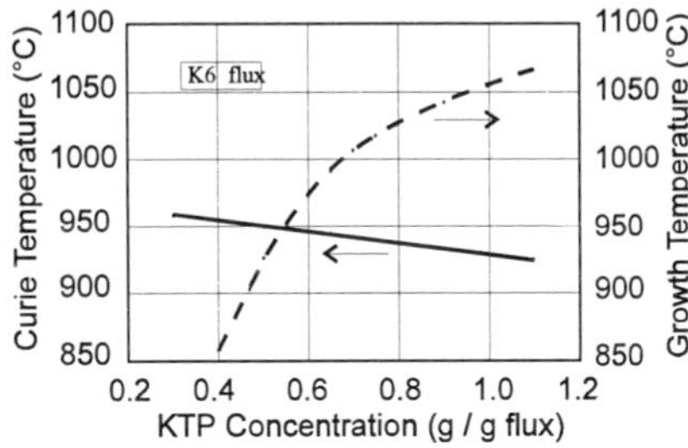

FIG. 2. T_c and growth temperature versus the KTP concentration in the flux.

We have suggested earlier[5] that TSSG on [100]-oriented seeds may yield large single-sector KTP crystals with an additional benefit of planar growth interface, i.e. maximum transverse optical uniformity for X-cut elements. This includes noncritically phasematched OPO elements and electrooptic Q-switches. The latter are not very sensitive to optical uniformity along the laser beam propagation direction[6]. In contrast, nonuniformity of n_x may cause phase mismatch and consequent reduction in the OPO process efficiency. Variation in n_x may be associated with the crystal's gradual compositional changes during growth, since the flux becomes increasingly richer in [K] as the growth temperature decreases. Figure 2 shows the Curie point (T_c) variation of KTP crystals as a function of growth temperature together with the solubility curve. The remarkable T_c increase with growth temperature lowering indicates that the [K] content is very sensitive to the flux composition. We have investigated the KTP stoichiometry using both electron microprobe analysis and T_c measurements on samples synthesized by solid state reactions from powders of variable initial compositions. The results show that KTP solidifies in a relatively broad "homogeneity" range, which does not comprise the stoichiometric composition. Its general chemical formula can be given by $K_{1\pm x}Ti_{1\pm y}PO_{5\pm z}$, where $0 \leq x,y,z \leq 0.02$. Extra potassium ions may fill in the titanium vacancies or occupy interstitial positions and a corresponding optical inhomogeneity may be induced in the crystal. Indeed, a gradual variation in the refractive indices from the seed area to the periphery has been observed during growth of large KTP crystals[7]. The experimental variation of n together with the known mass balance of the material allows to estimate that n_z increases at a rate of 2×10^{-6} °C^{-1}, while $n_{x,y}$ decrease at a rate of 10^{-7} °C^{-1}.

3. Performance of optical elements

If l is the typical crystal dimension, in the case of submerged seed, or volumetric growth, $dn_x/dl \sim l^2$. In the case of one-dimensional growth (TSSG with pulling) $dn_x/dl=const$, or is independent of the pulled crystal length. Naturally, the smaller are the lateral dimensions of the TSSG-pulled crystal the smaller is the refractive index gradient. In practical terms, large optically uniform crystals can be pulled from very large crucibles. We have fabricated long (25-35 mm) OPO elements from such crystals and tested their nonlinear optical performance. High conversion efficiencies of over 30% have been obtained for the eye-safe 1.57 μ signal frequency excited by a Q-switched Nd:YAG (1.06 μ) laser. Use of three 20 mm long elements in a ring resonator has allowed to obtain a 43% conversion efficiency for 7 ns pulses (12 Hz repetition rate) at a 200 MW/cm^2 power density. Apparently, small (and constant) dn_x/dl in the x-direction cause only slight mismatch, well within the acceptance limit of the noncritically phasematched OPO process. It is noteworthy that no refractive index gradient exists in the z-y plane. This explains the high quality of KTP switches obtained approaching an extinction ratio of 300:1 and, unlike $LiNbO_3$, showing no signs of photorefractive damage even at high peak power operation. Crystals grown at lower temperatures have also smaller concentrations of [K] vacancies and, therefore, smaller amounts of defects associated with gray-tracking. As a consequence, we have succeeded to produce very stable gray-track resistant frequency doublers (1064 nm into green) exhibiting a long-term low IR absorption and a conversion efficiency of over 70% at 200 MW/cm^2 power densities.

4. Conclusion

Adequate choice of the seeding and pulling directions as well as thorough control of the crystal chemical composition and stoichiometry can yield optically uniform and almost defect-free KTP crystals.

References:

[1] L. K. Cheng and J. D. Bierlein, *Ferroelectrics* 142 (1995) 209.
[2] B. Boulanger, M. M. Fejer, R. Blachman and P. F. Bordui, *Appl. Phys. Lett.* 65 (1994) 2401.
[3] J. T. Murray, N. Peyghambarian and R. C. Powell, *Optical Materials* 4 (1994) 55.
[4] G. M. Loiacono, T. F. McGee and G. Kostecky, *J. Crystal Growth* 104 (1990) 389.
[5] N. Angert, L. Kaplun, M. Tseitlin, E. Yashchin and M. Roth, *J. Crystal Growth* 137 (1994) 116.
[6] C. A. Ebbers and S. P. Velsko, *Appl. Phys. Lett.* 67 (1995) 593.
[7] T. Sasaki, A. Miyamoto, A. Yokotani and S. Nakai, *J. Crystal Growth* 128 (1993) 950.

PART 8

LASERS IN MEDICINE AND BIOLOGY

Chairperson: *B. Ehrenberg, Israel*

EXPERT EVALUATION OF RETINAL PATHOLOGIES USING DIGITAL IMAGE ANALYSIS

N.Yu.Ilyasova, V.G.Baranov, A.V.Ustinov, V.V.Kotlyar, V.A.Soifer

Image Processing Systems Institute, Russian Academy of Sciences,

E.Farberov

NNT Neural Networks Technologies Ltd, Bnei-Brak 51264, Israel

Abstract

We present an expert computerized system for evaluating the probability of developing eye-diseases, the efficiency of treatment and dynamical monitoring. The aim of the development is dual: both scientific research and application of the software complex in clinical practice for conducting early diagnostics, prognostication of development and selection of an optimal treatment for a variety of severe eye-diseases, such as diabetic retinopathy, retina exfoliation, glaucoma. The system is aimed at quantification of pathological changes in a live microvascular system (conjunctiva and retina). It includes original techniques for evaluating morphological parameters of retina microvessels and a software for analyzing bio-microscopic images.

1. Introduction

Fundus is a unique part of the human body where the vascular system is fully accessible to non-invasive observation. The state of retina vascular system is also a most important indicator and prognostication factor with diabetic retinopathy. Visible ophthalmologic changes in retina vessels produce an integral characteristic of retina haemodynamics. In future, the system for automated diagnostics of diabetic retinopathy may be extended to allow automatie diagnosing of other retina diseases. The international market of ophthalmologic equipment offers a wide range of systems for acquisition of extra-high quality digital retina images (Carl Zeiss Jena GmbH, Topcon Imagenet, Ophthalmic Imaging Systems Inc., "SAPI" Ecom, St-Petersburg). Note, however, that the software of the majority of the above systems involves the most common-used techniques for image preprocessing, quality enhancement and marking. In the meantime, the automated system for analyzing retina images we discuss in this paper features a wider and more universal software, allows individual diagnostics stages to be automated and quantitative monitoring of pathological retinal changes to be conducted.

2. Diagnostic parameters of pathological retina pattern

Among the earliest signs of retina vessels affection caused by such diseases as diabetes, artery hypertension, atherosclerosis are the changed ratio of artery-to-vein diameter, local changes in vessel diameters, excessive vessel tortuosity, and others. Measuring the vessel diameter is an important component of early diagnostics and monitoring of the treatment efficiency of retina diseases. But the practical quantification of the state of retina vessels is a challenge: one needs to get high-resolution digital images and reduce the error in measuring the retina vessel diameter. Accurately measuring the diameter is a topical problem since the image sampling affects the measurement process since fairly small image elements are to be evaluated. Because of this, many methods for evaluating the diameter through the generation of the cross-section profile [1,2] introduce great calculation errors. Besides, the familiar methods for evaluating the vessel cross-section prevents the tracing of all vessel diameter variations along the retina vessel route, in the meantime an uneven narrowing of vessel's diameter is an important characteristic of blood flow. In Ref. [3] we describe how to enhance the accuracy of measuring the vessel diameter via constructing various parametric models of vessel approximation. Among other clinically significant indicators that considerably contribute to the construction of an expert estimate of pathology degree and the probability of eye-disease development are statistical vessel parameters, such as ratio of artery-to-vein diameter, diameter irregularity (local artery spasm, clear-cut vein changes), vessel tortuosity, vessel bifurcation angles. It has been substantiated in medicine that it is possible to analyze the state of local haemodynamics using the results of measuring the diameter of arteries and veins, and their ratio in a corresponding segment. The described system for quantification of medical-diagnostic signs of pathology generates a set of initial geometric vessel characteristics in the process of tracing. These include the vessel length and the local parameters vector that characterizes every separate vessel point and is composed of the following elements: coordinates of the route scanning point at each tracing step, local vessel width, direction towards the next point of vessel tracing, the number of vessel bifurcations at the given point. The mathematical model of a vessel fragment is defined by the following functions: $x=x(t)$, $y=y(t)$, $r=r(t)$, $0<t<Lv$, where $x(t)$, $y(t)$ are differentiable functions that describe a center line called the route, $r(t)$ is the vessel thickness function (distance between the route and the vessel boundary, counted along the perpendicular to the vessel), t is the distance from the route initial point, measured along the route; L_v is the route length.

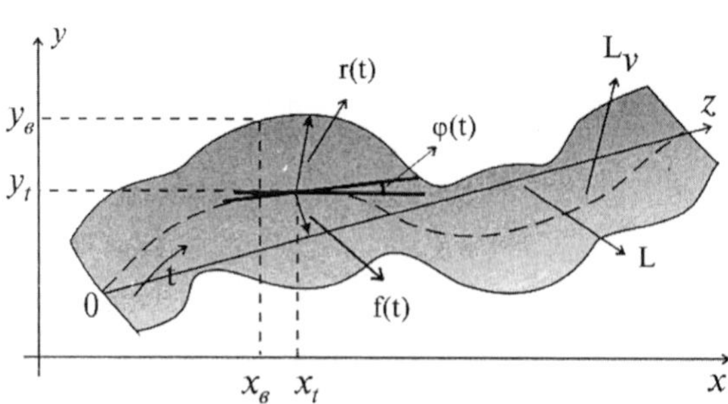

Fig. 1 Mathematical model of the vessel.

The above characteristics uniquely define the route direction function at every point $\varphi(t)$; the local height function $f(t)$, specified by the distance from the route current point to its projection onto the segment L connecting the initial and final points of the route, configuration of the vessel boundaries called walls. These are local characteristics calculated directly in the vascular system image. The set of local parameters accumulated at every tracing step are then used for evaluating the following diagnostic parameters of the vascular system: 1) linear haemodynamics parameter (local vessel diameter); 2) the average vessel diameter on a selected segment; 3) linearity (route deviation from the straight line); 4) beadwise shape (characterizes irregularity of vessel thickness on a measured segment with no regard to the character of changes); 5) route oscillation frequency; 6) route oscillation amplitude; 7) vessel thickness oscillation frequency (defines how many times the vessel wall changes its direction over a unit of length); 8) vessel thickness oscillation amplitude (defines to what degree the wall direction deviates from straight line); 9) route tortuosity (product of amplitude by frequency); 10) vessel wall tortuosity; 11) vessel bifurcation angles.

3. Description of the system for automated analysis of retina images

The developed software complex allows the ophthalmologist-doctor not only to quantify diagnostic parameters of patho-morphological retina pattern, but also to produce a prognosed value of the integral estimate of eye pathology degree. The prognostication is made using regression formulas for every type of pathology, with their coefficients previously derived on the basis of the learning sampling. At the learning stage, it is necessary to have a set of retina images for which the diagnosis has already been made using other diagnostic techniques. When dealing with the learning sampling, the doctor enters the type of pathology and its expert estimate in points (i.e. expressed numerically). It is desirable that the sampling volume be 50-100 measurements. The developed automated system includes: 1) programs for calculating diagnostic signs using digital analysis of retina image; 2) programs for studying dynamics of pathological regions on sequential retina images; 3) data base on calculated diagnostic signs, retina images and clinical information on the patient; 4) a program for automated calculation of pathology degree using expert estimates. The user's shell is intended for MS Windows 98. The system's data base stores the learning sampling and information about the patients. In the course of processing, vessel fragments are selected in the image and the results of their processing are entered into the table. It is possible to assign the results to one or several user-defined vessel groups. The vessel groups are introduced to make diagnostics more accurate, since when deriving the prognosis from the learning sampling it is only information about the chosen vessel group that is used. The results of segment processing contain the segment route and parameters of that segment. The graphic user's interface allows simultaneously viewing the image under analysis (including zoomed fragment), data on the patient and values of calculated diagnostic signs, diagram of vessel caliber variations along the selected segment (see Fig. 2).

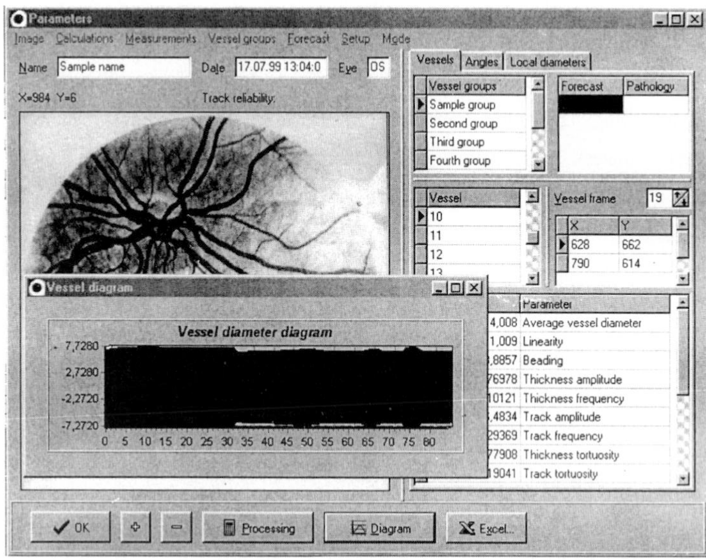

Fig. 2. Graphic user's interface in the diagnostic mode, the vessel diameter diagram.

4. Conclusions

The system may be successfully used in general therapeutical, obstetric and ophthalmologic practice for early diagnostics and evaluating the efficiency of treatment of vascular system diseases. As distinct from competing European analogs, the system enables the analysis of sub-clinical morphological changes on the basis of the original software to be done. The presented system possesses a wider and more universal software packet for image processing, allows diagnostics stages to be automated and quantitative monitoring of pathological retina changes to be conducted. The system features the use of expert system elements: data base on diagnostic signs, correlation-regression analysis with discard of unreliable data, expert-estimate-based prognostication.

5. References

[1] S.L.Brantchevsky, Yu.V.Vasiliev, A.B.Durasov, N.Yu.Iliasova, A.V.Ustinov. Method for the distinguishing and quantitative evaluation of the elements of pathological patterns in the retina (pathology of microciculation). Proceedings SPIE, vol. 2363, pp. 236-242, 1994.

[2] Schwaitzer, S. Guentheretal., International Ophthalmology, vol.16, pp. 251-257, 1992.

[3] S.L. Branchevsky, A.B. Durasov, N.Yu.Iliasova, A.V.Ustinov. Methods for estimating geometric parameters of retinal vessels using diagnostic images of fundus. Proceedings SPIE, vol.3348, pp. 316-325. (1998)

LASER SCANNING THIRD-HARMONIC GENERATION MICROSCOPY IN BIOLOGY

Dvir Yelin and Yaron Silberberg

Department of Physics of Complex Systems, Weizmann Institute of Science, Rehovot 76100, Israel.

Abstract

A laser scanning microscope using third-harmonic generation as a probe is shown to produce high-resolution images of transparent biological specimens. Third harmonic light is generated by a tightly focused short-pulse laser beam and collected point-by-point to form a digital image. Demonstrations with two biological samples are presented. Live neurons in a cell culture are imaged with clear and detailed images, including organelles at the threshold of optical resolution. Internal organelles of yeast cells are also imaged, demonstrating the ability of the technique for cellular and intracellular imaging.

Introduction

In laser scanning microscopy [1], an image is formed by scanning the sample point-by-point with a tightly focused laser beam. The scattered laser light, or, more commonly, fluorescence induced by the laser light is detected and collected to form a digital image. Utilizing nonlinear optical effects, such as two-photon [2] and three-photon [3] fluorescence, significantly improved the depth resolution and reduced the background noise. We have recently proposed and demonstrated a novel nonlinear scanning laser microscope that uses Third-Harmonic Generation (THG) to characterize transparent specimen [4].

In THG microscopy, third harmonic light is generated at the focal point of a tightly focused short-pulse laser beam. When the medium at the focal point is homogenous, the third harmonic waves generated before and after the focal point interfere destructively, resulting in zero net THG [5]. However, when there are inhomogeneities near the focal point, the symmetry along the optical axis breaks and measurable amount of third harmonic is generated. Due to its nonlinear nature, the third harmonic light is generated only in a close proximity to the focal point. Therefore, high lateral resolution can be obtained, allowing THG microscopy to construct three-dimensional images of transparent samples. Since all materials have non-vanishing third-order susceptibilities, THG microscopy can be utilized as a general-purpose microscopy technique. Recently, imaging of live biological samples, as well as imaging of other types of samples was reported by Squier and coworkers [6, 7].

We report here a study of laser scanning THG microscopy for biological imaging. We show that the inhomogeneity inherent to most biological specimen, and, in particular, the internal structures of various cells lead to generation of good quality THG images without any preconditioning such as labeling or staining that might induce undesirable effects in the live cell.

2. Experiment

As an imaging platform, we have used a Zeiss Axiovert-135 microscope, which was modified into a THG microscope. The laser source provides 130 fs pulses at a wavelength of 1.5 μm at a repetition rate of 80 MHz. The laser beam is coupled through one of the microscope ports and is focused into the sample by the microscope objective. The focal point is scanned in the x-y plane using two optical scanners, and along the z-axis using the motorized stage of the microscope. The third harmonic light at the wavelength of 0.5 μm is collected by the original condenser (NA=0.63) and measured by a photomultiplier tube after filtering out the fundamental wavelength using a band-pass interference filter (Center wavelength 500 nm, FWHM = 25 nm). The current generated by the photomultiplier was amplified, digitized and fed into a computer, which synchronizes the scanning process and the data collection.

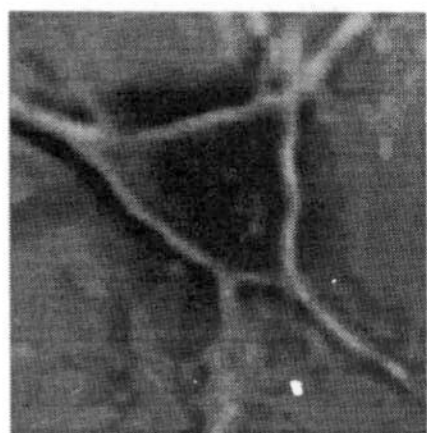
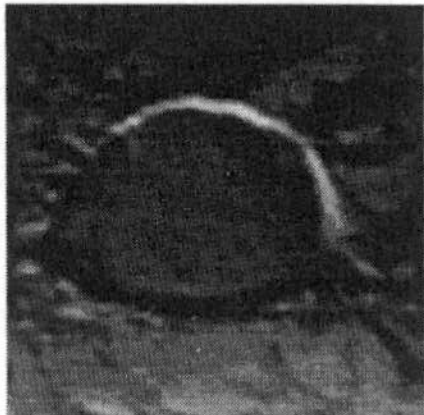

Fig 1. THG images of neurons in a cell culture. The size of the cell's soma is about 15 μm.

We first present THG images of single neurons. The study of single neurons and small networks of neurons in the central nervous system is one of the main tools of neurobiology. High-resolution, non-destructive imaging of the live neurons and their structures without any staining or markers can be very helpful. Shown in Fig. 1 are THG images of two neurons, taken with NA=0.6 (Left) and NA=1.4oil (Right) microscope objectives. The neuron's soma (cell body), the dendrites and the axon can be seen clearly. The plane in which these images were taken was chosen to be very close to the bottom of the cell.

The shadowing effect seen in both images is caused probably by the varying separation between the lower membrane of the neuron and the glass substrate.

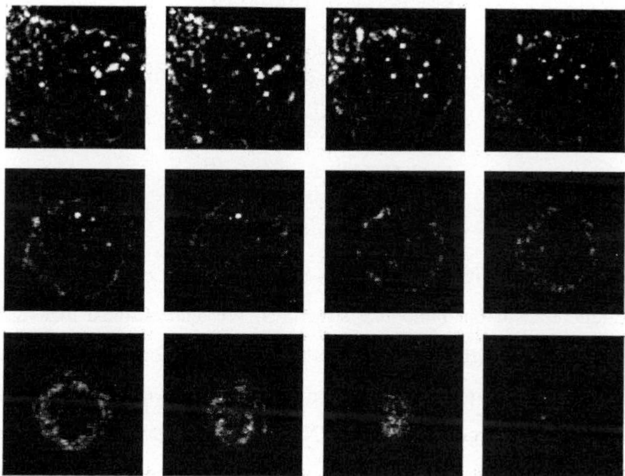

Fig 2. Sectioning of live neurons in a cell culture. Each image is a horizontal section of the neuron's soma. The sections are separated by 0.5 μm, where the top-left section is closer to the glass substrate and the bottom-right section is the top of the cell. The dimensions of each image are 20 x 20 μm.

The ability of the microscope to obtain three-dimensional images of *live* neuron soma is demonstrated in Fig. 2. Twelve consecutive horizontal sections along the axial direction (z-axis), separated 0.5-μm apart, are shown. The top-left section was taken from the bottom of the soma, while the bottom right image is from its very top. The neurons were grown on a layer of Glia cells on a thin cover glass. A strong THG signal from the Glia cells can be seen at the lower sections. Note that very strong THG signals from certain organelles within the cell saturate the detector. A large variety of organelles in the soma can be observed. The nucleus is the dark round region inside the soma. The nucleolus is easily observed inside the nucleus. We could not observe any kind of change or damage to the cell even after tens of scans, although the effect of the laser beam on a living cell should be studied more carefully.

Yeast cells (3-4 μm in diameter) in the middle of the budding process are imaged in Fig 3. Internal organelles and the cell's membrane are clearly noticed. We note that some organelles in these yeast cells appear much brighter than the rest of the structure.

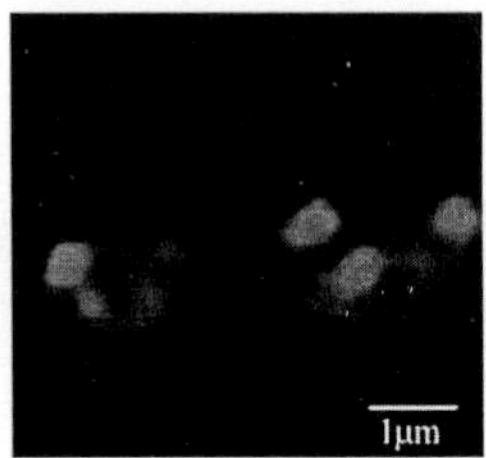

Fig 3. THG images of yeast cell.

3. Conclusions

In conclusion, laser scanning THG microscopy is shown to yield clear three-dimensional images of cellular and sub-cellular transparent biological structures. We demonstrated that this method could be used to identify even minute biological organelles, at the threshold of optical resolution.

High repetition rate short-pulse laser is used with peak intensities that are slightly below the ionization damage threshold of organic molecules. No observable damage to the inspected samples was observed. We believe that the results presented here further demonstrate that laser scanning THG microscopy is a promising general-purpose imaging technique with many possible application in biological imaging.

This research was supported by the Israeli Ministry of Science and Technology and by the Minerva Foundation.

References:

[1] T. Wilson, *Confocal microscopy*, (Academic, London, 1990).

[2] W. Denk, J. H. Stricker and W. W. Webb, "Two-photon laser scanning fluorescence microscopy," Science **248**, 73-76 (1990).

[3] M. Schrader, K. Bahlmann and S. W. Hell, "Three-photon-excitation microscopy: theory, experiment and applications," Optik **104**, 116-124 (1997).

[4] Y. Barad, H. Eisenberg, M. Horowitz and Y. Silberberg, "Nonlinear scanning laser microscopy by third-harmonic generation," Appl. Phys. Lett. **70**, 922-924 (1997).

[5] R. Boyd, *Nonlinear Optics*, (Academic, New York, 1992).

[6] M. Müller, J. Sqier, K. R. Wilson and G. J. Brakenhoff, "3D-microscopy of transparent objects using third-harmonic generation," J. Microsc **191**, 266-274 (1998).

[7] J. A. Squier, M. Muller, G. J. Brakenhoff and K. R. Wilson, "Third harmonic generation microscopy," Opt. Express **3**, 315-324 (1998).

PART 9

THIN FILMS, COATINGS AND MICRO-LITHOGRAPHY

Chairpersons: *K. Rabinovitch, Israel; Y. Shimony, Israel; J. Shamir, Israel*

METHODS TO REDUCE OPTICAL COATING AND OTHER PROCESS DEVELOPMENT TIME AND COST WHILE INCREASING PROCESS STABILITY

Ronald R. Willey

Consultant,
13039 Cedar St., Charlevoix, MI 49720, USA, rwilley@freeway.net.

Abstract

The development of optical coating processes to meet a set of performance requirements can involve dozens of process parameters which need to be controlled and optimized. The Design Of Experiments (DOE) methodology will be described along with its application to the development of optical coating process parameters. DOE can be used beneficially not only in optical coatings but also in most processes such as for a product, a service, a task, and others.

1. Introduction

Optical coating processes require the control of a great many process parameters. At many times and places in the past, coating development has been far less than efficient and not always successful. Some of the World's industries, such as automobile manufacturing, have used and evolved process development and refinement tools which provide a maximum amount of process understanding and control with a minimum number of tests or experiments. "User friendly" software tools and texts like that of Schmidt and Launsby[1] are now available which make process optimization and stabilization more certain and less time consuming. The DOE methodology offers a systematic approach to process development and optimization. A key characteristic is that it gives a maximum amount of information and insight in process development for a minimum number of test runs. As a result, the time and cost of process development are reduced and the probability of success is increased.

In any process development, it is advisable to have the process flow diagram in mind and preferable on paper. We should then consider all of the parameters which can cause variations and effect the results. Some of these can be controlled; some cannot be controlled and are therefore noise; and some are the subject of further experimentation. There may be a great number of parameters that could be the subject of experiments. In such case, there are screening procedures in DOE methodology to determine, with a minimum number of experiments, which are the vital few that have the major influence on the results. These few parameters are then the subject of detailed experimentation and optimization.

2. The Problem and Solution

There has been a tendency in process development in the optical coating industry and others to vary one variable parameter of a process at a time to find the desired maximum or minimum result for that variable and then to do the same with the next variable, etc., etc. One problem with this approach is that it requires many test runs if there are many parameters to be optimized. The second problem is that this approach may not find the optimum result for the combined parameters which have been varied. Figure 1 illustrates this effect. If variable B were first optimized with respect to the desired result Y and then the optimization of A were started from that point, it would appear that the optimum point in A and B was at point X. However, the real optimum is at point Z. With the proper statistical sampling techniques of DOE, it is possible to much more closely locate the true optimum at

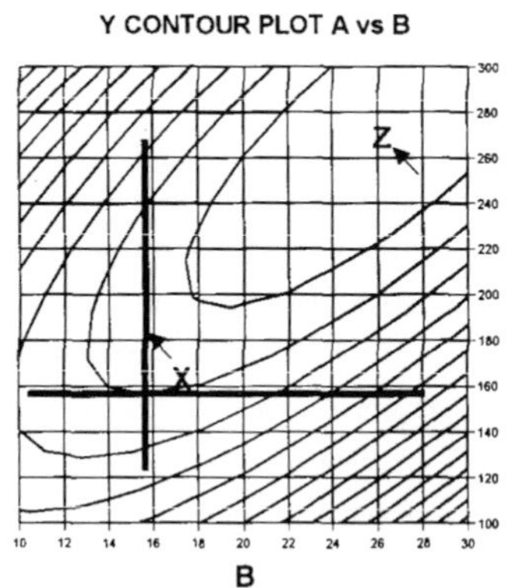

Figure 1. Contour plot of result Y as a function of variables A an B.

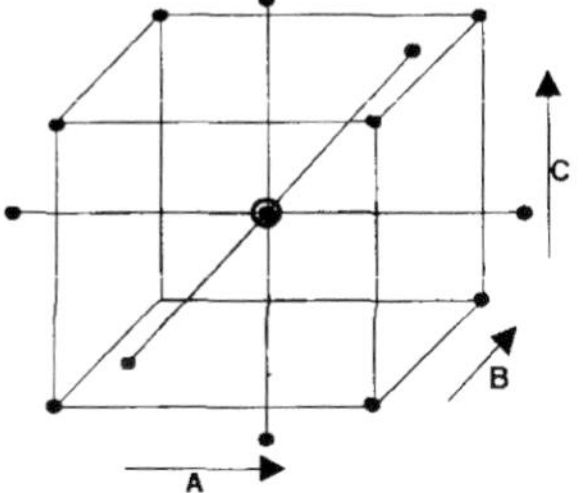

Figure 2. CCD sampling scheme for 3 variables. Dots are sample points.

point Z by only 5 sample points. For example, these might be data at points in the four corners of Fig. 1 and at the center point.

The case of Fig. 1 is for only two variables. Typically there are three or more critical parameters to be optimized. When there are three variables, it is still possible to show the distribution of sampling points graphically. Figure 3 shows the positioning of sampling points for one of the preferred DOE configurations known as the Box-Wilson or Central Composite Design (CCD)[1]. This is very efficient and flexible for second order modeling. Such a design also has rotatability so that the predicted response can be estimated with equal variance regardless of the direction from the center of the design.

Another frequently used design is the Box-Behnken Design shown in Fig. 3. These are potentially more efficient than CCD's for three factors (variables) and three levels. They allow estimation of linear and quadratic effects and all 2-way interactions. When the number of factors is greater than 4, the CCD would be more efficient. In both of these designs, the central or axial point is sampled 3 or more times to measure the repeatability of the data and allow the estimation of the standard deviation.

Experiments are then conducted at the conditions of each of the sample points and the results are recorded. The results are then processed in the DOE software[2] to fit (least squares) the data to a model for linear and quadratic effects and 2-way interactions. The model of the results can then be displayed in 2 and 3-dimensional graphics to aid visualizations of the process behavior. With the aid of these graphics, it is usually possible to find the values of each variable which will give the optimum process results.

3. An Example Case

We will illustrate the use of DOE with a fictitious example. Please note that this example is NOT TAKEN FROM REAL DATA, it is only for illustration purposes. We will imagine that we want to optimize the deposition of titania (TiO_2) using ion assisted deposition (IAD) with oxygen (O_2) and argon (Ar). The two major results desired are to have absorption (k) less than .001 and the spectral shift with humidity of less than 2 nm. There may be other results of secondary interest such as index of refraction and hardness, but adjusting for these could only be considered if the range of variables which satisfy the first two requirements leave some latitude to choose the best index and/or hardness results within that range.

We will further imagine that we have a conventional optical batch coater with heaters, electron beam gun (e-gun), ion gun, etc. What are the most important variables that could effect the desired results? From experience, we may know that some of these are: temperature, pressure, deposition rate, O_2/Ar mixture, ion current, ion voltage, etc. Our experience may allow us to bypass the screening experiments to determine which variables are important. The controls at our disposal are: deposition rate by e-gun power, ion gun current, temperature by heaters, and pressure and O_2/Ar mixture by mass flow controllers of the gasses to the ion gun. The flow is measured in Standard Cubic Centimeters per Minute (SCCM). Let us say that we know by experience that the temperature which would give the best results is the highest value that the equipment can provide. Therefore we will run the experiments at that temperature, eliminating it as a variable. Similarly, we know that the process runs most rapidly (and therefore economically) at the highest ion current that the source can provide. Again, we

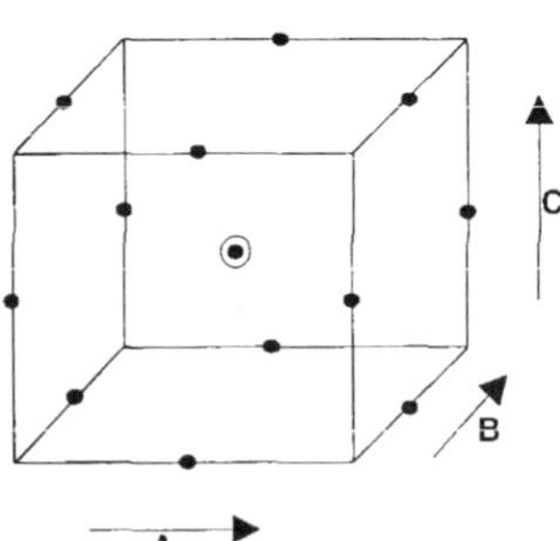

Figure 3. **Box-Behnken Design sample point scheme.**

Column #	1	2	3		RESULTS
Row #	RATE-A/S	Ar-SCCM	O2-SCCM		k x 1000
1	2	20	40		1.25
2	2	60	40		0.75
3	10	20	40		2.5
4	10	60	40		2
5	2	40	20		4
6	2	40	60		0
7	10	40	20		9
8	10	40	60		0
9	6	20	20		7.5
10	6	20	60		0
11	6	60	20		5.5
12	6	60	60		0
13	6	40	40		1.5
14	6	40	40		1.3
15	6	40	40		1.6

Figure 4. Design sheet for experimental points in Box-Behnken Design and k x 1000 results of those experiments.

will run at that current and eliminate another variable. The process pressure and O_2/Ar mixture will be affected by the flow rates of both gasses, but the pressure will also be influenced by the deposition rate due to the gettering of the reactive TiO_2. As a result of all of the above considerations, we conclude that the three independent variables that we will optimize are: Rate of deposition (measured in Angstoms per second (A/S) by a crystal monitor), Ar-SCCM, and O_2-SCCM.

We need to next decide what would be a reasonable range over which to experiment with each of these variables. Due to our experience with the coating chamber and equipment to be used, we choose 20 to 60 SCCM for both gasses and 2 to 10 A/S for the range of the rate.

FACTOR	COEF	P(2 TAIL)	TOL	LOW	HIGH	EXPER	ACTIVE
Constant	1.466667	0.002116					
RATE-A/S	0.9375	0.001747		2	10	6	X
Ar-SCCM	-0.375	0.05941		20	60	40	X
O2-SCCM	-3.25	4.48E-06		20	60	40	X
AB	0	1					X
AC	-1.25	0.002272					X
BC	0.5	0.070593					X
AA	0.079167	0.74169					X
BB	0.079167	0.74169					X
CC	1.704167	0.000666					X
R Sq	0.991461						
Adj R Sq	0.976092						
Std Error	0.436559						
F	64.50843		PRED Y			1.4667	
Sig F	0.000123						

Figure 5. Results of the analysis of the data in Fig. 4.

We can now enter these choices in an appropriate DOE software program such as DOEKISS[2] to calculate the points to be sampled which satisfy the Box-Behnken Design (in this case). *Figure 4* shows such a design sheet. We then need to choose which interactions of the variables will be included in the model to which the results will be fit. In this case, we will include all linear and quadratic interactions (as seen in Fig. 5). When the 15 experiments have been run, we enter the results into the RESULTS column of Fig. 4 and then analyze the matrix with the software.

Figure 5 shows coefficients of the model (derived from the experiments) and all of the statistical detail such as the standard error of the least squares data fit. In the EXPER column, we can enter specific values for the three independent variables and use the software to compute the predicted value (Y) based on the model. We can see here that this point at the center of the parameter space sampled (6,40,40) does not meet our need for a k-value (x 1000) less than 1. There are many more details which can be gleaned from Fig. 5, but they are beyond the scope of this paper.

The DOEKISS software facilitates the display of the results from the fit of the data to the model in a variety of graphs. Our first choice in a case such as this is usually to use a three dimensional plot of each type of result (such as "Humidity Shift" and "k x 1000") with respect to each of the independent variables taken two at a time. There would be three such plots per result if we were to view the "cube" from each of the three axes. Our general choice is to examine each of these at the plane containing the center point of the design. . We see such a plot in Fig. 6 for the k-value as a function of O_2 flow rate and deposition rate. In Fig. 7 for the Humidity Shift as a function of both gas flow rates, it can be seen that the lowest shifts occur at the lowest gas flows

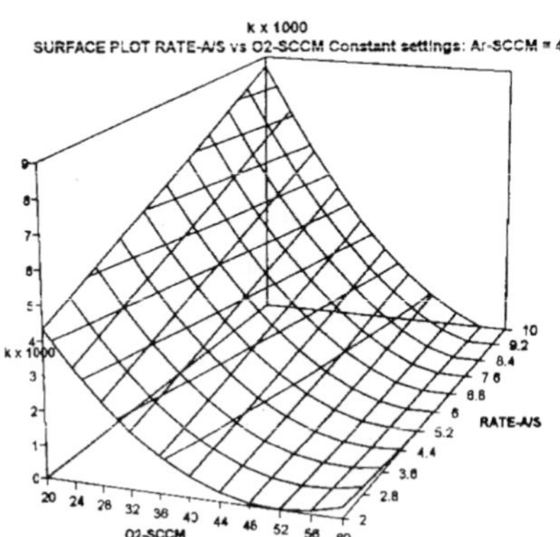

Figure 6. Surface plot of k-value versus O_2 flow and deposition rate.

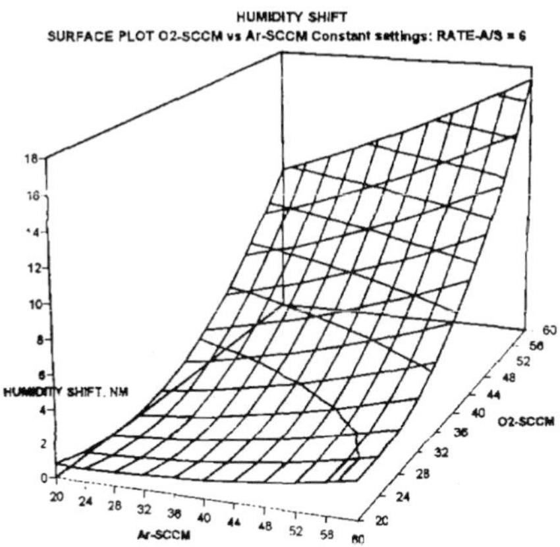

Figure 7. Humidity shift versus both gas flows.

(which result in the lowest chamber pressures). In Fig. 8 we see a similar plot for the k-value versus both gas flows. In this case, the lowest k-values are found at the highest gas flows. It can immediately be seen from Figs. 7 and 8 that the two results desired (of low humidity shift and low k-value) are in conflict, in that they "pull in opposite directions." The question to be resolved is whether there is a set of values within the range of variables where both results can be achieved.

We approach this question by using another graphic presentation, the Contour Plot. This is just a view from directly overhead of any of these plots, like a topographic map. In this case we choose to look down from the deposition rate axis to see the effects of the two gas flows on the k-value and the humidity shift. If there is an overlap between the regions of gas flow within some range of deposition rates that satisfy both of our objectives for these two results, we can solve the problem. Figure 9 is the contour plot of the humidity shift with gas flows at a Rate of 2 A/S. The lighter region meets the requirement of less than 2 nm shift. Figure 10 is the contour plot of the

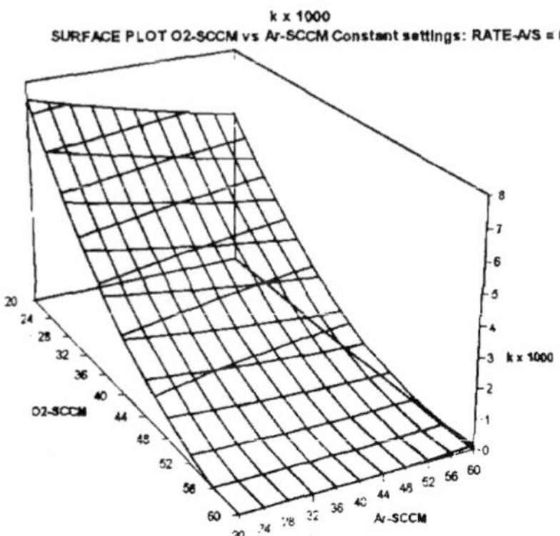

Figure 8. K-value versus both gas flows.

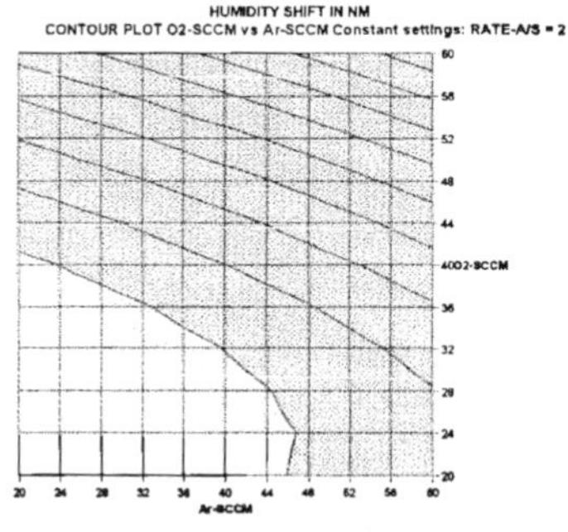

Figure 9. Contour plot of the humidity shift with both gas flows showing acceptable region in white.

k-value with gas flows at a Rate of 2 A/S. The light region is where k meets the requirement to be less than .001. It can be seen that the two white areas of Figs. 9 and 10 only overlap at a very small area in the vicinity of where the Ar flow is 20 and the O_2 flow is 41 SCCM. It was found from similar contour plots (not shown) of O_2 flow versus Rate at an Ar flow of 20 SCCM that there is no region of overlap at deposition rates greater than 2.5 A/S. We can conclude at this point that our result goals are only likely to be met in the variable region below 2.5 A/S in deposition rate, less than 24 SCCM of Ar flow, and greater then 41 SCCM of O_2 flow.

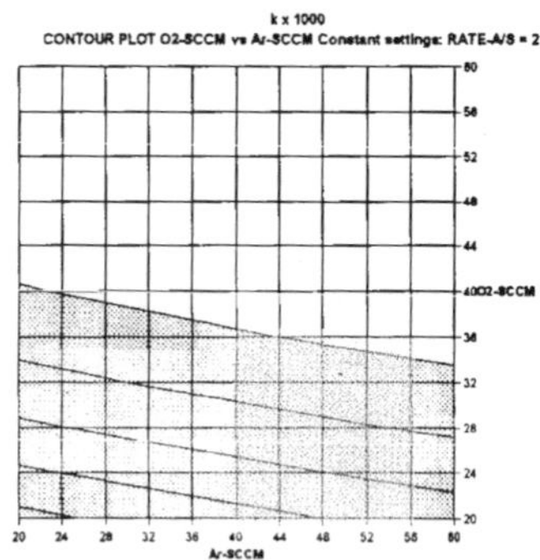

Figure 10. Contour plot of the k-value with gas flows at a Rate of 2 A/S showing acceptable region in white.

We might confirm these conclusions but an experimental test at these predicted values of the variables. If the results were satisfactory, we could consider that we are done with the DOE. If the results were not entirely satisfactory, we could use the additional set of data in the DOEKISS program as Historical Data to compute a further refined model which could even more nearly coincide with the experiments in the region of interest. However, in this particular case, we can see in Fig. 4 that the first experiment was essentially very near this variable point (2,20,40) and gave a result of k x 1000 = 1.25. Therefore, this particular would not be expected to add significantly to the data base. It appears that a test run at a rate of 2 A/S with 16 SCCM of Ar and 43 SCCM of O_2 would be more likely to satisfy the requirements while confirming the predictions. The data from such a test run could also be reprocessed as described above.

4. Summary

We have demonstrated the usefulness of the DOE methodology in finding the characteristics of a process and the optimum parameters to achieve desired results. The methodology is systematic and based on solid statistical concepts. The practical use of this tool or system is not dependent on deep understanding of the details of statistical mathematics. This like the fact that one can drive an automobile without being an experienced mechanic. The tools are "user friendly". We have shown some of the available graphics which aid in process visualization and how they might be used to gain insight and make process decisions. This methodology gleans to most information practical from a minimum number of tests. This has proved to be a great aid to efficient and successful process development.

5. References

1. S. R. Schmidt and R. G. Launsby, ***Understanding Industrial Designed Experiments***, Air Academy Press, Colorado Springs, CO, USA (1994).

2. ***DOEKISS*** Software, Digital Computations, Inc. and Air Academy Associates, LLC, Colorado Springs, CO, USA (1997).

Paper presented at the 11th International Conference on Electro Optics and Microelectronics, Tel Aviv, Nov. 9-11, 1999. To appear in Annals of the Israel Physical Soc., Vol . 14.

Low Temperature Nanoscopic Kinetics of Plasma-Assisted Si-microcrystallization in a-Si:H for Photovoltaic Applications

Yu. L. Khait [a], R. Weil [a], R. Beserman [a], F. Edelman [a], W.Beyer [b] and B.Rech [b]

a) Solid State Institute, Technion – Israel Institute of Technology, 32000 Haifa, Israel

b) Institut für Schicht und Ionentechnik, Forschungzentrum Jülich GmbH D-52425 Jülich, Germany

Abstract

A nanoscopic kinetic model of controlled plasma-assisted Si microcrystallite formation (PAμCF) in a-Si:H films at low temperatures is presented. It is shown that PAμCF can be induced by fluxes of energetic (heavy/light) plasma ions impinging on the film surface whose mass, energy, and flux control the nucleation rate, size, and density of Si microcrystallization.

Hydrogenated amorphous silicon (a-Si:H) degrades due to prolonged exposure to light (Staebler - Wronski effect [1]). The presence of Si-microcrystallites (μC) in a-Si:H stabilizes the material [2,3]. The mechanism of the Si-μC formation in the films is not clear [4,5]. In this paper we propose a novel nanoscopic kinetic model of controlled plasma-assisted Si microcrystallite formation (PAμCF) in a-Si:H films at low temperatures. The considerable enhancement of the Si crystallization by the plasma-solid interaction (PSI) is caused mainly by energetic plasma ions (EPI) impinging on the film surface. The EPI's generate in the film a great amount of nanometer short-lived (picosecond) hot spots (SLHS) of high energy density, which initiate high rates of atomic rearrangement towards the crystalline state. The EPI fluxes, the nucleation density and rates produced by heavy (e.g.Ar_2^+) and light (e.g. H_2^+) plasma ions are found. The proposed model combines results obtained in previously successful applications of nanoscopic electron-related kinetic models for the crystallization of a-Si:H and a-Si films [6-8,11],plasma-solid interaction (PSI) and plasma deposition [9,10]. The PSI which promotes the low temperature PAμCF, is associated with bombardment of the film surface by EPI's accelerated in the near surface electrical field to relatively high energies ε_i=20-100 eV [9,10]. The crystallization is associated with $\nu_N \approx$ 5-10 diffusion-like jumps of each involved Si atom [6,8]. The expectation time (per Si atom) of such a jump is [6,8,11]

$$t_d \approx 10^{-13} \cdot \exp\left(\Delta E / kT_S \right) \qquad (1)$$

where the activation energy $\Delta E \approx 3\text{eV} >> kT_S$. The flux $J_{is} \approx 0.25 a_i n_g u_{is}$ of a positive EPI of energy ε_i=20-100 eV bombarding the film surface [9,10] enhances considerably the crystallization rate. Here a_i is the ionization degree in the plasma gas of particle density $n_g = P_g (kT_g)^{-1}$, u_{is} is the EPI velocity.

First, we discuss phenomena generated in a-Si:H by a heavy EPI (e.g. Ar^+ ion) of energy ε_{is}=20-100 eV . The EPI is stopped during $t_f=10^{-14}$-10^{-13} s in the material at a distance R $\approx(3\text{-}6)10^{-8}$ cm The EPI energy ε_{is} is released within a small near-surface volume $V_f \approx 2R_f^3 = N_f\Omega_0$ containing a small number N_f of atoms (i.e. $N_f \approx 30$). (Here Ω_0 is the volume per atom). As a result, the EPI creates in V_f an initial fireball of very high energy density (per atom) $\varepsilon_f \approx \varepsilon_{is} N_f^{-1}$, e.g. $\varepsilon_f \approx$ 1-3 eV [9,10]. A part W_f of the energy is consumed in breaking the interatomic bonds, electron excitations, formation of mobile electrons etc. The fireball time evolution leads to the formation of the picosecond nanometer short-lived hot spot (SLHS) [9,10] which includes the following stages [9], fig. (1):

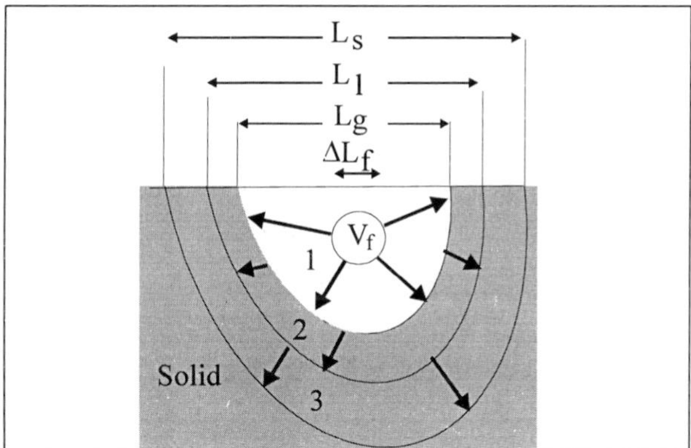

FIG. 1 Space scales of the fireball ΔL_f and phases of the SLHS formed by an EPI in a-Si:H: 1) The quasi-gas stage (QGS) , L_g . 2) Quasi-liquid stage (QLS), L_L.3) Hot solid state stage (HSSS), L_S. Arrows stand for the inward-directed transitional fluxes of energy, momentum and electric charge.

i) The quasi-gas stage (QGS) of duration $t_g \approx 10^{-13}$-10^{-12} s and average energy density (per atom) $\overline{\varepsilon}_g \geq kT_B$ is formed in the subsurface volume V_g containing N_g atoms; T_B is the boiling point.

(ii) The quasi-liquid stage (QLS) of duration $t_L \approx 10^{-13}$-10^{-12} s and average energy density (per atom) $\bar{\varepsilon}_L \geq kT_M$ is formed in the volume V_L; T_M is the melting point.

(iii) The hot solid state stage (HSSS) of duration $t_S \approx 10^{-12}$ s and average energy density (per atom) $\bar{\varepsilon}_S < kT_M$ is formed in the volume V_S containing N_S atoms. A part of the EPI energy ε_{is} is consumed in electronic excitations in the SLHS which promote Si crystallization. A single SLHS affects $3N_S \approx N_S \approx 10^3$ atoms for $\varepsilon_{is} \approx$ 30-50 eV. The corresponding number of valence electrons in the SLHS is $N_e \approx \gamma N_S$ (γ is the valence). Some of these electrons can be excited and become mobile during the SLHS lifetime $\Delta\tau$. The high energy density in the SLHS produces very large rates $K_d \approx t_d^{-1}$ (eq. (1)) of the atomic reconstruction in every SLHS. The formation of Si crystallites in the a-Si:H film results from the cumulative effect of a great number $g_N = J_{is} \cdot S \cdot t^{obs}$ of SLHS's formed by EPI flux $J_{is} \approx a_i n_g u_i$ impinging on the film surface area S during t^{obs}. The same surface area $L_S^2 \approx 10^{-13}$cm^2 of a single SLHS is affected by successive EPI's at the frequency $\omega \approx J_{is} L_S^2$. Hence one finds (for the PSI time $t^{obs} \approx 10^3$ s, flux $J_{is} \approx 10^{18}$cm^{-2}s^{-1}, $S \approx 1$ cm^2 and $L_S^2 \approx 10^{-13}$cm^2) $g_N = J_{is} \cdot S \cdot t^{obs} \approx 10^{21}$ SLHS/cm^2, $\omega_S \approx 10^6$ SLHS/(s L_S^2) and $\omega_S \cdot t^{obs} \approx 10^9$ in the same SLHS subsurface volume $V \approx L_S^3$ during t^{obs}.

We estimate the total probabilities of the diffusion-like atomic jump P_g, P_L and P_S, in the course of the QGS, QLS, and the HSSS using eq. (1) and making a 'pessimistic' assumption that $\Delta E \approx 3$ eV and neglecting the possible assistance of mobile electrons. Then the probability that one of the N_x atoms involved in the x- stage (x=g,L or s) of the SLHS experiences a diffusion-like jump is

$$P_x(\tilde{T}_x) = {t_x N_x}\Big/{t_d(\tilde{T}_x)} = t_x N_x t_{0x}^{-1} \exp\left(-\Delta E\Big/k\tilde{T}_x\right) \qquad (2)$$

where $\tilde{T}_x$ is the effective kinetic temperature for the x-th stage. Although $P_g(\tilde{T}_g > T_B)$ is much higher than $P_L(\tilde{T}_L < \tilde{T}_g)$ and $P_S(\tilde{T}_S < \tilde{T}_L)$, it is hard to expect it to dominate the atomic ordering jumps to more ordered positions since T_g is too high. The formation of Si crystallites seems to be more probable during the QLS and HSSS. From eq. (2) one finds (for heavy plasma ion bombardment) $P_L(\tilde{T} = T_M(\mathrm{Si})) \approx 2\cdot 10^{-6}$ and $P_S(\tilde{T}_S = 0.75T_M(\mathrm{Si})) \approx 3\cdot 10^{-9}$ whereas $P_g(\tilde{T}_g = T_B(\mathrm{Si})) \approx 2\cdot 10^{-2}$. Here we assumed relatively low effective temperatures $\tilde{T}_g \approx T_B(\mathrm{Si}) \approx 3.6\cdot 10^3\mathrm{K}^\circ$, $\tilde{T}_L \approx T_M(\mathrm{Si}) \approx 1.7\cdot 10^3\mathrm{K}^\circ$ which are near the lower energy density limits for the QGS and QLS. The total number of atomic displacements in the same SLHS region during the observation time t^{obs}=10^3-10^4 s is estimated by the expression $P_x(\tilde{T}_x)\cdot\omega_S\cdot t^{obs} \approx P_x(\tilde{T}_x)\cdot J_{is}\cdot L_x^2\cdot t^{obs}$. For heavy ions $\omega_S \cdot t^{obs} \approx 10^9$ and therefore heavy EPI's produce a high density of Si crystal nuclei. This leads to the formation of the microcrystallite phase. Light ion (e.g. H_2^+) bombardment produces Si crystalline nuclei at a rate and density a few orders of magnitude lower compared to that of heavy ions, since they form a fireball and SLHS of a much lower energy density, e.g. $\varepsilon_{fH} \approx 0.15\text{-}0.1$ eV $\leq T_M$(Si) . Thus, light EPI's produce a relatively low crystalline nuclei density, which is a few orders of magnitude smaller than those related to a heavy EPI. Consequently, light EPI's form microcrystals separated by distances of a few orders of magnitude larger than the radius of an individual nucleus.

In conclusion, it is shown why amorphous silicon can crystallize at low temperatures in a plasma, and that the size and density of the crystallites can be controlled by the ratio of heavy/light ions in the plasma.

Acknowledgements:

This research was supported by a grant from the National Council for Research and Development, Israel and the BMWi, Germany.

References:

[1] D.L. Staebler and C.R. Wronski, Appl. Phys. Lett. **39,** 292 (1977); H. Fritzsche, Mat. Res. Soc. Symp. Proc. **467**, 19 (1997).

[2] S. Guha et al., J. Appl. Phys. **52**, 859 (1981); L. Yang and L.F. Chen, Mat. Res. Soc. Symp. Proc. **336**, 669 (1994); D.V. Tsu et al., Appl. Phys. Lett. **71**, 131 (1997).

[3] Plasma Properties, Deposition and Etching, Ed's. J.J. Pouch and S.A. Alterovitz (Trens Tech. Pub., Switzerland, 1993).

[4] K. Saito et.al., Appl. Phys. Lett. **71** , 3403 (1997); R. Weil et. al., submitted for publication

[5] K. Panda et al., J. Appl. Phys. **85**, 1900 (1999); T. Takgi et al.,Thin Solid Films **345** , **75** (1999)

[6] Yu. L. Khait and R. Weil, J. Appl. Phys. **78**, 6504 (1995)

[7] I. Abdulhalim, et al., Appl. Phys. Lett. **55**, 1180 (1987)

[8] Yu. L. Khait, R. Beserman, Phys. Rev. B **33**, 2983 (1986).

[9] Yu. L. Khait, in ref. [3];Yu. L. Khait, in Plasma Processing and Synthesis of Materials, Ed's. Szekely, D. Apelian (Amer. Mat. Res. Soc. 1984), **V.30;** Yu. L. Khait, in Plasma Chemistry and Technology (Technomic Publishing Co., Basel 1985)

[10] Yu. L. Khait, A. Inspektor, R. Avni, Thin Solid Films **72**, 249 (1980)

[11] Yu. L. Khait, Monograph, Kinetics and Applications of Atomic Diffusion in Solids, Nanoscopic Electron-Affected Stochastic Dynamics (SCITEC Pub Switzerland, 1997).

NEW CHALCOGENIDE PHOTORESISTS FOR IR MICRO-OPTIC FABRICATION BY THE THERMAL REFLOW PROCESS

Salman Noach[1], Michael Manevich[1], Matvey Klebanov[2], Victor Lyubin[2] and Naftali Eisenberg[1]

[1] *Jerusalem College of Technology, POB 16031, Jerusalem 91160, Israel.*
Tel: 02-6751144, Fax: 02-6751244, e-mail: salman@mail.jct.ac.il

[2] *Department of Physics, Ben Gurion University of the Negev, Beer-Sheva, Israel*

Abstract

A maximum sag limited to 1.3μm is achieved when using Chalcogenide (CG) films of the type AsS or AsSe for fabrication of I.R. microlens arrays applying the modified proximity method, the gray scale, or continuous tone photolithography with classical UV exposure sources. Using the thermal reflow method one can extend this range. Therefore the development of a CG photoresist with lower melting temperature was necessary. A new material based on a modification of the CG structure by the introduction of iodine atoms was developed. An $AsSeI_{0.1}$ composition was obtained as the optimum structure, giving the maximum decrease in melting temperature, without appreciable loss of photosensitive properties. AFM measurements of an $AsSeI_{0.1}$ microlens made by the thermal reflow method were performed during the reflow process. The gain in the maximum sag of the microlens compared with the modified proximity method, gray scale or continuous tone lithography is about 50%.

1. Introduction

Microlens and microlens arrays can be found in an increasing number of optoelectronic applications, such as optical communication and computing, CCD cameras, faxes, imaging systems and I.R. technology [1-4]. Microlens arrays have been fabricated by a variety of techniques, including distributed index planar techniques [5], resin thermal reflow [6], laser beam ablation [7], laser beam exposure [8] and the photo expansion method [9]. The most common and widely used technique is photolithography in which the photosensitive material that is deposited on the substrate is exposed through a mask by tailored light distribution or by thermal reflow methods. Also, 3D-structures having the form of the microlens arrays are generated in a photoresist material [6,10], and then transferred by anisotropic etching into a transparent robust material [11], transparent in the spectral range for which the microlens arrays are intended. Anisotropic etching is not a simple process; therefore its elimination would be very desirable in microlens fabrication. Maximum sag limited to 1.3μm is achieved when using CG films of the type As-S or As-Se with gray scale or continuous tone lithography with classical UV exposure sources [12,13]. The reason for this is the photodarkening of the CG film upon exposure. In order to overcome this maximum thickness limit one can extend this range using the thermal reflow method. Therefore the development of a CG photoresist with a lower melting temperature was necessary, since the melting temperature of the As-S and As-Se type films is above 360°C. This new material is based on a modification of the CG structure by the introduction of iodine atoms. This results in a lowering of the melting temperature. An $AsSeI_{0.1}$ composition was obtained as the optimum material, giving the maximum decrease in melting temperature to 220°C without appreciable loss of photosensitive properties. AFM measurements of an $AsSeI_{0.1}$ microlens made by the thermal reflow method were performed during the reflow process.

2. Properties of Chalcogenide Photoresists and photodarkening effect

Chalcogenide photoresists are usually prepared from binary As containing CG glasses (As-S, As-Se). Some properties of these inorganic CG photoresists, especially important for IR microlens array fabrication, are reviewed in reference [12] .In conventional methods of photolithographic fabrication of microlens arrays using CG [13,14], the maximum sag is limited to 1.3μm using classical UV exposure sources. The reason for this is due to the photodarkening effect of the CG. This effect is identified by the change in the transmission curve of the material upon exposure (as can be seen from Fig 1) which accompanies the photoinduced structural transformations in the CG films. Curve No.1 shows the transmittance of the deposited film. Curve No.2.shows the transmittance following laser illumination and curve No. 3 shows the reversibility of the photodarkening effect upon annealing close to 170^0C.

figure 1. The photodarkening effect in AsSe Chalcogenide with thickens of 0.8μm

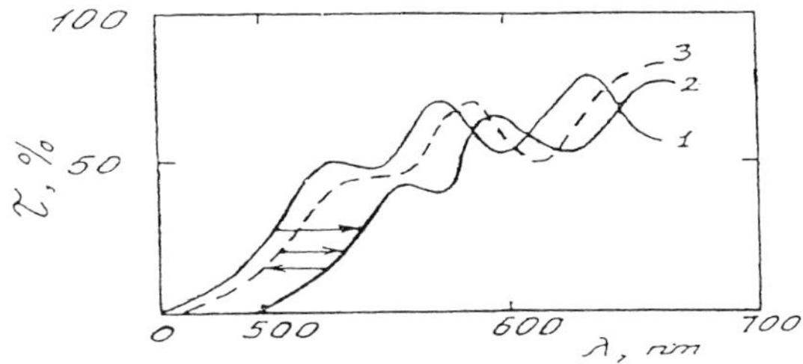

Analysis of this effect is described in Ref. [15,16]. This effect does not present a problem when low sag values are desired (below 1.3μm) but due to this effect there is a self-limitation process that limits the thickness that can be exposed. Using the classical UV sources for exposure the maximum thickness that can be exposed was no more than 1.3μm.

3. Thermal reflow method for microlense fabrication using $AsSeI_{0.1}$Chalcogenide

The thermal reflow method can avoid exposure problems. Using a binary mask containing holes or slits, islands of 3D binary shape can be formed that can be then transformed to 3D plano convex microlenses. This is done by heating the material close to the melting point, causing reflow and achieving the desired 3D shape. The thermal reflow method isn't applicable to conventional CG compounds since the melting temperature of AsS type films is above 360°C and is not practical in microoptic fabrication processes. A program was instituted which resulted in the development of several CG photoresists with lower melting temperatures. These materials are based on a small modification of the CG structure by the introduction of iodine atoms. This results in lowering the melt temperature [17] and also in degradation of the photoresist properties. Optimization results in the choice of the composition $AsSeI_{0.1}$, a compromise between the improvement due to lower melting temperature, and degradation of contrast of dissolution K (for negative photoresists $K = v_{non\ exp.} / v_{exp.}$, where v is the rate of dissolution) as shown in Table 1.

Glass composition	AsSe	$AsSeI_{0.05}$	AsSe $I_{0.1}$	$AsSeI_{0.2}$	$AsSeI_{0.25}$
T_s softening (°C)	233	191	168	153	145
T_m melting (°C)	360		220		
K	5.8	3.1	3.0	2.4	1.7.

Table 1: Influence of Iodine concentration on T_m and K.

The T_s values were determined by the scratch disappearance method [18,19] at a heating rate of 10-12 °C/min, a rate usually used for relaxation measurements and the contrast values were obtained with a weak polychromatic source of illumination. T_m values are well above the measured T_s values. Conventional K values using materials of this kind is between 3-6. The optimized composition was chosen based on the lower limit of K.

5. Experimental

After determining the optimum composition of the CG to be $AsSeI_{0.1}$, the photoresist was deposited by vacuum thermal evaporation onto glass, with a thickness of 830nm. A contact binary photomask containing slits for cylindrical microlenses was used. Samples were exposed for 5 minutes using a 200W halogen lamp, and then developed in the negative photoresist mode in a monoethanolamine developer for 25 seconds in order to achieve 3D binary shapes. The thermal reflow procedure consisted of heating the binary shapes at intervals of 10°C from 170°C -250°C for 5 minutes at each interval. After a short relaxation time, AFM topography measurements tracked the formation of the plano convex shape. Step heating of this particular sample resulted in a lens, with a gain of 1.4 (from 820nm bulk to 1130nm sag) with a small shrinkage of the initial diameter. Fig. 2 shows the formation of an $AsSeI_{0.1}$ cylindrical microlens made by the thermal reflow method with step heating.

bulk

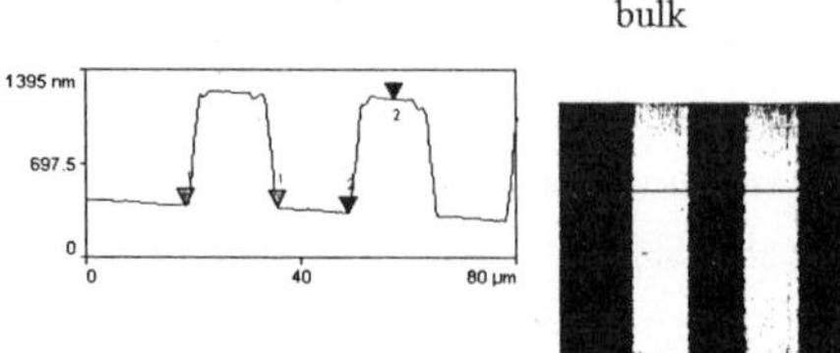

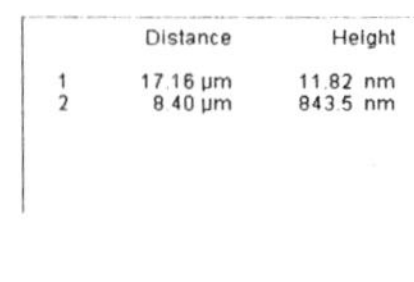

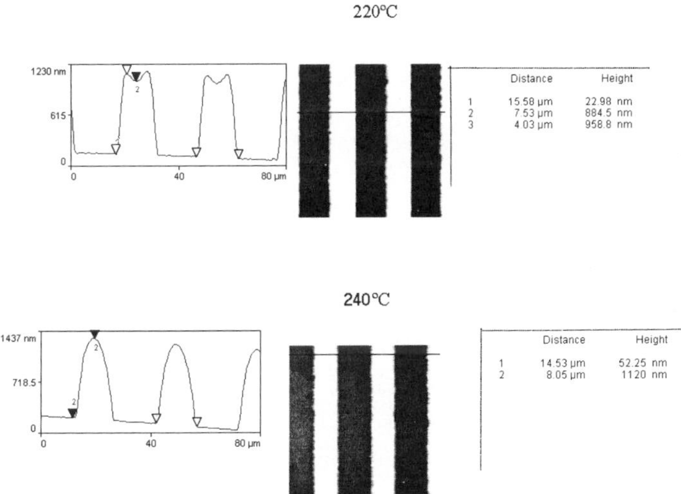

Figure 2: The formation of the $AsSeI_{0.1}$ microlens at various temperatures using the thermal reflow method.

However direct heating to 220^0C resulted in a higher gain, 2, in the maximum sag of the microlens compared with the bulk. The measured sag is about 1.9µm and the base width 13.7µm. The original thickness of the CG photoresist before thermal reflow was 0.95µm. Higher gain is obtained in a one step treatment.

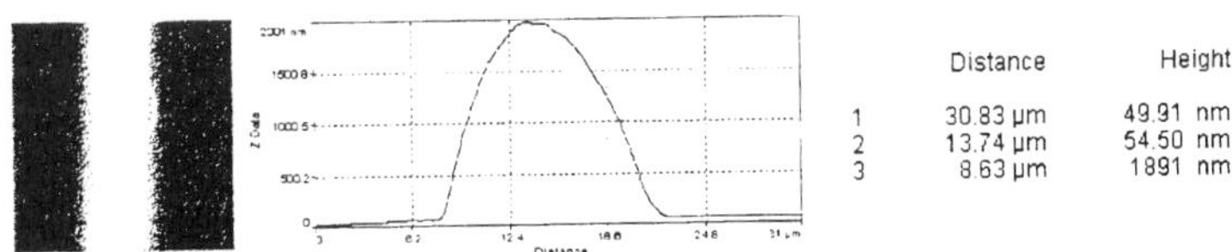

Figure 3: An $AsSeI_{0.1}$ microlens fabricated by the thermal reflow method in one step.

6. Conclusion

$AsSeI_{0.1}$ CG glassy films are shown to be a suitable material for fabrication of I.R. microlens arrays. Using the photolithography and thermal reflow methods the maximum sag limit caused by the photodarkening effect is eliminated. The gain in the maximum sag of the microlens compared with the modified proximity method and gray scale or continuous tone lithography is about 50%.

Acknowledgment

This work was supported by a grant from the Israeli Ministry of Science for generic research for refractive microoptics.

References

1. H. Nishihara and T. Suchara, in **Progress in Optics,** Vol. 24, p.1, North-Holland, Amsterdam, 1987.
2. R. H. Bellman, N. F. Borrelli, L. G. Mann, and J. M. Quintal, Proc. SPIE **1544**, 209 (1991).
3. T. R. Jay, M. B. Stern, and R. E. Knowlden, Proc. SPIE **1751**, 236 (1992).
4. M. W. Farn, M. B. Stern, W. B. Veldkamp, and S. S. Medeiros, Optics Letters **18**, 1214 (1993).
5. M. Oikawa, Appl.Opt. **21**, 1052 (1982).
6. Z. D. Popovic, R. A. Sprague, and G. A. N. Connell, Appl. Opt. **27**, 1281 (1988).
7. M. Terao, K. Shigematsu, M. Ojima, Y. Taniguchi, S. Horigome, and S. Yonezawa, J. Appl. Phys. **50**, 6881 (1979).
8. H. Hisakuni and K. Tanaka, Optics Letters **20**, 958 (1995).
9. S. Ramachandran, J.C. Pepper, D.J.Brady and S.G. Bishop. Journal of Lightwave technology, **15**,8, 1371(1997)
10. M. Manevich and N. P. Eisenberg, Proc. SPIE **2426**, 242 (1994).
11. E. J. Gratix, Proc. SPIE **1992**, 266 (1993).
12. S. Noach, M. Manevich, M. Klebanov, V. Lyubin and N.P. Eisenberg, Proc SPIE **3778,** 151 (1999).
13. N.P.Eisenberg, M.Manevich, M.Klebanov, S.Shtutina and V.LyubinV. Proc.SPIE **2426**, 235 (1995).
14. N.P.Eisenberg, M.Manevich, M.Klebanov, V.Lyubin and S.Shtutina. J. Non-Cryst. Sol. **198-200**, 766 (1996).
15. B. T. Kolomiets and V. M. Lyubin, Mat. Res. Bull. **13**, 1343 (1978).
16. K. Tanaka, J. Non-Cryst. Solids **35&36**, 1023 (1980).
17. Z.U.Borisova. Glassy semiconductors, Plenum Press, New-York 1981.
18. M.Wobst. J.Non-Cryst.Sol. **11**, 255 (1972).
19. B.T.Kolomiets, S.S.Lantratova, V.M.Lyubin and V.P.Shilo. Sov. Phys. Solid State **21**, 594 (1979).

PHOTOINDUCED ANISOTROPY OF CONDUCTIVITY IN CHALCOGENIDE AMORPHOUS FILMS

V.Lyubin and M.Klebanov,

Department of Physics, Ben-Gurion University of the Negev, Beer-Sheva 84105, Israel

Abstract

Reversible photoinduced optical anisotropy in amorphous $As_{50}Se_{50}$ films was shown to be accompanied by the reversible anisotropy of photoconductivity. The conclusion is made that the microanisotropic species presented in the amorphous chalcogenide film determine not only the light absorption process but also the process of transport of the non-equilibrium charge carriers.

1. Introduction

A phenomenon of photoinduced optical anisotropy in chalcogenide amorphous films, manifested as linear dichroism and birefringence has attracted attention of many researchers in recent years[1-6]. This phenomenon is interesting not only as an unusual physical effect but also due to prospects of its application in electrooptics[7,8]. Photoinduced anisotropy gradually appears during the course of irradiation of samples by linearly polarized light and can be reoriented when the electrical vector of exciting light is changed to the orthogonal one. Different approaches to explanation of PA have been developed[1,2,4]. A new model of PA in thin films based on polarized light orientation of randomly oriented valence-alteration pairs has been proposed recently [3,5,6].

In this paper we report new results demonstrating that photoinduced optical anisotropy induced in the amorphous $As_{50}Se_{50}$ films by linearly polarized laser beam is accompanied by the anisotropy of photoconductivity.

2. Experimental

Amorphous $As_{50}Se_{50}$ films were prepared by standard thermal vacuum evaporation at a pressure $3x10^{-6}$ Torr. The thickness of films was in the range of 300 - 2000 nm. Study of photoinduced optical anisotropy was carried out using an experimental set-up with two linearly polarized laser beams (inducing and probing beams) illuminating the same area of the film. A low power He-Ne laser beam (~ 0.4 mW/cm^2 at λ = 633 nm) was applied as a probing beam, while as a photoinducing beam we used either a He-Ne laser beam (~2.75 W/cm^2, λ = 633 nm) or an Ar^+ ion laser beam (~ 0.3 W/cm^2, λ = 488 nm). The photoinduced anisotropy (dichroism Δ) was determined from the equation $\Delta = 2 (I_y - I_x) / (I_y + I_x)$, where I_y and I_x are correspondingly the intensities of the probing beam transmitted through the film for the parallel and perpendicular polarized components of the electric vector of the inducing beam.

For investigation of anisotropy of the photoconductivity we used both the film samples having two parallel Au electrodes with space ~0.5 mm between them and the samples in which three Au electrodes were deposited on the oxide glass substrate before forming the chalcogenide film (Fig.1).

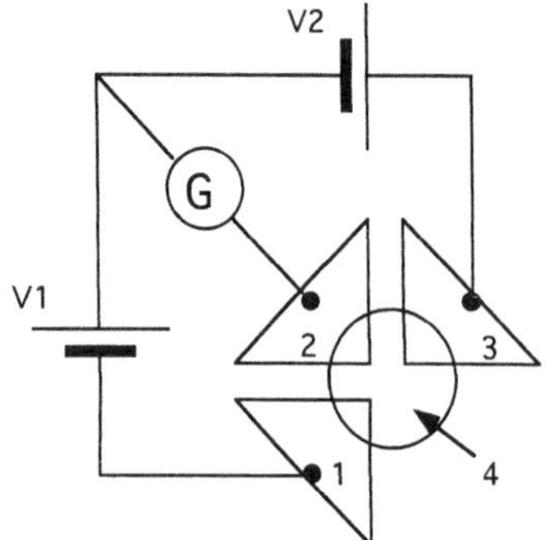

Fig.1. Scheme of electrodes and electrical circuit used. 4-position of the laser beam.

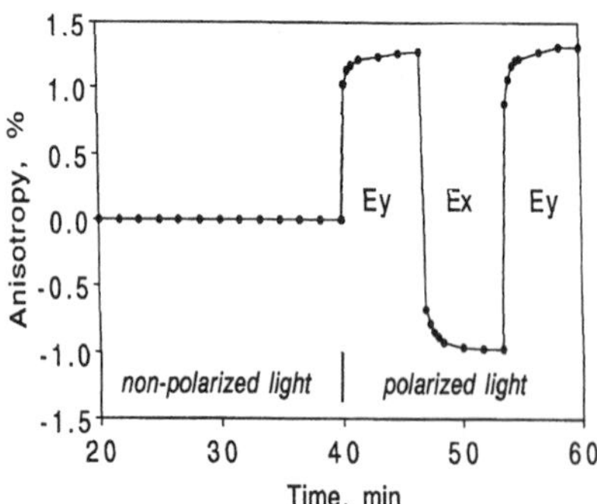

Fig.2. Kinetics of dichroism generation and reorientation under action of non-polarized and polarized light with x and y directions of electrical vector.

The electrical circuit shown in Fig.1 was adjusted by shifting the position of the illuminating light beam in such a manner that the nano-ampermeter G showed a zero current by illumination of the sample with non-polarized light. By irradiation with strong (inducing) linearly polarized light, the device G showed a current if the photoconductivity became anisotropic, as the electrical fields between electrodes 1-2 and 2-3 were directed either in parallel or perpendicular to the electrical vector of probing light. Application of these three-electrode samples can be considered as an analog of uses the lock-in measuring system and results in large increase of sensitivity.

3. Experimental results

Excitation of $As_{50}Se_{50}$ films by the linearly polarized light led to generation of linear photoinduced dichroism and the change of the inducing light polarization to the orthogonal one was accompanied by the dichroism reorientation as it is shown in Fig.2. These results are very identical to data obtained by different researchers[1-6].

The results of the study of photoconductivity in $As_{50}Se_{50}$ samples are demonstrated in Fig.3a, b. In the samples with two parallel electrodes, irradiation by non-polarized He-Ne laser light was accompanied by growth and next saturation of photocurrent (Fig.3a). Following irradiation by linearly polarized light with different electrical vector orientation led to modulation of photocurrent,

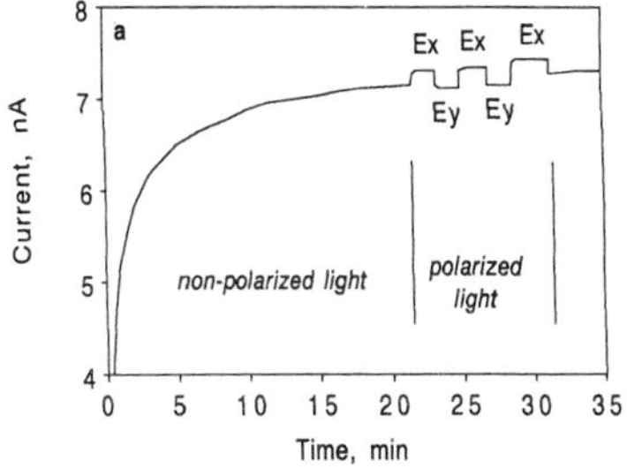

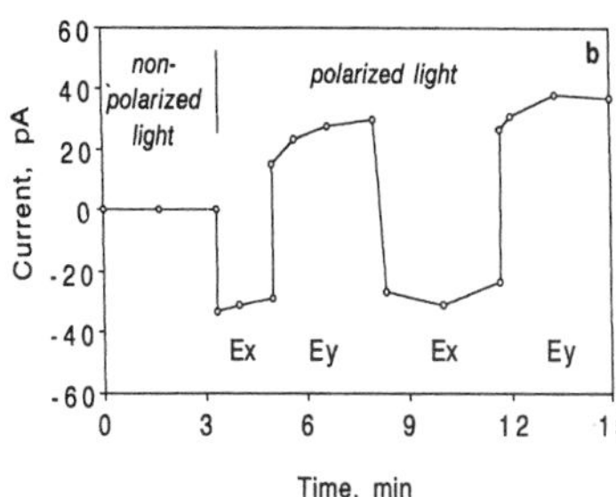

Fig.3. Kinetics of photocurrent induced by linearly polarized He-Ne laser beam with x and y directions of electrical vector in two-electrodes (a) and three-electrodes (b) samples.

speaking about the anisotropy of photoconductivity as it is shown in Fig.3a. More distinctly this result was seen from experimens using the samples with special electrodes and electrical circuit shown schematically in Fig.1. It can be seen from Fig.3b that the irradiation of such sample by the inducing linearly polarized light beam leads to appearance and growth of a photocurrent ΔI in the circuit. Changing the polarization direction of the inducing light was always accompanied by corresponding reversible jumps of the current. Kinetics of photocurrent is quite analogous to the kinetics of the optical dichroism changes shown in Fig.2.

4. Discussion

This research demonstrates that photoinduced optical anisotropy induced in the amorphous $As_{50}Se_{50}$ films by the linearly polarized laser beam is accompanied by the anisotropy of photoconductivity. To the best of our knowledge, this is the first reported case of the photoinduced electrical anisotropy observation in the chalcogenide films. Anisotropy of conductivity was shown to be reorientable: change of the inducing light polarization to the orthogonal one is accompanied by the change of the sign of electrical anisotropy.

It is known that photoconductivity $\sigma_{ph} = en_{ph}\mu$ is determined by both the concentration of photoinduced charge carriers n_{ph} and their mobility μ (e is the electronic charge). Both parameters n_{ph} and μ can be responsible for the observed anisotropy of photoconductivity. Moreover, the n_{ph} value in this case must be dependent on the light polarization as it is determined by the light absorbtion coefficient that is anisotropic. The effect of anisotropic n_{ph} value must be seen in the sandwich-type samples preliminary illuminated by linearly polarized light. However in our experiments on photoconductivity with the gap-type samples and without preliminary illumination with strong linearly polarized light, irradiation by the light of different polarizations but the same intensity generates the same n_{ph} and must not lead to the photoelectrical anisotropy. Such anisotropy can appear only due to different transport of charge carriers excited by the light with electrical vector parallel or perpendicular to the electrodes. Thus, the obtained data indicate that the microanisotropic species presented in the film (in particular, valence-alteration pairs) determine not only the light absorbtion process but also the process of transport of the non-equilibrium charge carriers (mobility of these carriers). In usual state, these microanisotropic species are oriented randomly and ensure the isotropic photoconductivity but under the polarized light excitation they are oriented either parallel or perpendicular to the electrodes creating anisotropic photoconductivity.

Photoinduced anisotropy of the photoconductivity in $As_{50}Se_{50}$ films, that posessed of strong photoresist effect, can be used for creation of microminiature low-power polarization-sensitive optoelectronic sensors.

5. Conclusion

The photoinduced optical anisotropy (dichroism and birefringence) in different chalcogenide amorphous films was investigated by many authors. In this paper we reported new results demonstrated that photoinduced optical anisotropy induced in the amorphous $As_{50}Se_{50}$ films by the linearly polarized laser beam is accompanied by anisotropy of photoconductivity. Anisotropy of photoconductivity was shown to be reorientable: change of the inducing light polarization to the orthogonal one is accompanied by the change of the sign of electrical anisotropy. The data obtained indicate that the microanisotropic species presented in the amorphous chalcogenide film (in particular, valence-alteration pairs) determine not only the light absorbtion process but also the process of transport of the non-equilibrium charge carriers (mobility of these carriers).

Bibliography

[1] V.M.Lyubin and V.K.Tikhomirov, J. Non-Cryst.Solids 114 (1989) 133.

[2] H. Fritzsche, Phys.Rev.B 52 (1995) 15854.

[3] V.K.Tikhomirov and S.R.Elliott, Phys.Rev.B 5 1 (1995) 5538.

[4] K.Tanaka, K.Ishida and N.Yoshida, Phys. Rev. B 54 (1996) 9190.

[5] V. Lyubin and M. Klebanov, Phys. Rev. B 53 (1996) 11924.

[6] V.Lyubin, M.Klebanov, V.Tikhomirov and G.Adriaenssens, J. Non-Cryst. Solids 198-200 (1996) 719.

[7] C.H.Kwak, S.Y.Park, H.M. Kim and E. H. Lee, Optics Commun. 88 (1992) 249.

[8] Joby Josef, F.J. Aranda, D.V.G.L.N.Rao, J.A.Akkara and M.Nakashima, Opt. Lett. 21 (1996) 1499.

TRANSMISSION ELECTRON MICROSCOPY OF DEFECTS IN $ZnGeP_2$

I. Dahan, G. Kimmel

Nuclear Research Center - Negev (NRCN), P.O. Box 9001, Beer-Sheva 84190, Israel.

R. Feldman, Y. Shimony

Rotem Industries, Rotem Industrial Park, Mishor-Yamin, M.B.O Arava, 86800, Israel.

Abstract

Coherently oriented cubic defects were recently reported to exist within the tetragonal lattice of zinc-germanium-phosphide, $ZnGeP_2$ (ZGP). X-ray diffraction utilizing whole pattern optimization (Rietveld Method) and line-profile-fitting (LPF) enabled to identify these defects. In the present research, transmission-electron-microscopy (TEM) was used for direct observation of these defects. The bright field and dark field TEM images of the ZGP foil, are indicate the existence of anti-phase domain boundaries (APB). Electron diffraction patterns clearly reveal two coherently oriented entities, one of a tetragonal symmetry in accordance with the chalcopyrite lattice, the other however of a cubic symmetry. This confirms our former finding of the existence of coherently oriented cubic defects in as-gown chalcopyrite ZGP crystal.

1. Introduction

A disordered zinc-blende structure was identified within some of the ordered chalcopyrite crystals belonging to the ternary I-III-VI_2 and II-IV-V_2 groups [1]. In most cases, this phenomenon is also accompanied with a solid-solid structural phase transition, which occurs at elevated temperatures [2]. Binsma et al. [3] have found that an order-disorder transition occurs only for chalcopyrite crystals of an axial ratio $c/a>1.95$. The ordered arrangement of the cation lattice of the chalcopyrites results in a tetragonal distortion δ, defined by $\delta=2-(c/a)$. Order-disorder transition has already been found in

$ZnSnP_2$, $ZnGeP_2$ (ZGP), and $CdSiP_2$ compounds [3]. In the case of $ZnSnP_2$, various degrees of ordering were obtained, depending on the cooling rate during the crystal growth [4]. This partly ordered crystal showed a mosaic structure, consisting of blocks, each having either the chalcopyrite or the zinc-blende structure [4]. In ZGP, the presence of a large number of zinc and germanium anti-sites is expected, since the material solidifies at 1027°C as a zinc-blende random alloy, and undergoes a phase transition to the chalcopyrite structure at 952°C [5]. Because the kinetics of the order-disorder transition is sluggish, some residual disorder may remain after cooling to below the transition temperature. Recently residual disorder was also observed in ZGP, by using x-ray diffraction [6]. It should be emphasized here, however, that although order-disorder transition has already been found in ZGP long ago by utilizing differential thermal analysis (DTA) [7], recent DTA measurements showed no such transition [8].

A wide optical absorption band near the band edge is usually found in ZGP crystals, which are grown either by the gradient freeze technique [9] or by the Czochralski method [10]. This band, extending from 0.7μm to about 2.5μm, is attributed to photo-ionization of a native defect-related acceptor labeled AL1 [11]. It was suggested that AL1 primarily displays itself as a component of a native defect complex, possibly V_{Zn}-V_P (zinc and phosphorus vacancies, respectively) [11]. Recent studies confirmed that the main defect in melt-grown ZGP is V_{Zn}, while V_P as the dominant donor exists in a highly compensated material [9]. The native point defects in the ZGP lattice have been already studied by various techniques such as by optical absorption [12], photoluminescence [13], EPR [14], ENDOR [15], and temperature-dependent Hall-effect [16]. Thermal annealing [17], electron beam, and γ-ray irradiation [18] treatments were successfully applied on ZGP to reduce this parasitic optical absorption in as-grown crystals.

Although correlation between the appearance of disordered cubic domains in melt grown ZGP crystal, to its residual absorption at the near-IR has already been confirmed [19], no direct observation of this structural defect has yet be demonstrated. In the present paper, we show our recent results on transmission electron microscopy (TEM) studies of ZGP. They provide the first direct observation of the disordered structural defects in melt grown ZGP, and provide a better understanding of their nature.

2. Experimental

ZGP single crystals for the present study were grown by horizontal gradient freezing [6]. Results obtained by powder x-ray diffraction were analyzed by peak-by-peak line profile fitting, and whole pattern optimization (Rietveld Method). For the TEM studies, a round ZGP sample was drilled out of a 2 mm thickness wafer by using an ultrasonic cutter. The sample thickness was then reduced to about 100 μm by utilizing Gatan precision grinder model 602. A 2 mm diameter dimple was then made in the center of the ZGP sample, decreasing its thickness to 5-10 μm. Finally a hole was drilled at the dimple center by ion milling. A PHILIPS CM200 STEM system was used to examine the ZGP samples.

3. Results and Discussion

Fig. 1(a) shows an X-ray powder diffraction pattern of a melt-grown ZGP crystal, which is apparently consistent with the chalcopyrite structure [5]. Rietveld refinement of this pattern, however, yielded a final agreement factor R = 0.23, indicating poor fitting. The fit was significantly improved by assuming that the crystal contains two constituents, one of the tetragonal chalcopyrite structure (assigned as α-ZGP), and the other of the zinc-blende cubic structure (assigned as β-ZGP). Both structures are related to each other by $a_{tet} \approx a_{cub}$, and $c_{tet} \approx 2a_{cub}$, with a and c being the lattice parameters. As a result of the Rietveld refinement, the R-factor was reduced to R_B=0.08 for the α-phase, and to R_B=0.10 for the β-phase. Based on the Rietveld refinement, the patterns of α and β phases were calculated, and shown in Figs. 1(b) and 1(c), respectively. Similar results were obtained with x-ray diffraction of crystalographically oriented ZGP single crystals, when the diffraction patterns were treated with line profile fitting. The diffraction pattern of the cubic phase is always obtained with broadened lines. Details of the x-ray diffraction analysis of melt-grown ZGP were recently published [19].

By using the TEM, small zones of irregularity were observed within the matrix of as-grown ZGP crystal. In fig. 2(a) the bright-field (BF) of such à typical zone is depicted, at 11.5K magnification. Fig. 2(b) shows the same defect at the same magnification in the dark field (DF). These two figures look the same, however, with a reverse contrast. In fig. 3 the same defect is shown at a magnification of 66K in both the bright and dark fields, which are characterized by typical fringe patterns. Such fringe patterns (Fig. 3) might be attributed to either stacking faults, planar precipitates with small misfit forms, bend contour, or anti-phase domain boundaries (APB) [20]. The possibility that a bend contour is responsible for the fringe pattern was excluded by a minute tilt of the sample, since this would cause the fringes to move. In addition, careful interpretation of the TEM patterns, figs. 2 and 3, leads to the conclusion that it arises from an APB defect, as it meets the demands for the identification of such a defect as APB [20]. Thus, it seems likely that the tetragonal matrix of ZGP contains an anti-phase domain boundary defect, also known as π-boundary [20]. Note that anti-phase domain boundaries usually occur in ordered materials. This defect appears when there is a change in the identity of the atom at a given lattice point but there is no atomic stacking change like those that occur at a stacking-faults [20].

In order to examine how the APB relates to the residual cubic phase as obtained by the XRD [19], electron diffraction images were taken from the vicinity of these domains. In fig. 4, a typical electron diffraction pattern is presented. This pattern was recorded from the same area appearing in fig.2. Two sets of dots can be observed within the pattern, coherently oriented with respect to each other. One set contains small and sharp dots, the other set is characterized by spread dots. The first set of sharp and small dots belongs to the tetragonal matrix of ZGP, since it fits well to the lattice parameters a=0.546 nm, and c= 1.071 nm [19]. The second pattern with spread dots fits to a cubic phase having a lattice parameter of a= 0.545 nm. Note that this result is in full agreement with our XRD results, including of the line broadening of the cubic phase. It thus turns out that these two structures definitely correspond to the ordered and disordered ZGP phases, respectively, which appear in Figs. 1(b) and 1(c). The zone axis of the electron diffraction patterns for both phases is <001>. Note that only in the <001> zone axis one can distinguish between the c- and a-axes of the ordered phase. The coherency between the two sets of electron diffraction patterns (Fig. 4) indicates that the two crystalline

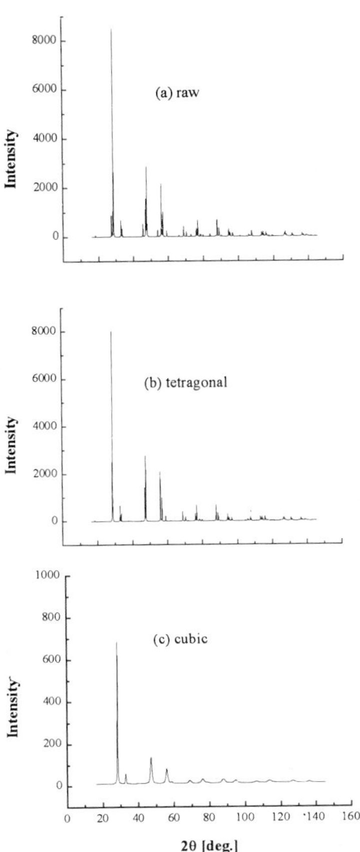

Fig. 1

X-ray powder diffraction pattern of melt-grown ZGP crystals; (a) The experimental pattern, also contains extra lines of germanium that was added to the ZGP powder for calibration. (b) and (c) are the tetragonal and cubic diffraction patterns, respectively, which were obtained by Rietveld refinement of pattern (a).

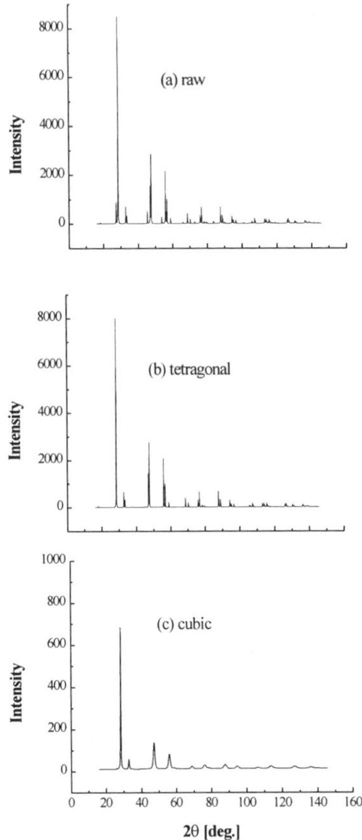

Fig. 1
X-ray powder diffraction pattern of melt-grown ZGP crystals; (a) The experimental pattern, also contains extra lines of germanium that was added to the ZGP powder for calibration. (b) and (c) are the tetragonal and cubic diffraction patterns, respectively, which were obtained by Rietveld refinement of pattern (a).

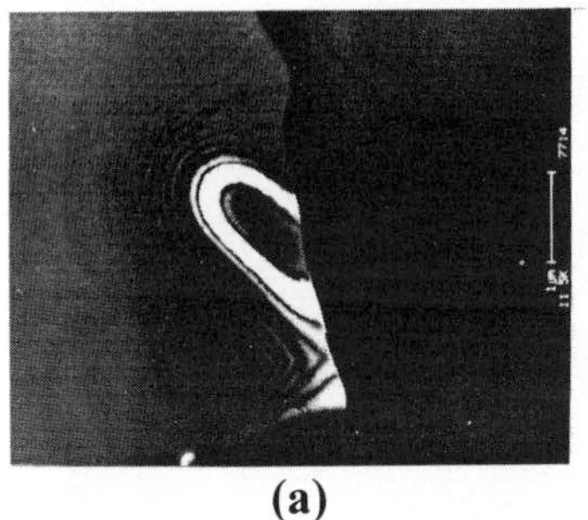

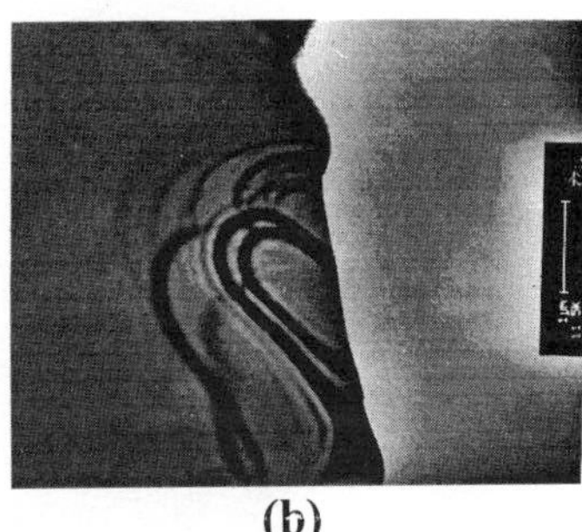

(a) (b)

Fig. 2: Bright-field (BF) (a) and dark-field (DF) (b) TEM images of anti-phase domain boundary in ZGP (Mag. 11.5K).

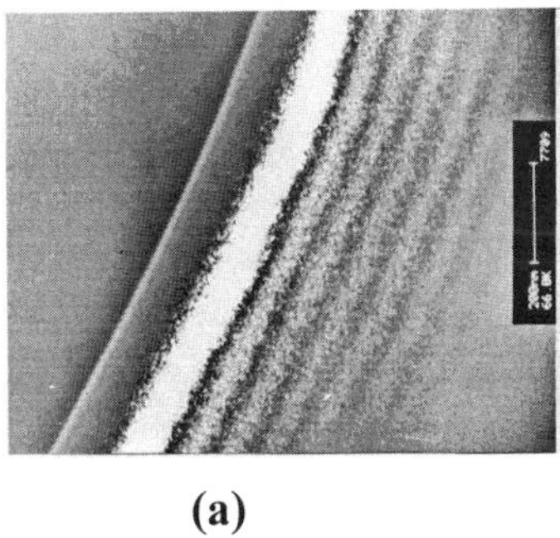

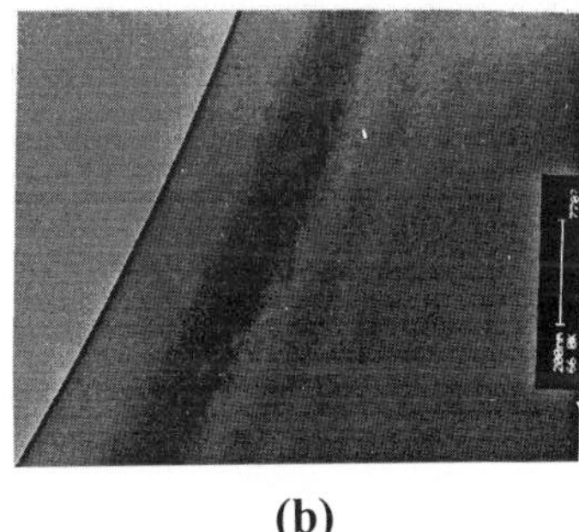

(a) (b)

Fig. 3: BF (a) and DF (b) TEM images of anti-phase domain boundary in ZGP (Mag. 66K), taken from the same area as in Fig. 2, respectively.

Fig. 4: Electron diffraction pattern from the anti-phase domain boundary regime in ZGP. Two different structures can be observed within the pattern, coherently oriented with respect to each other: a set of small dots corresponds to ordered tetragonal ZGP, and another set of spread dots which fits to disordered cubic ZGP. See text for details.

entities are mutually oriented. This result is compatible with the identification of the defect appearing in the TEM image as an APB. Similar phenomenon has already been observed at the phase transition between the NiAl and $NiAl_3$ phases [21].

As already mentioned, the optical absorption band in the near IR in ZGP is commonly assigned to native point defects. Such defects in the cation sub-lattice may arise either from non-stoichiometric evaporation of the constituent elements at elevated temperatures, or from disordering which occurs during solidification. In the present study, TEM and electron diffraction results reveal the appearance of anti-phase domain boundaries (APB) in the ordered ZGP matrix. It should be born in mind, however, that such boundaries are often associated with a variety of point defects, such as anti-site pairs, and vacancies. Thus, we tend to believe that the anti-phase domain boundaries revealed by the TEM and electron diffraction study is related to the creation of various point defects in ZGP.

4. Summary

Transmission-electron-microscopy and electron diffraction was used for direct observation of anti-phase domain boundaries in chalcopyrite ZGP. The structural defect thus observed correlates to our former results, based on x-ray diffraction, on the existence of a cubic structure embedded within the tetragonal matrix of ZGP. The anti-phase domain boundaries are coherently oriented relative to the tetragonal matrix, and might be associated with various point defects, which have already been found to exist in melt-grown ZGP.

References:

[1] A.A. Viapolin, E.O. Osmanov, and V.D. Prochukhan, Izv. Acad. Nauk SSSR, Neorg. Mater., 8(5), (1972) 947.

[2] L. Shay and J. H. Wernick, in: ***"Ternary Chalcopyrite Semiconductors: Growth, Electronic Properties, and Applications"***, Pergamon Press, New-York, N.Y., 1975.

[3] J.J.M. Binsma, L.J. Giling, and J. Bloem, Phys. Stat. Sol. (a) 63, (1981) 595.

[4] A.A. Vaipolin, N.A. Goryunova, L.I. Kleshchinskii, G.V. Loshakova, and E.O. Osmanov, Phys. Stat. Sol. 29 (1968) 435.

[5] M.H. Rakowsky, W.K. Kuhn, W.J. Lauderdale, L.E. Halliburton, G.J. Edwards, M.P. Scripsick, P.G. Schunemann, T.M. Pollak, M.C. Ohmer, and F.K. Hopkins, Appl. Phys. Lett. 64(13) 28 March 1994, 1615.

[6] Y. Shimony, G. Kimmel, O. Raz, and M.P. Dariel, J. Crystal Growth 198/199 (1999) 583.

[7] K. Masumoto, S. Isomura, W. Goto, J. Phys. Chem. Solids 27, (1966), pp. 1939.

[8] S. Fiechter, R.H. Çastleberry, M. Angelov, K.J. Bachmann, in ***"Infrared Applications of Semiconductors – Materials, Processing and Devices"***, M.O.manasreh, T.M. Myers, and F.H. Julien (eds.), MRS Symposium Proceedings, Vol. 450, MRS 1996 Fall Meeting, Boston, 2-6.12.1996, pp. 315.

[9] F. K. Hopkins, Laser Focus World, July 1995, pp. 87.

[10] D.F. Bliss, M. Harris, J. Horrigan, W.M. Higgins, A. Armington, and J.A. Adamski, J. Crystal Growth, 137 (1994) 1945.

[11] H.M. Hobgood, T. Henningsen, R.N. Thomas, R.H. Hopkins, M.C. Ohmer, W.C. Mitchel, D.W. Fischer, S.M. Hegde, and F.K. Hopkins, J. Appl. Phys. 73(8) 15 April 1993, 4030.

[12] L.S. Gorban,V.V. Grishchuk, and I.J. Tregub, Sov. Phys. Semicon. 18 (1984) 567.

[13] G.K. Averkieva, V.S. Grigoreva, I.A. maltseva, V. D. Prochukhan, and Yu. V. Rud, Phys. Stat. Solidi A39 (1977) 453.

[14] M.H. Rakowsky, W.K. Kuhn, W.J. Lauderdale, L.E. Halliburton, G.J. Edwards, M.P. Scripsick, P.G. Schunemann, T.M. Pollak, M.C. Ohmer, and F.K. Hopkins, Appl. Phys. Lett. 64(13), March 28, 1994, 1615.

[15] L.E. Halliburton, G.J. Edwards, M.P. Scripsick, M.H. Rakowsky, P.G. Schunemann, and T.M. Pollak, Appl. Phys. Lett. 66 (1995) 2670.

[16] A. Sodeika, A.Z. Silevicius, Z. Januskevicius, and A. Sakalas, Phys. Status Solidi , A69 (1982) 491.

[17] Y.V. Rud and R.V. Mesagutova, Sov. Tech. Phys. Lett. 7(2), February 1981, 72.

[18] M.H. Rakowsky, W.J. Lauderdale, R.A. Mantz, R. Pandey, and P.J. Drevinsky, D.C. Jacobson, D.E. Luzzi, T.F. Heinz, M. Iwaki (Eds.), Beam-Solid Interactions for Materials Synthesis and Characterization Symposium, Pittsburgh, PA, USA, (1995) 735.

[19] Y. Shimory, O. Raz, G. Kimmel, and M.P. Dariel, Optical Materials 13(1) (1999) 101.

[20] J.W. Edington, "***Practical electron microscopy in material science***", Van Nostrand Reinhold Company, New York, N.Y., 1976.

[21] B.W. Williams and C.B. Carter, "***Transmission Electron Microscopy***", Planum Press, New-York, N.Y., 1996.

STUDY OF THE OPTICAL BEHAVIOR OF p-$CuInSe_2$ FILMS BY PHOTOACOUSTIC SPECTROSCOPY

K.T.Ramakrishna Reddy*, M.A.Slifkin† and A.M.Weiss†

**Department of Physics, Sri Venkateswara University, Tirupati - 517 502, India*
†Department of Electronics, Jerusalem College of Technology, Jerusalem, Israel

Abstract

Polycrystalline and single phase p-$CuInSe_2$ thin films were grown by selenization of sputtered Cu/In multilayer at a temperature of 500°C. The samples were characterized using photoacoustic spectroscopy (PAS) to evaluate the optical properties of the layers. The layers showed a threefold optical structure with three high energy transitions which is the characteristic behavior of the material, evaluated for the first time using this method. The PA spectrum is reported and discussed.

Keywords: $CuInSe_2$ thin films, Selenization, Photoacoustic spectroscopy, Energy band gap

1. Introduction

$CuInSe_2$ is a member of the ternary chalcopyrite compounds, well known for its photovoltaic properties. [1] This material has been synthesized both in crystal form and in thin film form using a variety of techniques. [2, 3] Although this material has been extensively studied, the available data on the non-radiative processes involved in $CuInSe_2$ is very meager. [4] Furthermore, there is little or no detailed analysis of photoacoustic spectra of $CuInSe_2$ synthesized by selenization in the literature. The photoacoustic process can be briefly described in terms of a localized temperature rise within the sample produced by a periodically modulated light beam, resulting in a pressure wave being generated by the sample and transmitted to a sensitive acoustic detector. In the present study, a simple photoacoustic spectrometer was used to evaluate optical properties such as absorption coefficient, energy band gap and the optical structure of $CuInSe_2$ layers grown by a two step process. We believe that this is the first application of PAS to the study of this material.

2. Experimental

Thin films of $CuInSe_2$ were prepared by selenization of ultrathin Cu-In multilayer precursors. Alternate layers of Cu and In were grown by magnetron sputtering using the respective 5N pure targets in the presence of high purity argon, resulting in about 1700 bilayers with a film thickness of 1.0μm. These layers were selenized using high purity selenium in a closed graphite box, which was placed on the top of a strip heater in an annealing chamber. The chamber was evacuated to less than 10^{-2} Torr, and the layers were annealed at a temperature of 500°C. The design of the photoacoustic cell and the details of the optical arrangement used in the present study are described elsewhere. [5] The photoacoustic spectra were corrected for the spectral distribution of the optical system, the cell and the microphone by normalizing the response of the specimen to that of a fine powder of carbon black. All the spectra were recorded at room temperature using a modulation frequency of 112 Hz, corresponding to a thermal diffusion length of 120μm for $CuInSe_2$.

3. Results and Discussion

The as-grown $CuInSe_2$ films were strongly adherent to the substrate and blackish in appearance. Electrical measurements showed p-type conductivity in the layers. The structural analysis of the sputtered Cu-In precursor layers revealed a homogeneous $Cu_{11}In_9$ phase without any secondary phases. The synthesized $CuInSe_2$ films were uniform, and the crystal planes were strongly oriented along (112). The surface topology indicated triangular faceted crystals with a crystallite size of 2.0 μm. The detailed analysis of these layers was reported earlier. [6] The films showed a Cu/In ratio of 0.94 with an elemental composition of Cu=23.8 at. %, In=25.3 at. % and Se=50.9 at. %.

The normalized photoacoustic spectrum of $CuInSe_2$ layers synthesized in the present study is shown in Fig. 1. The steep increase in the PAS signal indicates the direct nature of the fundamental transition. The shoulder that appears close to the band gap is characteristic of p-type material. The absorption coefficient, α, was calculated using the formula [7]

$$\alpha = \frac{q^2 + q(2-q^2)^{1/2}}{\mu_s(1-q^2)} \qquad (1)$$

where q is the normalized photoacoustic amplitude and μ_s is the thermal diffusion length of the material. This formula is valid when the value of q is less than or equal to 0.2, which was the case in the present study. The calculated absorption coefficient is greater than 10^4 cm^{-1} for energies above the fundamental absorption band edge.

Analysis of the absorption coefficient shows that the increase in α for the energy range $h\nu <$ 1.10 eV follows the relation

$$\alpha_1 = (A_1/h\nu)(h\nu - E_{g1})^{1/2} \qquad (2)$$

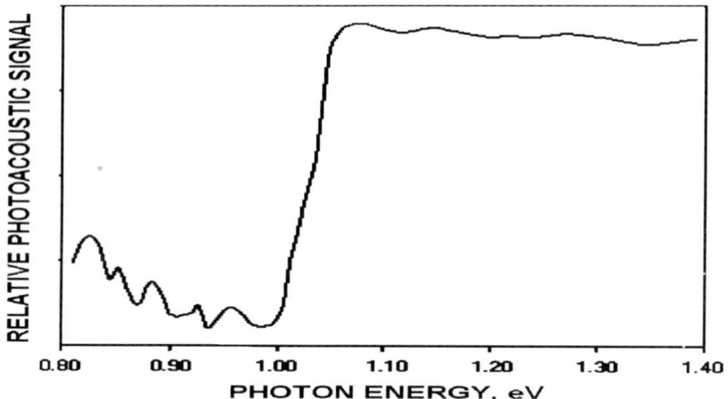

FIG. 1: Photoacoustic spectrum of $CuInSe_2$ films grown by a two-step process.

where Eg1 = 1.01 eV. However, when α_1 was calculated using A_1 and E_{g1} for energies higher than 1.10 eV, it was found that α_1 becomes considerably smaller than the absorption coefficient, α, measured experimentally, indicating the existence of an additional absorption processes. The analysis of the additional absorption coefficient, $\alpha_2 = \alpha - \alpha_1$, in the photon energy range 1.10 eV $< h\nu <$ 1.22 eV, showed that its dependence on $h\nu$ can be described by the relation

$$\alpha_2 = (\frac{A_2}{h\nu})(h\nu - E_{g2})^{3/2} \tag{3}$$

with E_{g2} = 1.07 eV, indicating that it is an indirect and forbidden band gap. When α_2 is calculated beyond $h\nu$ = 1.22 eV, the sum of α_1 and α_2 is still smaller than the value of α, indicating the presence of another higher energy transition. In the region beyond 1.22 eV, α_3 $(\alpha - \alpha_1 - \alpha_2)$ followed the relation

$$\alpha_3 = \frac{A_3}{h\nu}(h\nu - E_{g3})^{3/2} \tag{4}$$

with E_{g3} = 1.27 eV.

Fig. 2 shows the variation of $(\alpha_1 h\nu)^2$, $(\alpha_2 h\nu)^{2/3}$ and $(\alpha_3 h\nu)^{2/3}$ versus $h\nu$. From the present study, three energy band gaps have been determined for $CuInSe_2$ layers. They are E_{g1} = 1.01 eV, E_{g2} = 1.07 eV, E_{g3} = 1.27 eV. These values correspond to the three characteristic energy band

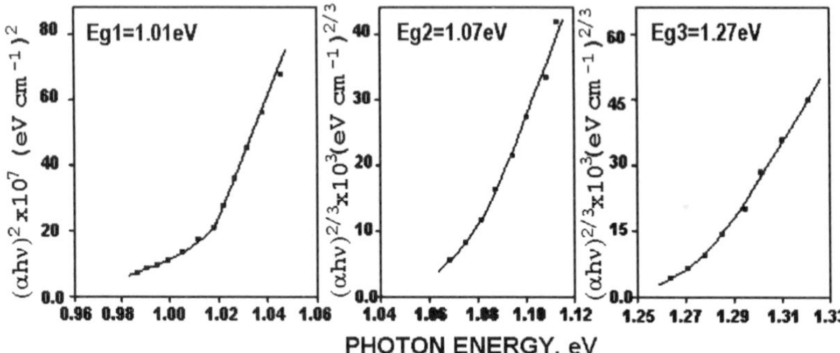

FIG. 2: Left: $(\alpha h\nu)^2$; **center and right:** $(\alpha h\nu)^{2/3}$; versus photon energy $h\nu$, for three energy bands of interest. The best fit energy gap is shown on each plot.

gaps observed in chalcopyrite compounds and are in agreement with the reported data on both single crystals and thin films. [8]-[10] The first band gap corresponds to the fundamental transition between the bands while the latter two are attributed to the crystal field and spin-orbit splitting in this material. The absorption spectrum in the sub-band gap region showed five peaks, probably due to the defect states present in $CuInSe_2$ layer. A detailed study of these sub-band gap states and the influence of the growth parameters on these states is in progress.

4. Conclusions

The present study demonstrates the potential of photoacoustic spectroscopy in the analysis of the optical properties of thin layers. The different optical parameters of $CuInSe_2$ films evaluated in the present study showed that these films have a band structure comparable with that of single crystals. The threefold optical structure was observed for the first time using the PAS technique in $CuInSe_2$ layers grown by a two-step process.

Acknowledgements: One of the authors (KTRR) wishes to thank the Indian Ministry of Science and Technology and the Israeli Ministry of Science for providing financial assistance under the Visiting Fellowship Program. The assistance of Dr.M.J.Carter, Dr.R.W.Miles and Prof.P.K.Dutta of the School of Engineering, University of Northumbria, Newcastle, U.K. in the preparation and characterization of the samples is greatly acknowledged. The PAS was built with a grant from the Israeli Ministry of Science.

References

[1] K. Mitchell, in *Solar Cells and their Applications*, Ed. L.D. Partain, John Wiley and Sons Inc., New York (1995).

[2] S.R.Kumar, R.B.Gore and R.K.Pandey, "Preparation and properties of a Cu-In alloy precursor for $CuInSe_2$ films," *Solar Energy Mater. and Solar Cells* **26**, 149 (1992).

[3] M.E.Calixto and P.J.Sebastian, "$CuInSe_2$ thin films formed by selenization of Cu-In precursors," *J. Mater. Sci.* **33**, 339 (1998).

[4] K. Yoshino, T.Shimizu, A.Fukuyama, K.Maeda, P.J.Fons, A.Yamada, S.Niki and T.Ikari, "Temperature dependence of photoacoustic spectra in $CuInSe_2$ thin films grown by molecular beam epitaxy," *Solar Energy Mater. and Solar Cells* **50**, 127 (1998).

[5] M.A.Slifkin, L.Lurie and A.M.Weiss, "Photoacoustic spectroscopy of photovoltaic materials," *SPIE* **3110**, 481 (1997).

[6] K.T.Ramakrishna Reddy, I.Forbes, R.W.Miles, M.J.Carter and P.K.Dutta, "Growth of high-quality $CuInSe_2$ films by selenising sputtered Cu-In bilayers using a closed graphite box," *Mater. Letts.* **37**, 57 (1998).

[7] J.Poulet, J.Chambron and R.Unterreiner, "Quantitative photoacoustic spectroscopy applied to thermally thick samples," *J. Appl. Phys.* **51**, 1738 (1980).

[8] J.L.Shay and J.H.Wernik, *The Ternary Chalcopyrite Semiconductors: Growth, Electronic properties and Applications*, Pergamon Press, New York (1975).

[9] W.Horig, H.Neumann, H.Sobatta, B.Schumann and G.Kuhn, "The optical properties of $CuInSe_2$ thin films," *Thin Solid Films* **48**, 67 (1978).

[10] D.Sridevi and K.V.Reddy, "Electrical and optical properties of flash evaporated $CuInSe_2$ thin films," *Indian J. Pure and Appl. Phys.* **24**, 392 (1986).

LINEAR HOLOGRAPHIC RECORDING IN AMORPHOUS As_2S_3

Michael Lisiansky, B. Spektor, J. Shamir, M. Klebanov and V. Lyubin

Series Editors, Annals of the Israel Physical Society

c/o Department of Electrical Engineering, Technion-Israel Institute of Technology, Haifa 32000, Israel

Abstract

An experimental study was performed to investigate the linearity of the phase recording in amorphous α-As_2S_3 films. In contrast to the previous results we demonstrate that such linearity is possible if proper precautions are taken. We found that the temperature, recording wavelength and a substrate quality are important factors for approaching linear relationship between exposure and the change in refractive index. Phase gratings were recorded by Ar^+ laser (458 nm, 488 nm, 514.5 nm) while resulted diffraction efficiency was monitored in real time by He-Ne laser. Linearity of recording was verified by simultaneous monitoring of the first- and second-order diffraction efficiency

1. Introduction

Most high efficiency optical elements are phase elements that operate only on the phase of an incident wave without affecting its amplitude. That is due the great interest exists to the development and implementation of diffractive optical elements (DOE) in which light intensity distribution can be directly and precisely recorded as spatial phase modulation. Materials with unique combination of parameters (large photoinduced change of refractive index, high photosensitivity, linear recording characteristics) are desirable for DOE fabrication. Amorphous arsenic trisulfide (α-As_2S_3) is an important candidate for this practical application. From earlier publication it is known [1,2] that high photosensitivity of this material goes with significant diffraction efficiency (DE) of phase holograms, but considerable non-linearity was indicated in their recording. Several reasons were suggested for the non-linearity of recording. The main objective of this work is to determine if the non-linearity observed earlier is an intrinsic property of As_2S_3 or suitable conditions can be found for linear recording.

2. Background and experimental procedure

As-deposited (non-annealed) As_2S_3 films of various thickness δ (0.3-4.2 μm) thermally evaporated on glass substrates were exposed at three different laser wavelengths (514.5 nm, 488 nm and 458 nm). The recording was monitored in real-time during exposure by a normally incident He-Ne laser (632.8 nm) beam about 1 mm in diameter. In contrast to most earlier publications the measurements were based on exposure with uniform sinusoidal intensity distribution generated by the superposition of two (almost) plane waves with aperture of 6 mm in diameter aligned symmetrically with respect to the normal to the sample surface. As a period of grating was $\Lambda \approx 20$ μm, the thin film approximation of hologram recording ($\Lambda/\delta > 5$) was kept during the experiments. The organization of experiment eliminates a possible influence of high light intensity gradients that exist in recording of phase gratings with narrow beams. Such precaution seems important in view of other processes that according to literature [3,4} could affect the refractive index modulation caused by the photoinduced structural change (PSC) - the main process. The most significant of these is related to relaxation structural changes (RSC) resulting from the spatial modulation of the material viscosity (photo-plastic effect) [3] and the recharging of localized states [4].

By our definition of linearity *recording is linear when proportionality exists between the exposure E and the change of refractive index* Δn. From this definition of linearity and sinusoidal intensity distribution of illumination $I(x) = I_b + I_0 \cos Kx$ follows [2,5]:

$$\Delta n(x,t) = pI_b t + pI_0 t \cos Kx \qquad (1)$$

where $I_b, I_0, K = 2\pi/\Lambda$ are constants, t is the exposure time. The parameter p represents the phase photo-sensitivity of the material and is constant for appropriate recording conditions. Since we view our recorded phase grating like a thin optical element that modulates a probe beam, we may describe it by its complex amplitude transfer function in the form,

$$T(x,t) \propto e^{jk\delta\, pI_0 t \cos Kx} \qquad (2)$$

where k is the probe beam wave number ($k = 2\pi/\lambda$), with λ representing the probe wavelength in the material. Expanding expression (2) into series we obtain a series Bessel functions. If we define the DE of the ν–order as the fraction of light intensity propagating in the corresponding diffracted wave by η_ν, we obtain:

$$\eta_\nu \propto \left[J_\nu(\phi)\right]^2 \qquad (3)$$

where J_ν is the ν-th order Bessel function of the first kind and $\phi = kpI_0 t\delta$.

3. Experimental results

Since earlier experiments [4,5,6,15] exhibited significant non-linearity for recording at 514.5 nm and 488 nm, the present experiments were extended to three wavelengths of the Ar^+ laser including 458 nm. To confirm the linearity of phase recording we introduced, in our experiment, the simultaneous monitoring of the first- and second order diffracted beams. If both diffraction orders satisfy Eq. (3), it is a strong indication of highly linear relationship since Bessel functions of different orders are mutually orthogonal. Moreover the second order Bessel function is related to the derivative of the first order one, leading to higher sensitivity in the detection of any non-linearity. The experimentally obtained results (Figs. 1 and 2) display the root square of the DE normalized to its maximum value, $\sqrt{\mathrm{DE}} = \sqrt{\eta_v / \eta_v(\max)}$, as a function of exposure with constant intensity. The dotted lines are appropriate Bessel functions fitted to the experimental curves. The detailed description of the fitting procedure is given in [5].

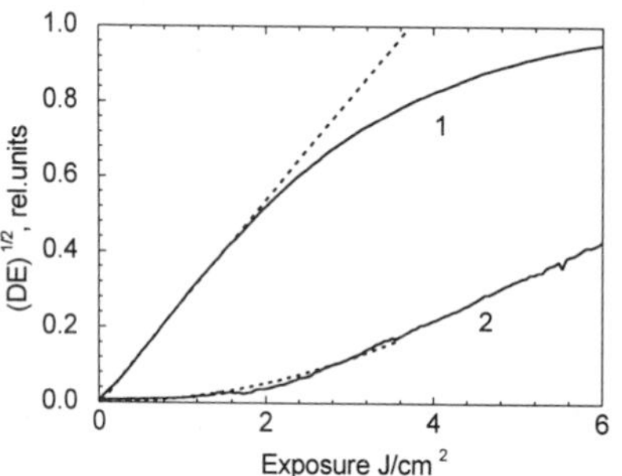

Fig. 1 Exposure characteristic of the thin film 0.3 μm recorded at 458 nm with light intensity 40 J/cm^2.

Basically, the results obtained at room temperature in films exposed by laser with wavelengths 488 nm and 514.5 nm are similar to earlier results [1,2]. But at short wavelength 458 nm where the rate of the PSC is strongest, the recording on films with thickness 0.3 and 1.1 μm is practically linear from zero exposure up to about 25% of the maximum value of the refractive index change (see Fig. 1). For longer wavelength there is a significant range of exposures where Eq. (3) for first-order DE is followed, provided the fitting curves are properly shifted from the origin as shown in Fig. 2a. Note that at higher exposures, where an apparent *quasi-linear* behavior was observed from the first diffraction order (curve 1, Fig. 2a), a significant non-linearity was still exhibited by the second diffraction order (curve 2). This indicates the higher sensitivity of the second order to non-linearity effects.

A qualitative explanation of our results is based on the microscopic model that there are two physical processes, PSC and RSC, going on in parallel. The non-linearity at low exposures and at long wavelength recording can be associated with a contradictory effect of the two processes on the changes of the refractive index. As result gratings produced by RSC effects (see Section 2) may be in opposite phase to the grating generated by PSC. A mathematical model describing the various competing processes may be written by modifying the parameter p in Eq. (3) as $p \rightarrow p - \beta(t, K, \delta, I)$, where $\beta(t, K, \delta, I)$ is a relaxation function which depends on time, intensity, layer thickness and the spatial modulation frequency. This relaxation function is most significant at low exposures where the PSC can be dynamically compensated by the RSC (p is of the order of β). As a result, recording linearity is violated.

However, when the relaxation process is completed ($\beta \to 0$), recording linearity is regained, but with a proper shift along the exposure axis, as indicated by our experimental results. Since the sensitivity parameter p increases strongly on wavelength, at short wavelength (458 nm) $p >> \beta$, leading to practically linear recording. At longer wavelength the competition between the PSC and the RSC is much more severe resulting in a significant non-linear region.

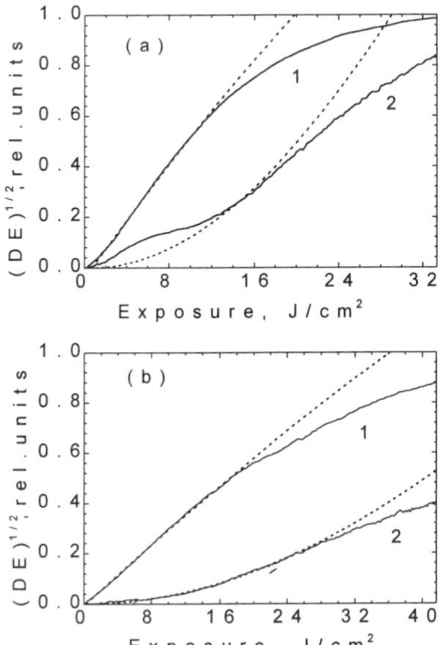

FIG. 2 Exposure characteristics recorded on the same film 1.0 μm at 514 nm with light intensity 160 mW/cm^2 at temperature: a)295 K; b)275 K.

Since both effects that lead to RSC are strongly temperature dependent [3,4], the overall RSC must be so too. If our hypothesis is correct the effects of RSC should decrease with decreasing temperature improving the linearity of the recording also at longer wavelength. Fig. 2 shows results of exposure by 514.5 nm obtained on the same film 1.0 μm thickness at two temperature 295 and 275 K. At 295 K recording is non-linear (Fig. 2a). But at 275 K (Fig. 2b) linear recording starts practically from zero exposure. This is indicated by an ideal matching between the experimental curves and theory, both for the first and the second order diffraction (curves 1 and 2, respectively), up to a diffraction efficiency of ~1.5%. It is worth noting that the substrate also effects on the linearity of recording. We succeeded to perform the linear recording at 514.5 nm only on films deposited on high quality optical glass. Thus proper conditions for linear recording in α-As_2S_3 in wide wavelength range were found and this open way to a various applications of the material as a light recording media.

Acknowledgement: This research was supported by the Israel Ministry of Science

References:

[1] A. Ozols, N. Nordman, O. Nordman, and R. Riihola, Phys. Rev. B 55, 14236 (1997).

[2] T. Suhara, H. Nishihara, and J. Kojama, Electron. Commun. Jpn. 59C, 116 (1976).

[3] D. K. Tagantsev and S. V. Nemilov, Sov. J. Glass Phys. and Chem. **15**, 220, (1989).

[4] A. V. Tyurin, A. Yu. Popov, V. E. Mandel', and V. M. Belous, Phys. Sol. State 38, 209, (1996).

[5] B. Spektor, M. Lisiansky, J. Shamir, M. Klebanov and V. Lyubin, J. Appl. Phys. (in press).

NEAR FIELD PHOTO-DARKENING OF CHALCOGENIDE GLASSY FILM

Yuval Isbi[1,2], Shmuel Sternklar[1], Er'el Granot[1], Victor Lyubin[3], Matvey Klebanov[3] and Aaron Lewis[2]

1 Non Linear Optics Group, Electrooptics Division, Soreq NRC , Yavne 81800, Israel
2 Division of Applied Physics, The Hebrew University of Jerusalem, Jerusalem, Israel
3 Department of Physics, Ben-Gurion University of the Negev, Beer-Sheva Israel

Abstract

Chalcogenide glassy film is suggested as a candidate for near-field nano-writing. Photo-induced changes in the transmission (photo-darkening) of Chalcogenide glassy film (AsSe), with cw and pulsed green light, was investigated. The sensitivity of this effect is ~100 times stronger with pulsed light as compared to cw illumination. A 25% reduction of transmission using one pulse, at an energy flux achievable in the near field, make this medium a good candidate for nano-writing. Nano-writing of 100nm-width lines is shown.

1. Introduction

In the quest for technologies to optically record small features, both for the microelectronic industry and for high-density data storage, near field optics is considered to be a promising technique [(1-3)]. Near field optics overcomes the diffraction limit imposed by conventional optics and has been used successfully to achieve sub-wavelength optical resolution in near field microscopy for the last fifteen years[(4-5)].

Near field nano-writing is based upon the fact that light passing through a small hole in an opaque screen with diameter ρ smaller than the wavelength λ, is effectively collimated until a distance $r \sim \rho$ [(6)]. Consequently, light that passes through a 100nm hole at the end of a metal-coated tapered fiber placed at a distance of 10-50nm above a surface, illuminates a 100nm spot of light on the surface. Controllable motion of the fiber tip and hence the light spot allows optical nano-writing with the proper choice of light, surface material and the interaction between them.

Chalcogenide glass is known to be a suitable material for optical recording[(8-9)]. Its high resolution makes it suitable for high-density data storage and building nano-structures. With different quantities of composites it may be used as a positive or negative photoresist material. Among its interesting features, we use the photo-induced change of its transmission spectrum, also known as photo-darkening (PD)[(8-9)]. The transmission spectrum in the visible is higher at longer wavelength and reaches its maximum close or in the IR. Light at short wavelengths that is absorbed changes the spectrum in a manner that "shifts" it towards longer wavelengths, so that the transmission in the visible is lowered and the area that was illuminated looks darker.

The amount of the PD depends, as we will show, on the total absorbed energy and on the power. This writing process is reversible and can be erased simply by heating. In this work we characterize an

AsSe glassy film by studying its PD behavior under cw 514.5nm and pulsed light (7ns) at 532nm. We will show that the PD sensitivity to pulsed light is larger by two orders of magnitude. We will argue that this light level is suitable for nano-writing using NSOM microscope. With our NSOM, we show first results of writing on this medium with resolution better than the diffraction limit.

2. Far field AsSe characterization

The studied Chalcogenide glassy film is a composite of arsenic and selenium ($As_{50}Se_{50}$) made with vacuum evaporation technique with a thickness of ~20μm. This glass was chosen because of its low sensitivity to 633nm light and high sensitivity to green light. The PD effect studied here is the change of the transmission at 633nm HeNe light after illumination with green light.

The set up of this experiment is shown in fig 1.

Fig. 1

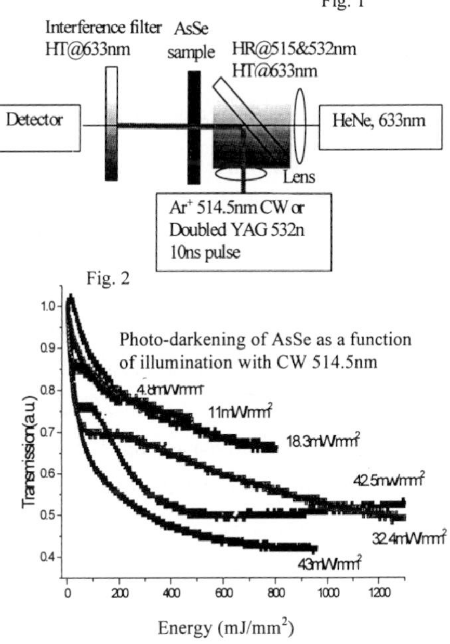

Fig. 2

Initially, cw (λ=514.5 nm) light from an Ar^+ laser illuminated a spot with a diameter of 3mm and intensities ranging from 5 to 50 mW/mm^2. The 633nm (HeNe laser) beam had a 0.5mm diameter and probed the transmission at the center of the green PD spot. The initial transmission was 50%. The intensity of the HeNe beam was kept low so that a secondary PD effect due to the weak 633nm probe light was negligible.

Fig.2 are plots of the photo-induced change in the transmission as a function of the total exposure energy for different cw light intensities.

From this plot one can see that an energy exposure of 100mJ/mm^2 reduces the transmission by 25%. Intensities higher than 50mW/mm^2 caused damage after some time and were not useful for PD.

With the same set-up we measured the PD due to pulsed light at 532nm. Fig. 3 (next page) plots the change in the transmission as a function of the total energy from additive pulses for different pulse energy. It is seen that the first pulse is responsible for the most dramatic change of the transmission. Fig.4 is a plot of the change in the transmission as a function of the first pulse energy.

One can see that a single pulse of 7ns duration with energy distribution less then 1mJ/mm^2 changes the transmission by 25%. The comparison between cw and pulsed light shows that the sensitivity of the PD effect to pulse light is more than 100 times larger. Because of this high sensitivity the material is suitable for high-density data storage written with pulsed lasers.

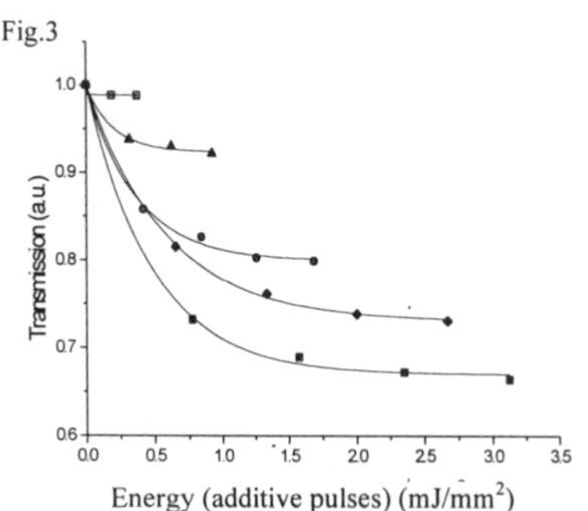

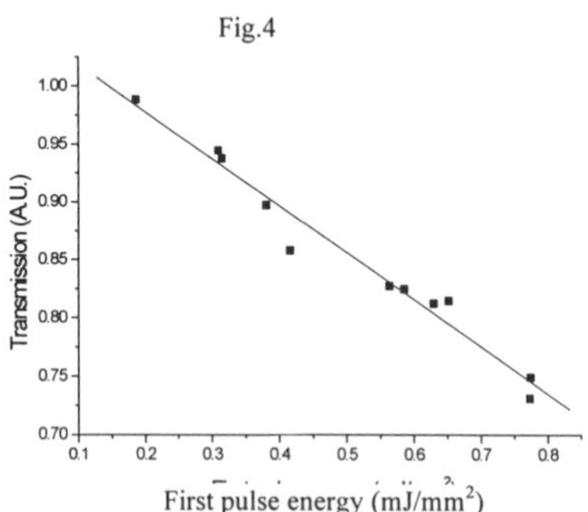

3. Near field sub-wavelength writing

We use a homemade near field microscope based on Nanonics (Jerusalem, Israel)(7) design. Cantilevered, bent and tapered, Cr/Al coated fiber tip, is held a few nm above the sample placed on a low profile piezo-electric 3D stage. Atomic force feedback regulates the distance between the tip and sample. Nano-writing is due to green light (cw or pulsed) that is launched through the fiber tip and creates a PD spot on the sample. PD patterns are creates with computer control movement of the sample in the xy plane. In the reading step he 633nm HeNe light is launched through the fiber tip and illuminates a sub-wavelength spot on the sample. The transmission signal from this spot is collected by a high-NA objective and focused on a low light level avalanche photo-diode (EG&G sq-151). The stage moves in a raster scan manner and the transmission image of the sample as well as topographic picture is build by the computer pixel by pixel. Fig. 5 is a schematic (not to scale) of the system.

The tip aperture diameter is ~100nm. The light throughput is 10pJ/pulse at 7ns pulse(10) and 10nW at cw. Which corresponds to an energy fluence of $1mJ/mm^2$ for pulsed light and $1W/mm^2$ for cw light, at the near field of the tip.

Following our analysis of the PD behavior, we chose the AsSe for sub-wavelength writing because of its high sensitivity at low light levels, especially for pulsed light, which is compatible with the near field tip's low light level throughput.

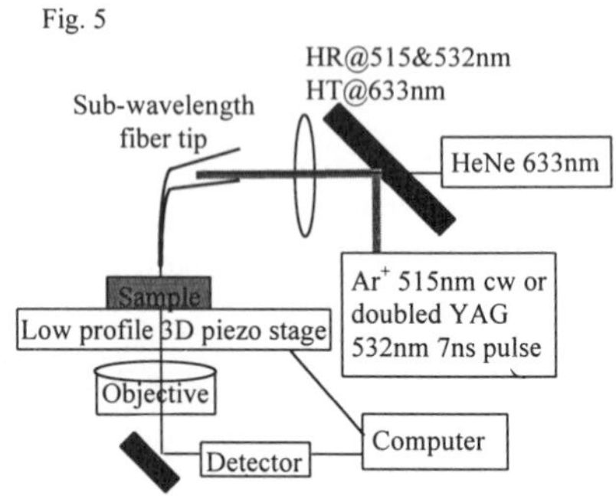

In this work we demonstrate sub-wavelength writing using cw illumination. CW light was launched through the tip, while a line 7μm long was repeatedly scanned. Each scan took 0.9sec and the total exposure time was 1min. The next Figure (fig 6) is the near field image of the written line. The change of the transmission is ~15% and there is no change in the topography. The readable line width is ~250nm. The line width is a convolution of the written width and the tip. This implies that the written width is ~125nm.

We have not yet succeeded in writing one point using cw illumination because of sample damage, possibly due to heating of the tip. Writing with pulsed light is obviously the next step.

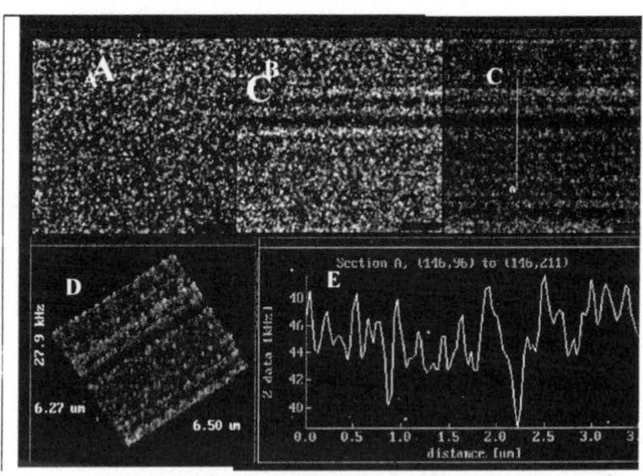

Fig 6. A: near field image before writing, B: image after writing line with cw light for 20sec, C: image after writing for 60sec, D: 3D image of C, E: line section in C.

4. Conclusion and future remarks

We proposed Chalcogenide glassy film as an optically writeable material with sub-wavelength resolution, by using near field techniques. We characterized the photo-induced reduction of the transmission in the visible, known as photo darkening (PD), for cw and pulsed light. Significantly higher sensitivity (x100) to pulsed light was demonstrated. Pulsed-light exposure may overcome slow writing speeds. Writing sub-wavelength lines with cw light is a very encouraging step towards the aim of reversible optical high density data storage. We show that, at least for cw light, the near field sub-wavelength behavior is similar to the far-field behavior. We thank the Israeli Ministry of Science for support under infrastructure project no. 8527-1-95.

5. References:

1: Rudman M., Lewis A., Mallul V., *Haviv V., et al*: J. Appl. Phys. 72 (9), 4379-4383 (1992)
2: Tsujioka T., Irie M.: J. Opt. Soc. Am. B 15 (3) 1140-1146 (1998)
3: Terris B.D., Mamin H.J., Rugar D. Appl. Phys. Lett. 68 (2) 141-143 (1996)
4: Lewis A., Isaacson M., Harootunian A., Muray A. : *Ultramicroscopy* 13 , 227-230 (1984)
5: Pohl D.W.: Scanning Near Field Optical Microscopy: Advances in Optical and Electron Microscopy Vol. 12 243-312 (Academic Press London 1991)
6: Leviatan Y. : J. Appl. Phys. 60, 1577 (86).
7: Lieberman K., Ben-Ami N., Lewis A. : Rev. Sci. Inst. 67 (10) 3567-3572 (1996)
8: Lyubin V., Klebanov M., Bar I., Rosenwaks S., *et al*: J. Vac. Sci. Tech. B 15(4) 823-827 (1997)
9: Lyubin V., Klebanov M., Bar I., Rosenwaks S., *et al.* : Applied Surf Science 106 502-506 (1996)
10: private communication with Klony Liberman from Nanonics (Jerusalem)

PART 10

DIFFRACTIVE OPTICS

Chairpersons: *A.A. Friesem, Israel; N. Davidson, Israel*

Annals of the Israel Physical Society, v. 14

ELECTRO-OPTICS and MICROELECTRONICS
Eds: Raphael LAVI and Ehud AZULAY

Analysis of A Diffractive Lenslet Array for Non-Uniformity Correction

Asa Fein

RAFAEL, Electro-OPTICS dept., POB 2250, Haifa 31021, Israel.

Abstract

Non-Uniformity in Focal Plan Arrays is of major concern in infrared imaging systems. One method for non-uniformity correction uses averaging of the scenery over the FPA in a uniform manner. This paper describes the theoretical and experimental analysis of A diffractive lenslet array was implemented for that purpose. This system was found to have reduced transmission and parasitic effects due to scattering from its diffractive elements that can limit its use in operational systems.

1. Introduction

The performance of Focal Plane Array (FPA) infrared imaging systems is strongly affected by the spatial non-uniformity in the photoresponse of the detector elements of the array. This non-uniformity results in a spatial pattern that can be viewed as a fixed pattern noise that reduce the resolving capabilities of the imaging system. To overcome this problem, the non-uniformity is electronically corrected. The common correction techniques use a uniform source such as a blackbody and image it onto the FPA. The individual response of each detector element is compared to the average response and corrected. This can be also done in real time by averaging the scenery over the FPA in a uniform manner. One method to achieve this uniform illumination is by using a lenslet array that causes a strong defocus of the scenery over the FPA [1].

2. System requirements and implementation

For a system to be operational, these conditions have to be met: 1. Strong defocus and uniform illumination over the FPA; 2. Close to 100%

transmission; 3. Minimal zero order transmission; 4. Only scenery radiation should reach the FPA.

The lenslet array consists of 1mmX1mm lenses with spherical profile and 3mm focal length which ensures a uniformly defocused scenery over the FPA. The array was etched by the Holo-Or Company, from ZnSe on a 1mm thick substrate. Each lens is made of 7 zones and each zone is etched in 14 steps. A close-up of few lenslets is shown in Fig. 1.

Fig. 1 A close-up of a 1mm square sided etched lenslets.

3. Optical performance

Measurements of the BFL and PSF verified that all lenses were diffraction limited and the BFL was as designed.

Calculation of the diffraction efficiency, using the usual procedure for multi-level binary optics [2,3], shows that an ideal lens, at the wavelength range 4-5μm, will have 97% 1st order efficiency and 0.4% zero order efficiency. The deviation from the ideal lens profile can be estimated from the zero order efficiency. Two methods were used for this purpose: 1) An IR camera and narrow band-pass filter were used and the zero order efficiency was found to be 3-4%; 2) A spectrometer with very narrow field of view was used to measure only the zero order transmission for a large wavelength range. The results, presented in Fig. 2, show clearly that the deviation from an ideal lens performance lie in errors in the zone depth that is around 10%.

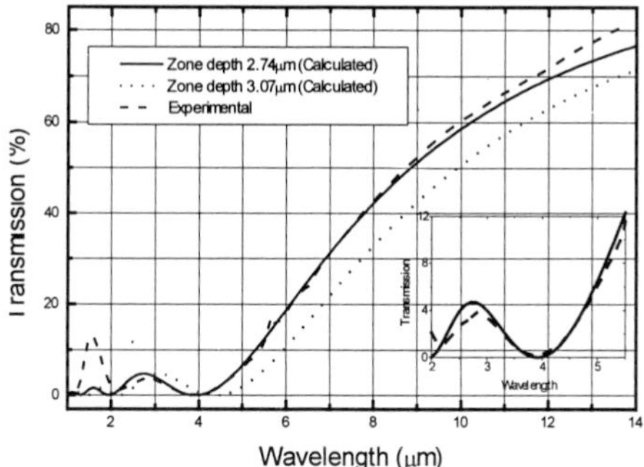

Fig. 2 Zero order diffraction efficiency measured and calculated for two zone depths, as a function of wavelength.

4. Functional analysis

The purpose of the measurements performed was to test the overall transmission, uniformity of illumination and scattering of radiation from internal parts onto the FPA.

Spatial uniformity was tested for both uniform and non-uniform backgrounds. In the first experiment a uniform blackbody was imaged through the lenslet array and it was found that the array did not introduce any non-uniformity. In the second experiment we imaged the outside scenery and performed non-uniformity correction, which was of good quality. The remaining concern with regard to illumination uniformity is the residual zero order transmission.

Overall transmission was measured by introducing an extended and hot (100^0C) blackbody in front of the array. As shown in Fig. 3., the transmission was measured as a function of the spatial angle subtended by the blackbody. One can see here that the transmission is angle dependent and, moreover, there is radiation that enters the FPA from outside its field of view. This is caused by scattering from the etched surfaces and is proved by the fact that this scattering is somewhat reduced when the array was anti reflection coated. This scattering explains the rather low transmission. Thus, although the transmission from a coated array was lower then expected but acceptable, the scattering is a serious problem for applications in operational systems.

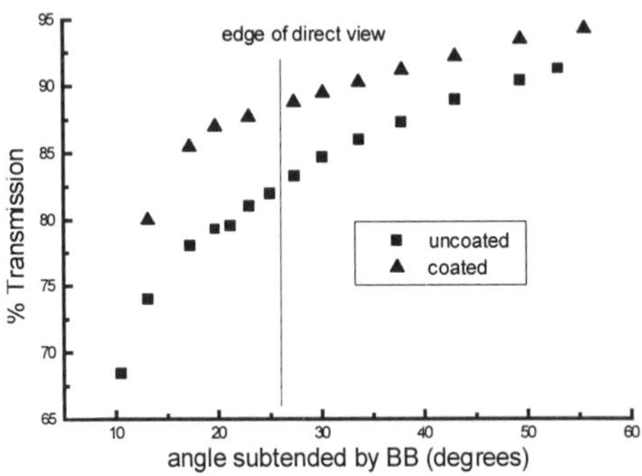

Fig. 3 Transmission of the lenslet array as a function of the spatial angle subtended by an extended blackbody.

5. Conclusions

A diffractive lenslet array for the purpose of non-uniformity correction was manufactured and analyzed. The analysis shows that: 1. The illumination over the FPA is uniform under nominal conditions; 2. The total transmission of the system is 85%; 3. This rather low transmission is caused by scattering that can bring about light scattered from the optical system onto the FPA; 4. Zero order transmission is on the order of 3%, which can be of a problem when there are strong sources in the background.

6. References

[1] D. Even-Sturlesi and Micha Oron, United States Patent 5,471,047 (1995).

[2] G.J. Swanson, "Binary Optics Technology: The theory and design of multi-level diffractive optical elements", Lincoln Laboratory Tech. Rep. **854** (1989).

[3] G.J. Swanson, "Binary Optics Technology: Theoretical limits on the diffraction efficiency of multi-level diffractive optical elements", Lincoln Laboratory Tech. Rep. **914** (1991).

11th International Meeting on Electro-Optics and Microelectronics in Israel, Tel-Aviv, Israel, November 1999.

DIFFRACTIVE OPTICS FOR 3-D OPTICAL METROLOGY

E. Hasman, and V. Kleiner

Faculty of Mechanical Engineering, Optical Engineering Group
Technion-Israel Institute of Technology, Haifa 32000, Israel

N. Davidson, S. Keren, and A.A. Friesem

Department of Physics of Complex Systems
Weizmann Institute of Science, Rehovot 76100, Israel

Abstract

A novel approach to light stripe triangulation configuration that allows parallel, fast, real-time 3-D surface metrology with a large depth-measuring range and high axial and lateral resolution is presented. The method is based on a color-coded arrangement that exploits polychromatic illumination and a cylindrical element that disperses the incident light along the axis. This leads to an increase of the depth-measuring range without any decrease in the axial or the lateral resolution. A 20-fold increase in the depth of focus was experimentally obtained, while diffraction-limited light stripes were completely maintained.

1. Introduction

The increasing need for non-contact three-dimensional profile measurements has already led to the development of several electro-optical measuring systems[1], based on structured light triangulation. The simplest of these projects a single point of light from the source onto the surface of a three-dimensional object. The single point triangulation approach is relatively inexpensive and capable of high resolution. However, the measurement of three-dimensional surfaces involves lengthy time-consuming axial and lateral scanning which is often impractical. A more complicated system involves the projection of a light stripe onto the surface of the object and a two-dimensional detector array. In this system, only one lateral scan in the direction perpendicular to the stripe, is needed, thereby reducing the overall measurement time[2]. Conventional light triangulation systems cannot simultaneously achieve a large depth measuring range and high horizontal resolution, so in general, there is a trade-off between range of depth and lateral resolution.

Here we propose and demonstrate a novel approach for light stripe triangulation configuration, which allows parallel, fast, real-time three-dimensional surface metrology with large depth measuring range and high axial and lateral resolutions. The method is based on a color-coded arrangement that exploits polychromatic illumination and cylindrical element that disperses the incident light along the axis, in order to increase the depth measuring range without any decrease in the axial and lateral resolutions[3].

2. The color coded approach

The principle of our color-coded approach is described here. The polychromatic light source can be either a white light source or an ultra-short pulsed laser, that produces relatively large spectral bandwidth. A collimating lens forms plane waves which are then focused by axially dispersing optics (ADO), such as cylindrical diffractive optical element or a combined diffractive-refractive optical element. The ADO forms a "rainbow" light sheet which consists of light stripes of different wavelengths at different distances from the lens. An object whose depth is smaller than the "rainbow" focal depth ΔF, intersects the "rainbow" light sheet, and the intersection profile of the object is then imaged with an off-axis configuration (at an angle θ from the illumination optical axis) to a two-dimensional array CCD detector. The detected data is processed by a computer to give the profile in virtually real-time. After a line is detected, a computer controlled stepper motor shifts the object to get another line, and so on, until a complete 3-D profile of the object is obtained. Although the "rainbow" light sheet is comprised of thin stripes (limited by the diffraction) of the individually focused wavelengths, each stripe is surrounded by background light of other wavelengths. Thus, the detected profile is relatively broad. This profile can be significantly narrowed by suppressing the background light and allowing only the light of the proper (focussed) wavelength to be detected at each object depth. This is achieved by inserting into the detection path an interference filter, whose transmitted wavelength is spatially non-uniform. With such a variable wavelength filter (VWF) optimal performance is obtained when the spread and specific wavelength location matches that of the light from the axially dispersing optics. When the match is exact, it is possible to obtain diffraction limited resolution over the entire depth of focus without any scanning.

3. Realization and experimental results

In order to test our approach, we designed and built a color coded triangulation arrangement. For the illumination source, we used a 75 Watt xenon arc lamp with a continuous spectrum ranging from 400nm to 700nm, and utilized a heat absorbing glass in order to reduce the IR radiation. The light emerging from the lamp was focused by a parabolic mirror onto a slit whose aperture was $40\,\mu m$ from where it was collected and formed a "rainbow" light sheet with 2-f lenses configuration. The first lens in the 2-f configuration was an achromatic lens with 500 mm focal length, and the second lens was the ADO. The ADO was a combination of a cylindrical refractive lens with focal length $f_r = 496mm$, and an adjacent quadratic phase diffractive cylindrical lens with focal length of $f_d = 1040mm$, both at $\lambda_o = 529nm$. The diffractive lens was recorded by using a computer generated mask, photolithographic techniques and reactive ion etching to form 16 binary level element on a fused silica substrate[4]. The combined ADO lens configuration had an aperture of $D = 9mm$ and it was designed to have an approximately linear dispersion, in order to match exactly the linear dispersion of the commercially available VWF: SCHOTT VERIL S 60. The measured and calculated combined focal length of the ADO at $\lambda_o = 529nm$, was $F_o = 326mm$ and the "rainbow" focal depth was $\Delta F = 48mm$. In comparison, focal depth for a conventional refractive lens is $\delta F \cong 4\lambda_o F_\#^2 \cong 2.7mm$ (80% of the maximal axial intensity), at a similar F-number of $F_\# = 36$. Therefore, the enlarging factor of the focal depth M_o, which is defined as the ratio between the "rainbow" focal depth of the ADO, ΔF, to that of a conventional lens focal depth δF, is $M_o = \Delta F / \delta F \cong 18$.

To test the experimental arrangement, we performed a series of measurements on a flat object, placed at an

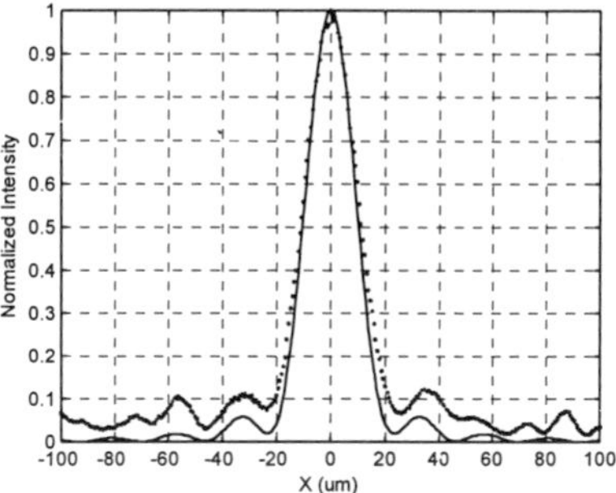

Fig.1(a)

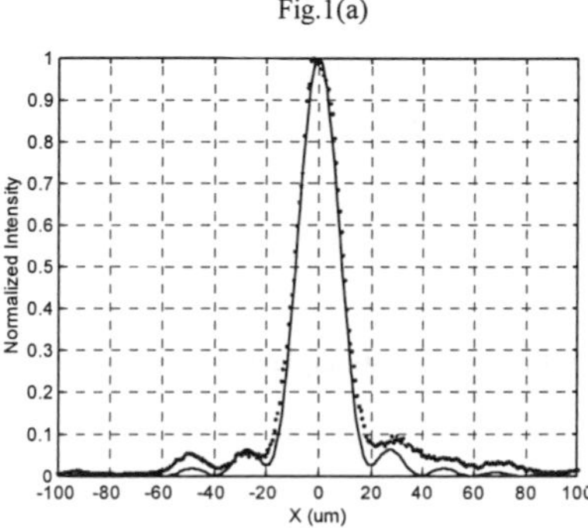

Fig. 1(b)

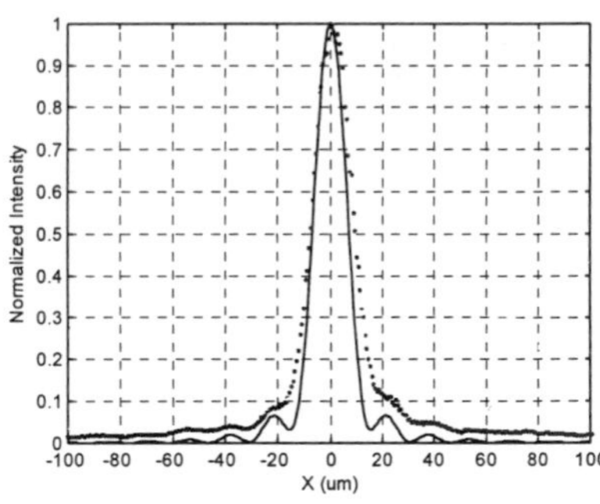

Fig. 1(c)

Fig.1:The intensity cross sections at three positions along the focal range; calculated intensity by solid curves, measured intensity by dotes, 1(a) for z=299mm; 1(b) for z=323mm; 1(c) for z=347mm;

angle, so as the entire focal range is included in the measurements. First, we measured the intensity cross-section distributions of the reflected light at the start, middle and end of the ADO focal range, (z=299mm, 323mm, and 347mm, respectively). Figures 1(a-c) show the measured and calculated intensity cross sections. The intensity for each wavelength was calculated using Fresnel diffraction integral multiplied by the VWF response and taking into account the dispersion relation of the ADO. The total intensity was obtained by integrating those intensities over the entire spectrum of the source. These results show that diffraction limited line width, of about $\Delta x \cong 20\mu m$ at full width half maximum, is maintained throughout the focal range with good agreement between experiment and theory. These indicate that near-perfect matching between the ADO and filter dispersions was indeed obtained.

Finally, in our experiments we used real time software for processing the data from the CCD camera and calculating the center of the detected stripes. An entire line cross section of the object's profile can obtained at video rates, to enable fully automatic and fast 3-D measurement of the overall surface of a three-dimensional object.

4. Concluding remarks

We presented a novel approach for rapidly determining the surfaces of three-dimensional objects. It is based on a color-coding and decoding arrangement that exploits polychromatic illumination and cylindrical, axially dispersing, optics to increase the depth measuring range without any decrease in the lateral and axial resolutions.

Acknowledgements:
This research was supported by B. and G. Greenberg Research Fund (Ottawa).

References:

[1] R.G. Dorsch, G. Hausler, and J.M. Herrmann, Appl. Opt. **33**, 1306 (1994).
[2] D.W. Manthey, K.N. Knapp II, and D. Lee, Opt. Eng. **33**, 3372 (1994).
[3] E. Hasman, S. Keren, N. Davidson, A.A. Friesem, Opt. Lett. **24**, 439 (1999)
[4] E. Hasman, N. Davidson, and A.A. Friesem, Opt. Lett. **16**, 423 (1991).

INTRACAVITY PHASE ELEMENTS FOR SPECIFIC LASER MODE SELECTION

Ram Oron, Nir Davidson, Asher A. Friesem

Weizmann Institute of Science,
Department of Physics of Complex Systems, Rehovot 76100, Israel

and

Erez Hasman

Technion - Israel Institute of Technology,
Faculty of Mechanical Engineering, Haifa 32000, Israel

Abstract

We present novel phase elements that can be incorporated into the laser resonator, so as to select a specific high-order transverse mode. Such elements are essentially lossless for the desired selected mode, but introduce high losses to other modes. Thus, the laser operates with a single high order mode that is stable and has a better output beam quality than with multi-mode operation. The design of the phase elements, as well as experimental results with Nd:YAG and CO_2 lasers, that typically operate with many transverse modes, are presented. The results reveal good mode selectivity, and a possibility for improving the laser output power with respect to the fundamental mode operation by more than 50 percent.

1. Introduction

In a laser resonator operating with many transverse modes, each mode leads to a different divergence, so the emerging output beam quality is relatively poor. To improve the quality, it is generally necessary to lower the Fresnel number, $N_F=a^2/\lambda L$, where a is the aperture radius, L is the resonator length and λ the operating wavelength. Typically, this is achieved by inserting an aperture inside the resonator so as to reduce the effective radius of the gain medium. The size of the aperture can be decreased until only the optimal fundamental TEM_{00} mode of Gaussian shape exists. Unfortunately, the introduction of the aperture results in a significant reduction of the output power, since only a small volume of the gain medium is exploited. Other means to improve the quality of the output beam involve the insertion of specialized mirrors and diffractive elements into the resonator [1-4].

In order to retain a high output power and yet improve the beam quality, we propose and demonstrate how newly designed intracavity phase elements can ensure that only a single high-order transverse mode, rather than the fundamental mode, is allowed to exist [5,6]. Such a high order mode exploits a relatively large volume of the gain medium, so the output power is relatively high. Yet, the single higher mode has a low divergence, and thereby a good beam quality. In the following we present the principles and design of these elements, along with representative experimental results for CO_2 and Nd:YAG lasers.

2. Mode Selection with Phase Elements

We begin by considering the field distribution inside a laser resonator. In cylindrical coordinates, each mode is characterized by a radial node index p and an angular node index l. Specifically, the field distribution $E(r, \theta)$ for a TEM_{pl} and TEM_{pl*} modes, can be expressed by [7]:

$$E_{pl}(r, \theta) = E_0 \rho^{l/2} L_p^l(\rho) \exp(-\rho/2) \cos(l\theta) = R_p^l(r) \cos(l\theta) , \tag{1}$$

$$E_{pl*}(r, \theta) = E_0 \rho^{l/2} L_p^l(\rho) \exp(-\rho/2) \exp(il\theta) = R_p^l(r) \exp(il\theta) , \tag{2}$$

where r and θ are the cylindrical coordinates, E_0 the magnitude of the field, $\rho=2r^2/w^2$ with w as the spot size of the Gaussian beam, L_p^l are the generalized Laguerre polynomials of order p and index l, and $R_p^l(r)$ includes the r dependence of E. Equations (1) and (2) indicate that with the exception of the fundamental mode TEM_{00}, the TEM_{pl} modes will either have spots or rings or both, whereas the TEM_{pl*} modes will have rings. Also, adjacent spots and adjacent rings have opposite phases (π phase shift).

Now, in order to select a specific single high order mode operation, we insert into the laser resonator either discontinuous phase elements (DPEs) or continuous spiral phase elements (SPEs). As shown in Fig. 1, the phase elements are inserted near the resonator mirrors. The phase elements are so designed, to have sharp phase discontinuities where the desired modes have low intensity. Thus, they introduce very low losses to the desired modes, but high losses to other modes, especially to the fundamental mode. This leads to very high mode discrimination, with the capabilities to obtain a single, well-defined and stable high-order mode operation. The high-order modes can exploit more of the gain medium, and thereby obtain relatively high output powers. The DPEs can be designed to select a high order mode in both the azimuthal and the radial direction, by introducing either an angular dependent phase shift or a radial dependent phase shift, whereas the SPEs select modes only in the azimuthal direction. Both types of elements may have no radial dependence, so they can be insensitive to axial displacement or thermal lensing. Moreover, since all parts of the desired mode distribution are in phase, the far field of the output beam intensity has a high central peak, albeit with some side-lobes.

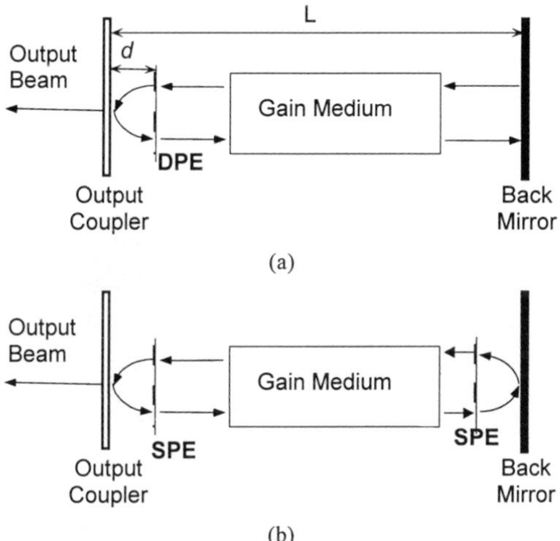

FIG. 1: Basic laser resonator configurations with phase elements:
(a) A DPE inserted next to the output coupler and (b) two SPEs inserted each next to a laser mirror.

3. Experimental Results

Several DPEs and SPEs were fabricated by advanced micro-lithographic techniques, and inserted into CO_2 and Nd:YAG lasers. Figure 2 shows the theoretical and experimental near-field intensity distributions of a TEM_{10} mode. Figure 2(a) shows the theoretical near-field pattern, which is composed of a central lobe and a ring, whereas Fig. 2(b) shows the corresponding experimental intensity distribution emerging from a CO_2 laser.

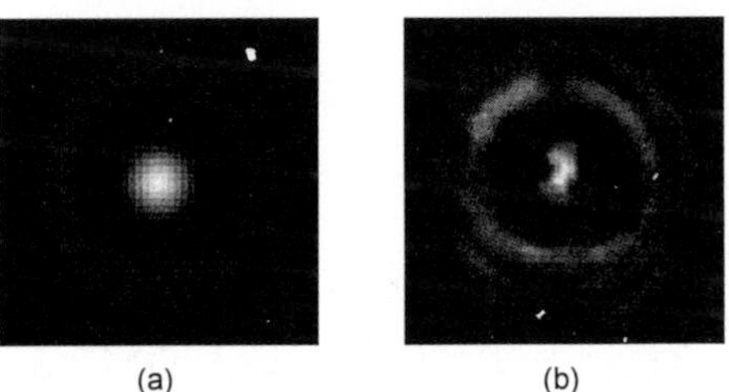

FIG. 2: Near-field intensity distributions of the TEM_{10} mode:
(a) Theoretical and (b) emerging from CO_2 laser with in which a DPE is inserted.

Figure 3 shows the experimental near- and far-fields intensity distributions from the Nd:YAG laser, in which we inserted SPEs designed to select the TEM_{01*} mode. Figure 3(a) depicts the near field intensity distribution pattern of the TEM_{01*} mode, with the doughnut shape and Fig. 3(b) shows the corresponding far-field intensity distribution pattern, which has a high central peak, indicating that all parts of the near field pattern are in phase. Here, an improvement of 50 percent in output power, with respect to the fundamental mode, was obtained.

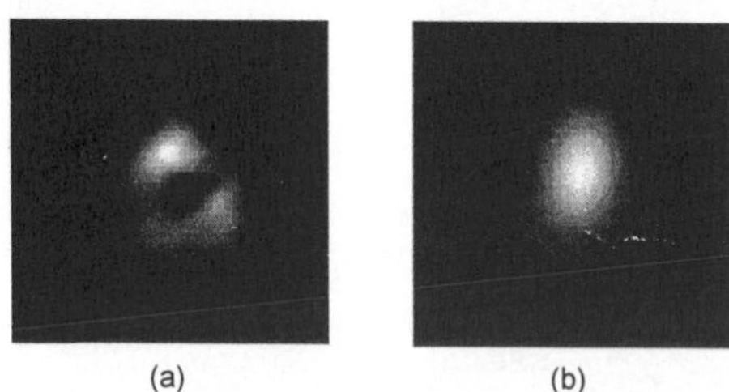

FIG. 3: Experimental Intensity distributions that emerge from a Nd:YAG laser in which SPEs were incorporated to select either the TEM_{01*} mode; (a) near-field; (b) far-field;

4. Conclusion

We presented a method, in which discontinuous phase elements are inserted into laser resonators, so as to obtain a stable operation with any selected single high order mode. As a result, more of the volume of the gain medium is exploited, thereby obtaining higher output powers. Specifically, we obtained an improvement of more than 50 percent in the output power with respect to the single fundamental mode TEM_{00} operation, by incorporating DPEs in Nd:YAG lasers. The method was illustrated with Nd:YAG lasers, but it is generic to many other lasers.

Acknowledgements:

This work was supported by Pamot Venture Capital Fund.

References:

[1] K. M. Abramski, H. J. Baker, A. D. Colly and D. R. Hall, *Appl. Phys. Lett.* **60**, 2469 (1992).

[2] P. A. Belanger, R. L. Lachance and C. Pare, *Opt. Lett.* **17**, 739 (1992).

[3] J. R. Leger, D. Chen and K. Dai, *Opt. Lett.* **19**, 1976 (1994).

[4] M. Piche and D. Cantin, *Opt. Lett.* **16**, 1135 (1991).

[5] R. Oron, Y. Danziger, N. Davidson, A.A. Friesem and E. Hasman, *Appl. Phys. Lett.* **74**, 1373 (1999)

[6] R. Oron, Y. Danziger, N. Davidson, A.A. Friesem and E. Hasman, *Opt. Commun.* **169**, 115 (1999).

[7] W. Koechner, *Solid State Laser Engineering*, Springer, Berlin Heidelberg, 4th edition (1996).

REFRACTIVE BEAM SHAPING ELEMENTS FOR COMPLICATETD INTENSITY DISTRIBUTIONS

Matthias Cumme and Ernst-Bernhard Kley

Friedrich Schiller University Jena, Max-Wien-Platz 1, 07743 Jena, Germany

Abstract

Refractive optical elements show high efficiency independent of the wavelength of the incident light. Therefore they are interesting for fabrication micro-optical lenses and beam shaping elements. Making such elements requires computation of smooth phase distributions which do not contain spiral phase dislocations. Using the iterative fourier transform algorithm (IFTA) and a stepwise modification of the desired signal, smooth phase profiles were computed for producing some complicated intensity distributions. Fabrication of designed elements with gray scale lithography is shown together with experimental results.

1. Introduction

Beam shaping elements have found rapidly growing application in fields like material processing, information processing, medical technique and measuring techniques.

Two types of phase elements are known, viz. diffractive and refractive elements. The phase distribution of diffractive elements is produced by applying mod $2\pi n$ with n an integer. Therefore the corresponding surface profile shows discontinuities. Evidently their optimum performance is achieved at a particular wavelength. On the other hand, the surface of refractive elements is smooth, so that their performance is nearly independent of the wavelength. The computation of phase elements for producing prescribed intensity distributions may lead to the occurrence of spiral phase dislocations (speckles) which are often undesired. With diffractive elements, these phase dislocations impair the signal quality, whereas phase distributions which contains dislocations cannot be produced at all with refractive elements.

2. Computation of refractive beam shaping elements using modified IFTA

Several methods are known for computation of phase elements. Analytical procedures based on the geometrical approach use the saddle-point method and yield distributions without phase dislocations [1], [2]. Therefore they look promising for the design of refractive phase elements, however, they are limited to separable or isotropic intensity distributions. A numerical procedure for arbitrary intensity distributions has been described by [3] .

The well known iterative procedure IFTA [4], [5] is based on wave-propagation theory and may be used for the design of elements which produce arbitrary amplitude distributions. However, the resulting phase distributions are not necessarily free from speckles. In order to avoid speckles, IFTA has been modified by introducing "soft" phase changes between iteration steps. Speckles which are still present may be removed by a final procedure [6]. The starting phase has to be chosen by the user.

In the present contribution, another modification of IFTA based on stepwise changes of the desired amplitude distribution is presented. Modified IFTA uses a sequence A_0, A_1,...,A_N of desired amplitude distributions. A_0 is nearly gaussian whereas A_N is the finally desired distribution (Fig.1).

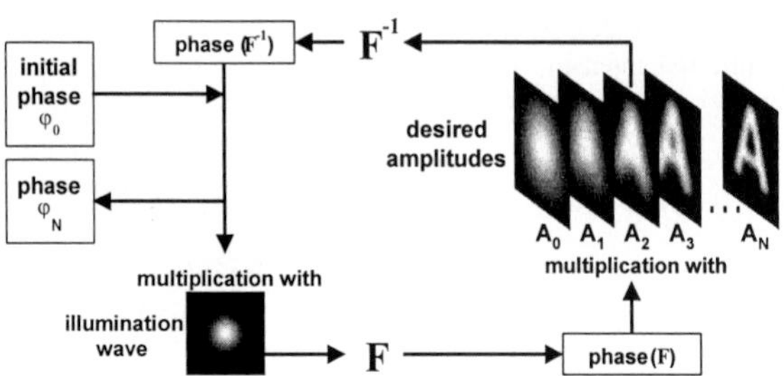

Figure 1: Modified IFTA.

Before IFTA is started, A_0 is generated. The dislocation free starting phase φ_0 for IFTA is calculated so as to produce A_0 in high approximation from the gaussian illumination wave. Since A_0 itself is gaussian-like, φ_0 can be calculated analytically. As soon as A_i has been approximated sufficiently by IFTA, it is replaced by A_{i+1} ($i = 0, 1,..., N-1$). A_i is calculated as follows: $A_i = |F^{-1} [F(A_N) g(\omega_i)]|$ where F denotes the Fourier-transform and $g(\omega_i)$ is a gaussian amplitude distribution with a waist ω_i. ω_0 is very small, $\omega_{i+1} > \omega_i$, and ω_N is sufficiently large for $g(\omega_i)$ to cover the essential features of A_N

3. Design, fabrication and optical performance of refractive beam shaping elements

Gray tone lithography is suited for producing continuous surface profiles [7]. As the accuracy of surface forming by gray-tone lithography depends on the element thickness, high accuracy requires thin elements. Evidently the required thickness will increase with the diameter of the illuminating beam. Given the beam

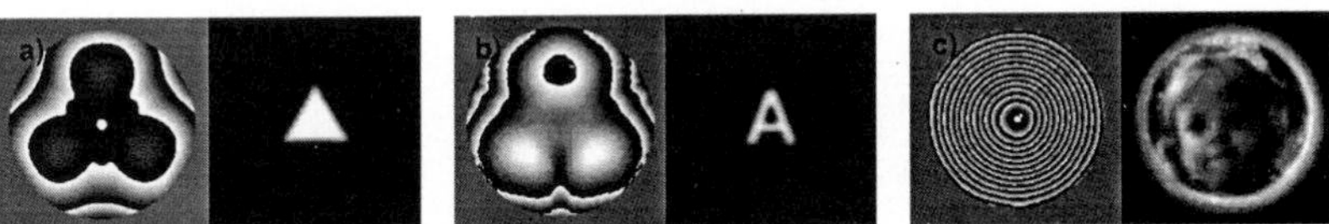

Figure 2: Computed phase distributions (left parts) and corresponding computed intensity distributions (right parts). The phase distributions were cut off where the normalized intensity was 0.002.

diameter, minimum thickness will be achieved if the element does not alter the beam divergence. According to these principles, different refractive elements were designed for shaping beams emerging from a single mode fibre with NA = 0.12. The elements were positioned 100 μm apart from the end of the fibre, the beam diameter was 20 μm at this point. Results are shown in Fig. 2. As to be seen, the corresponding height profile of the element of Fig. 2c will be much thicker than that of the other elements. The reason is that depicting the face shown in Fig. 2c requires an enlarged

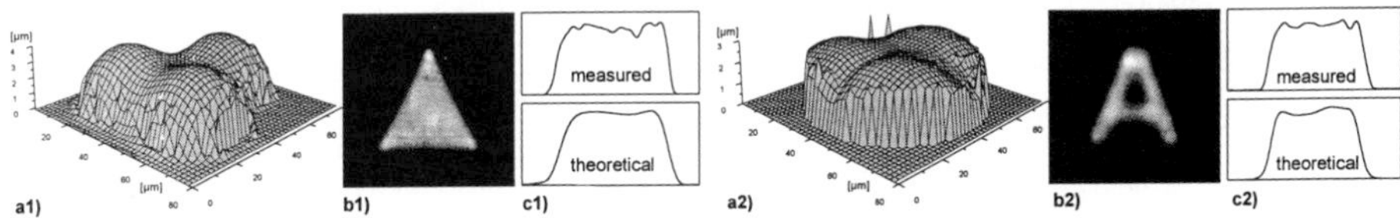

Figure 3: a1, a2) measured height profiles of fabricated phase elements according to Figs. 2a) and 2b); b1, b2) measured intensity distributions; c1, c2) intensity profiles along the dashed lines.

signal size in order to reproduce all details in spite of the diffraction determined resolution limit. It is planned to produce this element by pre-form techniques [8]. Elements produced according to Fig.2a and b are shown in Fig. 3 . They show more then 99 % conversion efficiency in the wavelength range 488-633 nm. The rms deviation of the measured from the computed intensity profiles was less then 10 %.

4. Conclusions

The presented modified IFTA is a well suited method for computing dislocation free phase elements which form some complicated intensity distributions. The fabricated elements show high efficiency and signal quality with an acceptance of a large wave length range.

Acknowledgements

The authors would like thank Mr H.-J. Fuchs, Mr. H. Schmidt and Mrs. W. Gräf for their co-operation in fabrication.

References:

[1] Bryndahl, O. (1974) "Geometrical transformations in optics", J. Opt. Soc. Am. **64,** 1092-1099.

[2] Aagedahl, H. Schmid, M. Egner, S. Müller-Ouade, J. Wyrowski, F. (1997) "Analytical beam shaping with application to laser-diode arrays", J. Opt. Soc. Am. **14,** 1549-1553.

[3] Arieli, Y. Eisenberg, N. Lewis, A. Glaser, I. (1997) "Geometrical transformation approach to two-dimensional beam shaping", Appl. Optics, **36,** 9129-9131.

[4] Gerchberg, R.W. and Saxton, W.O. (1972) "A practical algorithm for the determination of phase from image and diffraction plane pictures", Optik **35,** 237-246.

[5] Wyrowski, F. and Bryngdahl, O. (1988) "Iterative Fourier-transform algorithm applied to computer holography", J. Opt. Soc. Am. **5,** 1958-1966.

[6] Aagedahl, H. Schmid, M. Beth, T. Teiwes, S. Wyrowski, F. (1996) "Theory of speckles in diffractive optics and its application to beam shaping", Journal of Mod. Opt. **43,** 1409-1421.

[7] Däschner, W. Long, P. Stein, R. Wu, C. Lee, S. "One step lithography for mass production of multilevel diffractive optical elements using High Energy Beam Sensitive (HEBS) gray-level mask" Proc. SPIE Vol. 2689.

[8] Wittig, L.-Chr. Kley, E.-B. (1999) "Approximation of refractive micro-optical profiles by minimal surfaces", Proc. SPIE Vol. 3879 32-38.

HIGH-FREQUENCY CIRCULAR DIFFRACTION GRATINGS DIAMOND TURNED ON STEEP CURVED SURFACES

Michael A. Golub

Holo-Or Ltd., Kiryat Weizmann, P.O.B. 1051, Rehovot 76651, Israel.

Abstract

Method for design of tool-path for single point diamond turning of dense diffractive optical elements elaborated and experimentally tested. Compensation for finite radius and fixed offset angle of diamond tool is proposed. Blazed diffractive grooves with stable depth, opening angle and optimal orientation with respect to normal of curved surface are achieved. Shown that the limits of single point diamond turning can extended towards fabrication of high precision diffractive optics with a full groove width of few microns

1. Introduction

High-numerical-aperture refractive-diffractive lenses with aberration correction and beam shaping are required for coupling and transformation of light beam emerging from fibers, diode lasers as well as for wide angle imaging and uniform illumination. Proper circular blazed diffraction grooves feature period of few microns and reside on steep curved substrate surface with the slope varying within a wide range close to 90° degrees. Single point diamond turning have severe limitations on the period of grooves and zone-transfer region[1-3] due to finite radius and included angle of diamond. Half-radius tools minimize zone-transfer region but still leaves problems for groove period less than 200 μm. Reasonable opening angle and aspect ratio of dense diffraction grooves is not provided by above mentioned diamond tools. We used special sharp tools with small radii of diamond.

2. Creation of toolpath for sharp diamond tool

Classical "tool-nose" radius compensation does not work for micron-scale grooves with sharp tool. Fixed offset angle of the diamond tool raise problems to cut right grooves on substrate with varying normal. The main problem was to build unique tool path in CNC file for sharp tool diamond turning such a way that to keep required blazed shape and depth of fine diffractive grooves as well as their orientation with respect to a normal to substrate. Our constrain was the fixed offset angle αof of the tool on 2-axis lathe and quite a big included angle αt of the tool diamond required to avoid fragility of the sharp tool. Limits for two (inner and outer) boundary slopes of grooves slopes depend on the tool's offset and included angles are given by equations

$$\beta inn(\theta) = \left[\theta - \left(\alpha of - \frac{\alpha t}{2} - \varepsilon\right)\right] \qquad \beta out(\theta) = \left[\left(\alpha of + \frac{\alpha t}{2} + \varepsilon\right) - \theta\right] \qquad (1)$$

where θ - current angle of normal, counted from axis of revolution. It is clear from Eq. (1) that boundary there is limited range from θ_1 to θ_2 of angle θ for the achieving symmetrical orientation of grooves and opening angle β.

$$\theta 1 = \alpha of + \frac{\alpha t}{2} - \frac{\beta}{2} + \varepsilon \qquad \theta 2 = \alpha of - \frac{\alpha t}{2} + \frac{\beta}{2} - \varepsilon \qquad (2)$$

Grooves have to become asymmetrical out of range $[\theta_1 \;,\; \theta_2]$. Allowing for asymmetrical grooves with opening angle $\beta 0$ brings us to equations for really available inner and outer groove slopes $\beta curInn(\theta)$, $\beta curOut(\theta)$ being $\beta/2$ in the range $[\theta_1, \theta_2]$ and $\beta curInn\operatorname{Re}q(\theta)$, $\beta curOut\operatorname{Re}q(\theta)$ respectively,

$$\beta curInnReq(\theta) = \max\left(\operatorname{atan}\left(2\cdot\tan\left(\frac{\beta}{2}\right) - \tan(\beta out(\theta))\right), \beta 0 - \beta out(\theta)\right)$$

Asymmetricity of grooves shape is characterized by relative position of groove's depth within the period

$$\xi cur(\theta) = \text{if}\left(\theta<\theta 1, 1 - \frac{\tan(\beta out(\theta))}{2\cdot\tan\left(\frac{\beta}{2}\right)}, \text{if}\left(\theta<\theta 2, 0.5, \frac{\tan(\beta inn(\theta))}{2\cdot\tan\left(\frac{\beta}{2}\right)}\right)\right) \qquad hmax = \frac{b}{2\cdot\tan\left(\frac{\beta}{2}\right)}$$

Full symmetricity and maximum depth of groove hmax is achieved in the middle of sphere. In other places of substrate maximum depth is less and is described by equations (Ses Figs. 1-3).

$$hmaxCur(\theta) = \text{if}\left(\beta curInn(\theta) + \beta curOut(\theta) \geq \beta 0, hmax, \frac{b}{\tan\left(\beta curInn(\theta)\cdot\frac{\pi}{180}\right) + \tan\left(\beta curOut(\theta)\cdot\frac{\pi}{180}\right)}\right)$$

$$hmaxDelCur(\theta) = hmaxCur(\theta) + DelhmaxCur(\theta)$$

Graphs show that considerable (about 30%) reduction of grooves depth takes place

3. Compensation for finite tool radius and included angle of diamond

We found out corrected toolpath that ensures better symmetricity and depth of grooves:

$$hmaxCurC(\theta) = hmaxCur(\theta) - (DelhmaxCur(\theta) - DelhmaxCur(\alpha of)),$$

$$\beta curInnC(\theta) = \mathrm{if}\left[\theta<\theta 1, \beta curInn(\theta) - b\cdot\left(\frac{\cos\left(\beta curInn(\theta)\cdot\frac{\pi}{180}\right)}{hmaxCur(\theta)}\right)^2\cdot(DelhmaxCut(\alpha of) - DelhmaxCut(\theta)), \beta curInn(\theta)\right]$$

$$\beta curOutC(\theta) = \mathrm{if}\left[\theta>\theta 2, \beta curOut(\theta) - b\cdot\left(\frac{\cos\left(\beta curOut(\theta)\cdot\frac{\pi}{180}\right)}{hmaxCur(\theta)}\right)^2\cdot(DelhmaxCut(\alpha of) - DelhmaxCut(\theta)), \beta curOut(\theta)\right]$$

$$\xi curC(\theta) = \mathrm{if}\left(\theta<\theta 1, \frac{hmaxCurC(\theta)}{b}\cdot\tan\left(\beta curInnC(\theta)\cdot\frac{\pi}{180}\right), \mathrm{if}\left(\theta<\theta 2, 0.5, 1 - \frac{hmaxCurC(\theta)}{b}\cdot\tan\left(\beta curOutC(\theta)\cdot\frac{\pi}{180}\right)\right)\right)$$

where Rt is the tool tip radius,

$$DelhmaxCut(\theta) = -\left[\frac{Rt}{\sin\left[\frac{(\beta curInn(\theta)+\beta curOut(\theta))}{2}\cdot\frac{\pi}{180}\right]}\cdot\cos\left[\frac{(\beta curOut(\theta)-\beta curInn(\theta))}{2}\cdot\frac{\pi}{180}\right] - Rt\right]$$

Comparative results on depth of grooves correction are presented on the Figs 1-3. Parameters after correction are denoted by additional letter "C" in identifiers.

Experiment performed on the PRECITECH OPTIMUM lathe proofed design concept and resulted in quite good measured shape of grooves with aspect ratio about 0.5 with respect to normal to the substrate.

Author is grateful to Yuri Paskalov and Michael Levkovich (Holo-Or, Ltd.) for their qualified operation on the precision diamond-turning lathe.

Summary

- Possibility of single point diamond turning of sharp grooves was demonstrated
- Design concept for diamond tool path elaborated and proved
- Experiment performed on the PRECITECH OPTIMUM diamond turning lathe
- Blazed diffractive grooves with near-to-symmetrical shape, stable depth, opening angle and optimal orientation with respect to normal of curved surface are achieved
- Measured grooves feature aspect ratio about 0.5 for a period of few microns

References:

[1] Riedl M.J. Design example for the use of hybrid optical elements in the infrared.Appl. Opt., 1996, v.35, p.6833

[2] C. Gary Blough; M. Rossi; Stephen K. Mack; Robert L. Michaels. Single-point diamond turning and replication of visible and near-infrared diffractive optical elements, Appl. Opt., 1997,36, 4648

[3] Wender, David C. Diamond turning of aspheric infrared optical components (A). J. Opt. Soc. Am., 1980, v. 70, p.1055

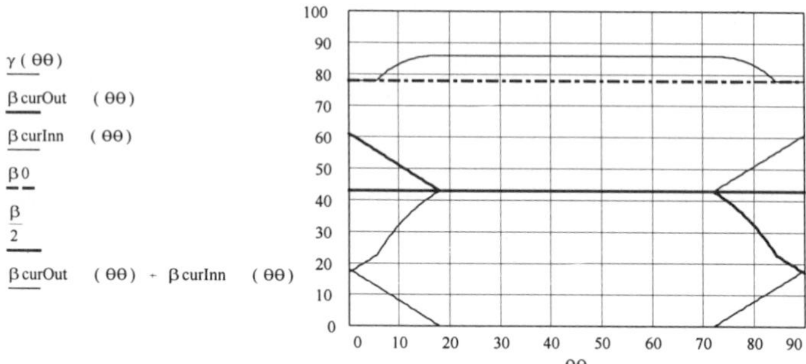

Fig.1. Boundary angles of grooves depending on the slope of normal to the surface

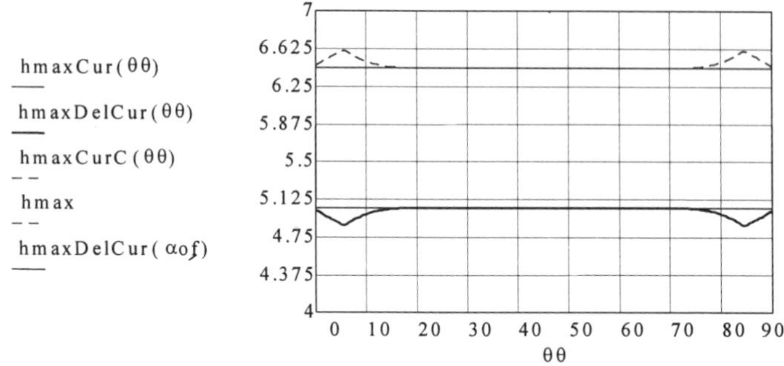

Fig.2. Variation of maximum depth of grooves depending on the slope of normal to the surface

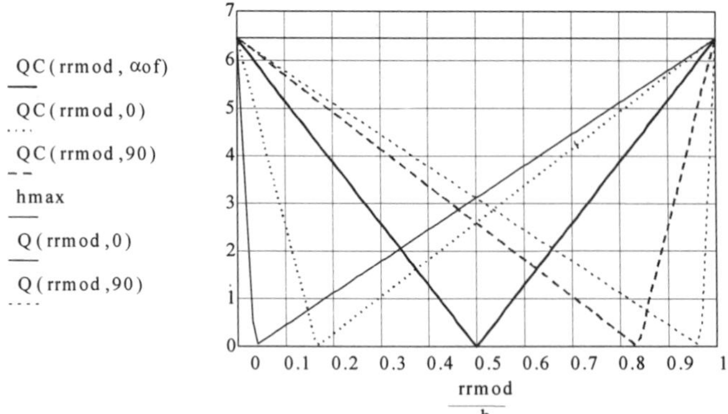

Fig.3. Variation of groove shape depending on the slope of normal to the surface

PREDICTION OF PERFORMANCE FOR DIFFRACTIVE OPTICAL ELEMENTS FROM MICRORELIEF PROFILE MEASUREMENTS

Michael A. Golub

Holo-Or Ltd., Kiryat Weizmann, P.O.B. 1051, Rehovot 76651, Israel.

Abstract

Method for reliable prediction of optical performance for diffractive optical elements by computer simulation of light diffraction is elaborated. Diffraction efficiency was numerically estimated just from data files of microrelief pattern received from profilometer scans of diffractive lens.

1. Introduction

Full optical performance check of diffractive optical elements (DOEs) might involve quite complicated measurements on optical bench. However regular measurements done in DOEs fabrication process are only profile scans of diffractive microrelief pattern in DOE substrate surface. We propose to use just the profile scan data for estimation of full DOEs performance. Simulation of DOE performance was done in works [2-3] but measured microrelief profiles were not considered. Based on relation between sampled and continuos versions of integral transforms we created algorithms and software for estimation of diffraction efficiency and point-spread-function of DOEs from their microrelief profile data files. Experimental results are given for blazed grating and diffractive lens.

2. Profile measurements for diffractive microrelief

Diffractive optical elements (DOEs) have microrelief profile with ring-type, linear or curvilinear zoned structure. Fabrication of diffractive element by lithographic and reactive etching technology introduces multilevel staircase approximation of the profile. Some errors of misalignment profile height and microrelief shape might also occur. Higher diffraction orders are appearing. Therefore diffractive lenses provide several focuses along with the useful focus. The point-spread function in the useful focal plain is superimposed with defocused images from other foci. This way contrast of the image and the energy concentration in the image are reduced compared with ideal diffractive lens. We propose to measure the profile, compare it with ideal one and estimate of higher diffraction orders and other performance. Profile measurements of diffractive optical elements were performed in Holo-Or on profilometer Form Talysurf Series 2, that originally deals only with smooth surfaces (planar, spherical, aspherical) as a reference. Thus we created algorithms and software that check agreement between measured profile and designed blazed staircase profile and estimates the proper deviation parameters. The comparative simulation data for ideal staircase profile and measured profile of diffractive lens is presented both on the graph Figs.1

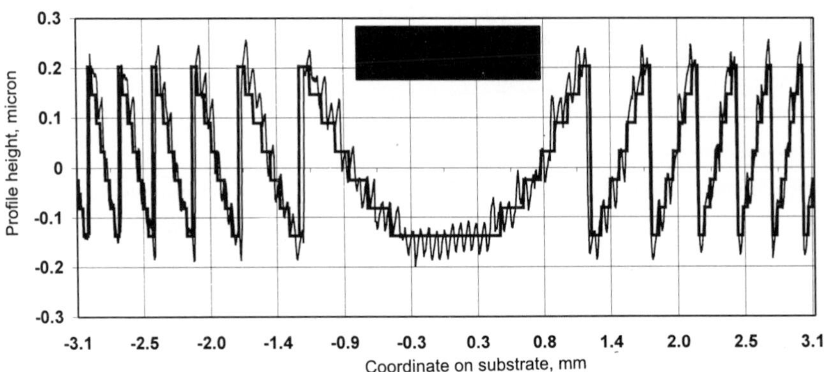

Fig.1. Ideal staircase profile and measured profile within 6 mm clear aperture of diffractive lens

3. Computer simulation of light diffraction on diffractive microrelief

Data from profilometer was converted, acquired to the computer and transferred to performance simulation program. Light propagation is modeled by scalar Kirchhoff integral [1], brought to 2-D or 1-D Fast Fourier Transform algorithm by special approximation[2,3]. Approximation considers compensation of sampling by adding zeros to the function in the DOE plane, extraction of proper part of focal plane, selection of dimensions of digital transforms. Software package offers windows-style user interface and supports profile scan data files conversion, running of simulation and estimation of diffraction efficiency and focal intensity distribution for measured profile as well as for designed profile. Computer simulation of 1-D diffraction grating diffraction based on groove profile scans resulted in estimation of expected efficiency in diffraction orders. Analytical solutions are available for comparison and verification for this important example. Ideal blazed profile theoretically expected to give the power 100% to the first diffraction order, while other orders do not receive at all the power. Staircase profile theoretically expected to give the power to the diffraction orders in accordance with the formula [4]:

$$\frac{\varepsilon_l}{\varepsilon} = \mathrm{sinc}^2\left(\frac{l}{M}\right)\left(\frac{\sin[\pi(l-1)]}{M\sin\left(\frac{\pi(l-1)}{M}\right)}\right)^2 \tag{1}$$

Table 1 presents the simulated power of diffraction orders ε_l in focal plane relative to the total light flux ε just after the plane of blazed grating for clear aperture 20 mm. It is seen that results of numerical simulation are in ağreement with analytical estimation. Dimension of FFT required for high-resolution simulation was quite high 524288 pixels. Special modifications of simulation software have been done to ensure working with such high dimension.

Table 1. Power of diffraction orders in focal plane (in % relative to power of incident beam) for blazed grating for clear aperture 15 mm, period 96 μm, number of levels 16.

Diffraction order No.	Profilometer profile	Ideal profile 16 levels	Analytical estimation
-17	0.10	0	0
-15	0.19	0.44	0.44
-9	0.58	0	0
-8	0.45	0	0
-7	0.82	0	0
-6	0.78	0	0
-5	1.05	0	0
-4	0.78	0	0
-3	1.17	0	0
-2	1.32	0	0
-1	1.83	0	0
0	1.55	0	0
1	80.01	98.58	98.72
2	0.22	0	0
3	0.04	0	0
4	0.03	0	0
5	0.19	0	0
6	0.03	0	0
7	0.02	0	0
8	0.01	0	0
9	0.04	0	0
17	0.34	0.34	0.34

Computer simulation of light propagation through a sample of diffractive lens gave clear prediction of reduction in focal encircled power (Fig. 2) due to etching depth and alignment error present in fabrication process as compared with designed performance. One string of radial symmetrical profile was just rotated (in software) to obtain 2-D bitmap file. Then 2-D simulation program was used. It is seen that the total power in the analyzed focal domain is only 81% for measured profile, while being 94% for the ideal staircase profile. The cause of difference is in the higher diffraction orders directed out of analyzed focal domain.

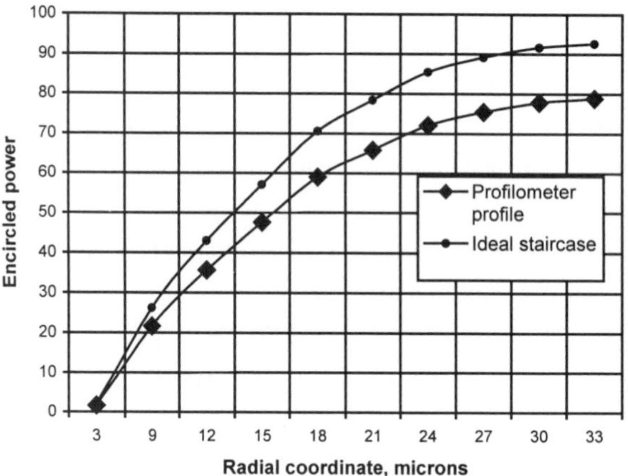

Fig. 4.7. Comparative results of computer simulation for measured and ideal profile.

Conclusions

- Profile data processing with computer program gives the possibility to analyze the details of steps and supports inspection of diffractive of staircase profile
- New method of computer simulation of performance relates the real microrelief profile data with intensities in diffraction orders and focal intensity distribution for prediction of DOE performance before optical bench measurements

References:

[1] J.W. Goodman. Introduction to Fourier Optics. New York, McGraw-Hill (1968)

[2] Golub M.A., Grossinger I.Diffractive optical elements for biomedical applications. Proceedings of the SPIE, 1997, v. 3199, pp.220-231.

[3] Duparre M., Golub M.A., Luedge B., et al. Paveliev V.S., Soifer V.A., Uspleniev G.V., Volotovskii S.G. Investigation of computer-generated diffractive beam shaper for flattening of single-modal CO_2-laser beam. Applied Optics, 1995, v.34, No.14, pp. 2489-2497.

[4] G.J. Swanson. Binary optics Technology: Theoretical limits on the diffraction efficiency of multilevel diffractive optical elements. Technical report 914, Lincoln Laboratory, Massachusetts Institute of Technology, Lexington, MA, 1991

CONTINUOS-PROFILE DIFFRACTION GRATINGS ON FUSED SILICA

Michael Levkovich, Helena Yanovsky

Holo-Or Ltd., Kiryat Weizmann, P.O.B. 1051, Rehovot 76114, Israel.

Abstract

The present work is dedicated to developing of single step "continuos profile" production technology for diffractive optical elements (DOEs). The profile depth, linearity and roughness are controlled by lithography parameters with an optimized reactive ion etching (RIE) process. An influence of the lithography on the DOEs quality is represented in the work.

1. Introduction

Binary mask fabrication of diffractive optical elements (DOEs) is complicated by several stages and features as alignment errors, relatively low resolution, cycle duration and etching errors. In the past time the interest to development of new technologies using different types of gray-scale masks for DOEs fabrication grown due to the advantages in the final blaze grating quality and in considerable reduction of the production duration and number stages[1,2]. Using gray-level masks may solve the problems[1]. An optimization of the blaze grating profile and its depth can be achieved by adjustment of such technological parameters as resist type, UV exposure, soft and hard bake regime. Today a regular production technology of the DOEs with using gray-scale masks includes contact printing photolithography and RIE. Usually engineers try to obtain a RIE process selectivity 1 (1:1) between resist and substrate, which would provide necessary grating depth equal to the resist one. It requests development of lithography and a RIE process to be optimized, but sometimes it is not easy to obtain[3]. In case when the RIE process is already optimal for a multilevel technology (it means optimal RIE uniformity, surface roughness and a fixed selectivity) it is reasonable to develop a suitable lithography process only without any change of RIE. In works of other authors lithography stage was not subjected to a precise analysis of lithography variations and usually the technology developing is allowed by a calibration of the RIE[3]. Our paper is dedicated to the development of gray level technology where an optimizing profile is adjusted on the lithography stage only while the RIE is kept unchanged as equal as in the case of multilevel technology.

2. Experimental investigation.

For the DOEs manufacturing on lithography step we used MJB 3 Karl Suss mask aligner with HBO 200 W/2 UV lamp. The fused silica optical elements were etched by Oxford Plasmalab RIE system. For measurements of the DOEs surface profile Talysurf series profilometer with 2 μm tip probe was used.

A regular lithography process for the DOE surface relief creation includes cleaning and drying of a substrate, coating of the substrate by resist (spinning method), soft bake, UV exposure, developing, rinsing, drying of the element and hard bake. The novolac-based S1800 Shipley resists series were applied for the investigation. All of the principal parameters have been varied in order to investigate their influence on the surface relief quality.

It is known that for continuos profile creation the soft bake parameters must be changed in order to increase a sensitivity of the resist layer to UV light through HEBS (high energy beam sensitive) mask, and to form a threshold of the profile depth[1]. There are two common methods for this resist coating preparing, namely decreasing of the soft bake duration in comparison with a regular one, and using of the regular soft bake in combination with UV pre-exposure of the resist without HEBS mask. An effect from these two methods can be estimated only in complex with the UV exposure through HEBS mask because of a strong connection between these parameters. The shorter soft bake time must be optimized very strictly because the heat treatment duration increasing or decreasing leads to the nonlinearity of the resist profile. For example the a shorter soft bake of less than regular 30 min at a definite temperature in combination with a UV exposure of few sec can give a qualitative blaze grating of 1-1.2 μm height. A change of the optimized soft bake duration and in UV exposure leads to the profile worsening in other words to the nonlinearity at different profile depths. The grating depth values of 1.2, 1.38, 2.36 μm for fused silica are compatible for the visible, UV light, and Nd-Yag laser respectively, but in this case the RIE selectivity must be less than a regular one for this kind of resist. Normally the selectivity exceeds 2.5 for S1828 Shipley resist in a fluorinated plasma mixture, which gives an excellent uniformity and surface roughness. Usually, for decreasing of the resist resistance to the plasma, a small oxygen addition to the gas mixture are used, but it could worse in turn the RIE process uniformity. As the experiments show the application of the soft bake in combination with UV pre-exposure (pre-exposure is UV exposure without using of HEBS mask) and exposure through the gray level mask can give a needed resist profile depth and linearity. A threshold of the resist grating depth can be adjusted easily by different UV exposures. Taking into account the selectivity, which equal to 4 about, after a long time hard bake, the optimal resist blaze grating height can be calculated. The resist height should be in the range of 0.32-0.35 μm for the visible and UV diapason for instance. The role of the pre-exposure is increasing of the resist sensitivity to the final UV exposure through the HEBS mask and. With the UV exposure dose increasing thickness of the exposed resist layer is increased also, that adjusts the final blaze grating height. There is one more important parameter for the blaze grating quality estimation, which defined as[4]

$$Q=1-x/T,$$

Where T is period of a blazed grating, x is the transition width or in other words the lateral distance between lower and higher points of the vertical wall of the blaze grating.

The developing conditions also influence on the blaze grating quality. The concentration of the developer in distillated water (351 Shipley developer was used) should be less than 1:3 for the grating linearity improvement. Duration of the developing is a very critical point because the overtime could cause a bend of the grating line, in the bottom of the grating, especially.

The hard bake, really, creates a stable microstructure of the resist patterns. The hard bake during several hours at 100 °C in oven seems to be optimal in order to achieve the stable selectivity of the RIE process.

3. Results on continuos profile blaze gratings.

It was found that resist, which has a higher content of a viscous component, has the best relief quality and result repeatability. A proper combination of the optimized soft bake and UV exposures of definite longevities could allow to achieve the needed profile form. With decreasing of period size the quality factor Q is increased in the resist gratings and in the fused quartz ones as well (Fig. 1, 2). It is important that up to period size of 15 μm the quality factor is high and its worst value is obtained at period of 15-30 μm and equal to 0.7-0.73 about (Fig. 2). This result is better than in case of spatially filtered halftone screens mask by comparison, where 0.7 quality factor was obtained at 50 μm grating period[4]. Other important issue is change of the blaze grating depth depending on the period size. With the period decreasing the grating height is monotonically decreased. This phenomenon took place on photolithography step and after RIE as well. In general, the blaze gratings defined by a very strict profile form. It is clear shown in the pictures, where the grating depth in the fused silica equal to 1.24 μm for period of 500 μm (Fig. 1), but for 15 μm period the depth is 0.9 μm already (Fig. 2).

4. Conclusions

In conclusion, we have developed the DOEs production technology with using HEBS-glass mask for different grating depths, which compatible to the visible, UV light, and Nd-Yag lasers. The developed gray level lithography technology allows using of a regular (used for multi-step technology) optimized RIE process. The advantages of the technology in comparison with the regular multi-step one are high precision form of the final profile, an essential reduction of the production cycle, reduction of the alignment and etching errors.

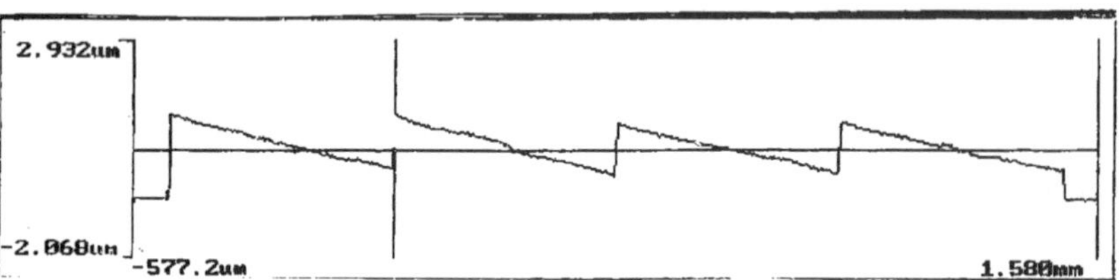

Fig. 1. Continuos profile relief in fused silica after RIE. Blaze gratings have period of 500 μm. Depth of the blaze gratings is 1.24 μm. Quality factor Q equal to 0.98 about.

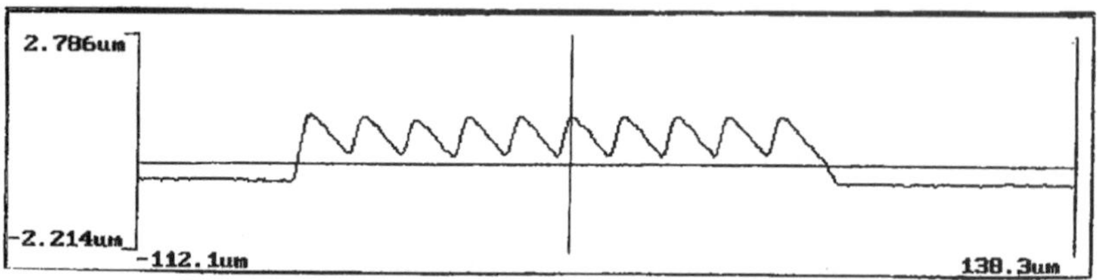

Fig. 2. Continuos profile relief in fused silica after RIE. Blaze gratings have period of 15 μm. Depth of the blaze gratings is 0.9 μm. Quality factor Q equal to 0.72 about.

Acknowledgements: We thank “Holo-Or Ltd.” chief scientist Dr. Michael Golub for scientific support.

References:

[1] W. Daschner, P. Long, R. Stein, C. Wu, and S. H. Lee. Cost-effective mass fabrication of multilevel diffractive optical elements by use of a single optical exposure with a gray-scale mask on high-energy beam sensitive glass. Applied Optics. Vol. 36, No 20. 10 July 1997.

[2] T. J. Suleski and C. O’Shea. Gray-scale masks for diffrative-optics fabrication: I. Commercial slide imagers. Applied Optics. Vol. 34, No. 32. 10 November 1995.

[3] M. Eisner, J. Schwider. Transferring resist microlenses into silicon by reactive ion etching. Optical Engineering. 35(10) 2979-2982. October 1996.

[4] D. C. O’Shea and W. S. Rockward. Gray-scale masks for diffractive-optics fabrication: II. Spatially filtered halftone screens. Applied Optics. Vol. 34, No 32. 10 November 1995.

PART 11

OPTICAL DEVICES, SWITCHING AND COMMUNICATION

Chairperson: *Z. Kotler, Israel*

ACTIVE GRATING WAVEGUIDE STRUCTURES BASED ON SEMICONDUCTORS MATERIALS

N. Dudovich, G. Levy-Yurista, D. Rosenblatt, A. Sharon and A. A. Friesem

Department of Physics of Complex Systems, Weizmann Institute of Science,. Rehovot 76100, Israel.

and

H.G. Weber, H. Engle and R. Steingrueber

Heinrich Hertz Institute, Berlin, Germany

Abstract

Under certain conditions, a tunable resonance phenomenon can occur in a waveguide grating structures that are formed with semiconductors materials. By controlling the material properties and the optical mode properties, the ratio between the tunable range of the wavelength and the absorption is maximized. The basic principles for actively tuning the resonance wavelength are described and numerical results are presented.

1. Introduction

Although diffraction of light from grating structures has been investigated for many years, the recent developments in electron beam and lithographic technologies led to a resurgence of investigations. This is particularly true for high-resolution gratings that have specific resonance anomalies. Such anomalies were first observed by Wood [1]. Subsequently they have been theoretically investigated extensively by a number of groups [2]-[5]. More recently the investigations have been expanded to resonance anomalies in grating waveguide structures (GWS), and included new theoretical and experimental developments [6]-[8].

GWS have multilayer configuration, the most basic of which is comprised of a substrate, a thin dielectric layer or semiconductor waveguide layer, and an additional transparent layer in which a grating is etched. The basic configuration is shown in Figure 1. When such a GWS is illuminated with an incident light beam, part of the beam is directly transmitted and part is diffracted and subsequently trapped in the waveguide layer. Some of the trapped light is then rediffracted outwards, so that it interferes destructively with the transmitted part of the light beam. At a specific wavelength and angular orientation of the incident beam, the structure "resonates"; namely, complete interference occurs and no light is transmitted. The spectral bandwidth of the resonance depends on geometrical parameters such as the grating depth and duty cycle, and thickness of the waveguide layer, and on optical parameters such as the refractive indices of the layers. The bandwidth can be designed to be very narrow, which is of interest for spectral filtering and modulation applications.

When the GWS are formed with semiconductors materials, it is possible to actively vary the resonance wavelength. This can be achieved by varying the index of refraction of the waveguide, which will lead to a different phase matching condition, and accordingly to a resonance wavelength shift.

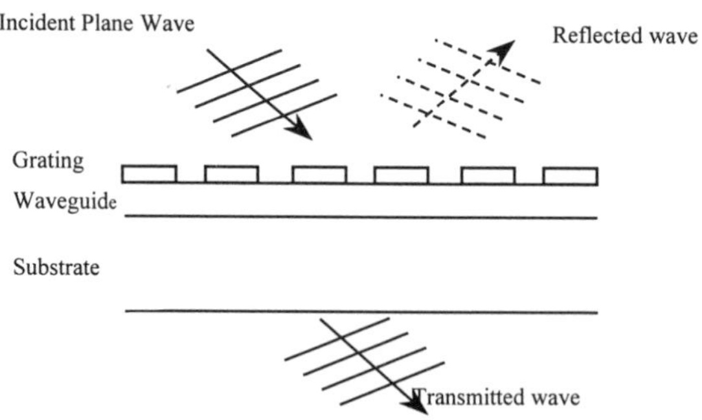

FIG.1: Basic configuration of a Grating-Waveguide Structure (GWS)

2. Active Tuning of the Resonance Wavelength

Several different effects that can lead to index of refraction variations in the semiconductor waveguide have been considered. These can be divided into two main subgroups: charge carrier related effects and electric field related effects. Specifically, either the injection of carriers or the application of an external electric field will cause changes in the absorption coefficient of the waveguide, which in turn induce changes in the refractive index via the Kramers-Kronig relations.

Charge carrier effects mainly include bandfilling, bandgap shrinkage and plasma effects, where each dominates at different band gap energies of the waveguide [9]. Electric field related effects mainly include the electrooptic effect that is characterized by a linear dependence of the change in the index of refraction on the applied electric field, and the electrorefractive effect that is characterized by a square law dependence on the electric field. These two effects have polarization dependence. The contribution of each effect depends on the difference between the photon energy and the bandgap energy, the applied electric field and the free carriers concentration.

By controlling the material properties (such as the energy bandgap and doping concentration) and the optical mode properties, the interaction between electrons and photons is maximized. We found that a large interaction will induce a high absorption in the waveguide cavity, thereby reducing the finesse. So, optimization is needed so that the ratio between the range of wavelength shift and the bandwidth is maximized. We performed some numerical calculations, and the results are presented in Figs. 2 and 3.

Figure 2 shows the calculated resonance shift $\Delta\lambda_r$ as function of the doping concentration for a GWS with a reverse voltage configuration, where a pn junction is located in the middle of the waveguide. As the doping concentrations and the applied voltage increases, the tunability range increases until the absorption limit is reached. When the bandwidth of the active GWS is designed to be equal to the absorption, then the maximal ratio between the tunability range and the bandwidth is calculated to be 8.6. For a GWS with a forward voltage configuration, higher values of $\Delta\lambda_r$ can be reached compared to one with a reverse voltage configuration. Since the active region does not depend on the free carriers concentration change (as in the reverse voltage case), the $\Delta\lambda_r$ will be dominated by a linear dependence. Figure 3 shows the calculated tunability range as function of the bandwidth for different values of maximal current. As evident, when the bandwidths are narrow, the tunability range is fixed since it is limited by the bandgap shrinkage effect [9]. When the bandwidth increases, so does the tunability range, until reaching a limiting value dictated by the maximal current. For a maximal current of 100 mAmp the maximal ratio between the tunability range and the bandwidth was calculated to be 10.

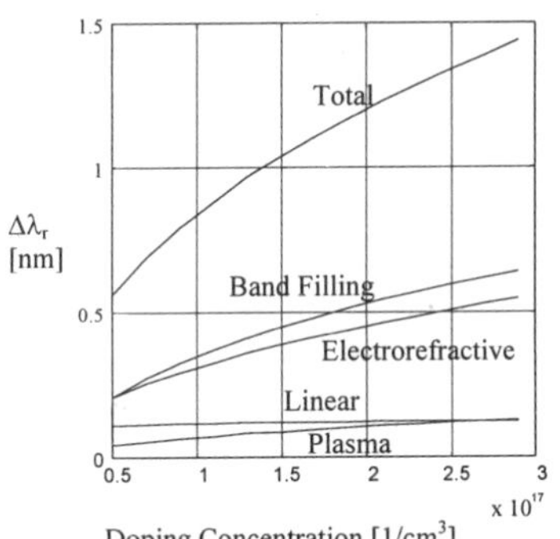

FIG. 2: The calculated tuning range $\Delta\lambda_r$ as a function of the doping concentration N_d for a PpnN configuration. Equal doping concentrations of the p and the n layers, reverse voltage of −5 V, bandgap wavelength of 1.3μm, photon wavelength of 1.55μm and waveguide thickness of 0.5μm.

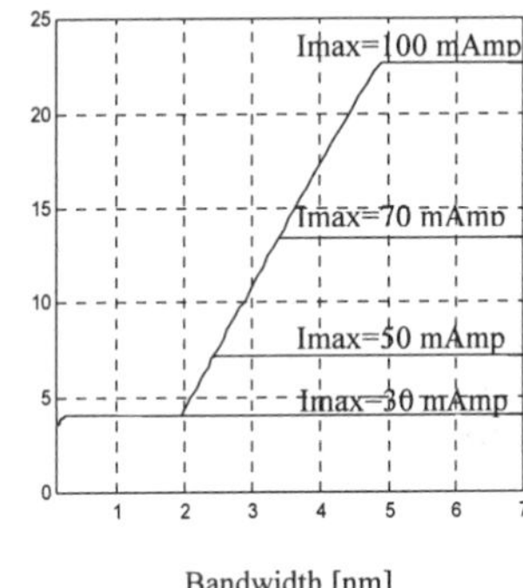

FIG. 3: The calculated tuning range $\Delta\lambda_r$ at 90% allowed resonance peak reflection, as function of the bandwidth for different maximum current values. The area of the GWS is 100*30μm^2

3. Concluding Remarks

The GWS exhibit high finesse, resulting in high sensitivity to changes in the resonance conditions, so that the resonance wavelength can be tuned. They can be designed with semiconductor materials such that the ratio between the tunability range and the resonance bandwidth is maximized. The results suggest that such GWS can serve as dynamic spectral filters, and fast optical switches, all of which can be useful in advanced signal processing and communications systems.

This work was supported in part by the German Ministry of Science and the Israeli Ministry of Science.

References:

[1] R. W. Wood, *Philosophical Magazine* **4**, 396 (1902).

[2] A. Hessel and A. A. Oliner, *Applied Optics* **4**, 1275, (1965).

[3] M. Neviere, "*Electromagnetic Theory of Gratings*", R. Petit(ed) Springer, Berlin, Chapt. 5, (1980).

[4] P. Sheng, R.S. Stepleman, and P.N. Sanda,, *Physics Review B* **26**, 2907, (1982).

[5] E. Popov, L. Mashev and D. Maystre, *Optica Acta* **32**, 607, (1986).

[6] A. Sharon, D. Rosenblatt, A. A. Friesem, H. G. Weber, H. Engel, and R. Steingrueber, *Optics Letters* **21**, 1564, (1996).

[7] A. Sharon, A. A. Friesem, and D. Rosenblatt, *Applied Physics Letters* **69**, 4154, (1997).

[8] D. Rosenblatt, A. Sharon and A. A. Friesem, *IEEE Journal of Quantum Electronics* **33**, 2038, (1997).

[9] B. R. Bennett, R. A. Soref and J. A. del Alamo, *IEEE Journal of Quantum Electronics* **26**, 113, (1990).

ANALYSIS OF GRATING ASSISTED COUPLERS

Nahum Izhaky and Amos Hardy

Tel Aviv University
Department of Electrical Engineering-Physical Electronics
Ramat Aviv, Tel Aviv 69978, ISRAEL

Waveguide gratings play a key role in the field of integrated photonics. They are used in many applications such as distributed feedback lasers, wavelength division multiplexing, optical switching, and guided wave coupling.[1-3] Recently, a unified approach to some coupled mode phenomena was presented,[4] based on an exact formulation, which is applicable to parallel waveguides with or without diffraction gratings. The model is a generalization of the coupled mode theory (CMT) for directional couplers.[5]

In this paper we present, after a brief introduction of the unified coupled mode formalism, its applicability for various cases. First we discuss forward coupling with and without gratings between two waveguides, and then we present the case of grating assisted backward coupling, also between two parallel waveguides. We show that the reduction of the full unified model (four-wave coupling problem) to a more commonly used models of two coupled equations is possible only in some specific cases. Finally, the influence of the grating parameters such as groove depth and duty cycle is considered. Chirped and parallel gratings as well as gratings with sinusoidal envelope periodicity are also addressed.

Let us consider, as an example, the structure depicted schematically in Fig. 1. It consists of two

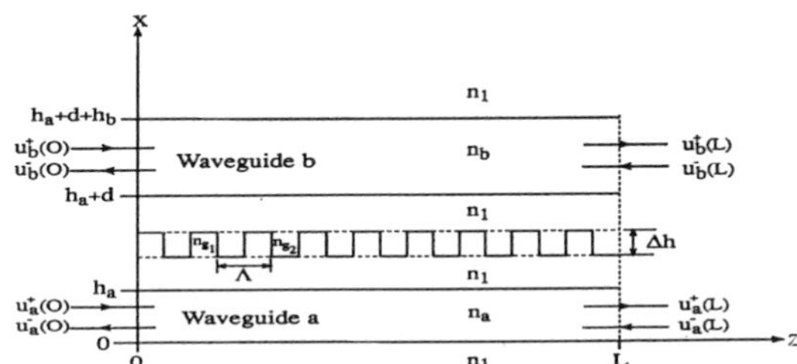

Fig. 1 Schematic illustration of two parallel waveguides (a and b), a rectangular grating between them, and the four involved waves.

parallel slab waveguides with core refractive indices of n_a and n_b, and cladding index of n_1. Their widths are h_a and h_b, and the distance between them is d. The grating period is Λ, with groove depth of Δh. The analysis of the above structure is based on the unified coupled mode formalism.[4] The model is derived directly from Maxwell's equations, and is based on the mode expansion conjecture. As a result, for most practical cases the expansion of the transverse fields is approximately satisfied by superposition of only four guided modes. The z-dependent amplitudes of the individual guided modes are denoted by $u_q^K(z)$, (q=a,b for waveguides a and b, respectively). The plus and minus superscripts stands for forward and backward propagation, respectively. Inserting that superposition into Maxwell's equations, one obtains a set of four coupled mode equations[4]

$$\frac{d\mathbf{U}(z)}{dz} = i\mathbf{M}(z)\mathbf{U}(z) \tag{1}$$

where $\mathbf{U}^T(z)$=[$u_a^+(z)$, $u_b^+(z)$, $u_a^-(z)$, $u_b^-(z)$], and **M**(z) is a 4x4 matrix consisting of propagation constants, coupling coefficients and other overlap integrals, as defined in Ref. 4. For the grating to assist coupling between the two waveguides, near wavelength λ_0, it should satisfy the Bragg condition

$$\beta_b(\lambda_0) \pm \beta_a(\lambda_0) \cong \frac{2\pi}{\Lambda} \tag{2}$$

where the plus and minus signs stands for backward and forward coupling, respectively. We used two analytical methods to solve Eqs. (1). The first technique takes the grating periodic dielectric constant as a Fourier series and considers only the dominant element for each interaction. Then, using the eigenvalues and the eigenvectors of the system it is solved analytically. The second technique considers the index profile exactly as it is, and use the transfer matrix method (TMM).[6] Both methods provide nearly identical results.

As an example for forward coupling, consider two nonidentical slab waveguides with core refractive indices of $n_a = 3.58$ and $n_b = 3.6$, and cladding index of $n_1 = 3.4$. Their thickness is 0.3 μm, and the distance between them is 0.4 μm. The grating period is Λ=58.26 μm, with groove depth of 0.1 μm. In this case of forward coupling, only two waves interact with each other. Hence two equations may suffice, and the full unified model may be reduced to a set of two coupled equations. In Fig. 2, we show the amplitudes $\left|u^+_{a,b}(z)\right|^2$ along the waveguides for initial conditions of $u^+_a(0) = 1$ and $u^+_b(0) = 0$. Dashed curves are without and solid curves are with the grating periodicity. As expected, the grating improves the power transfer considerably.

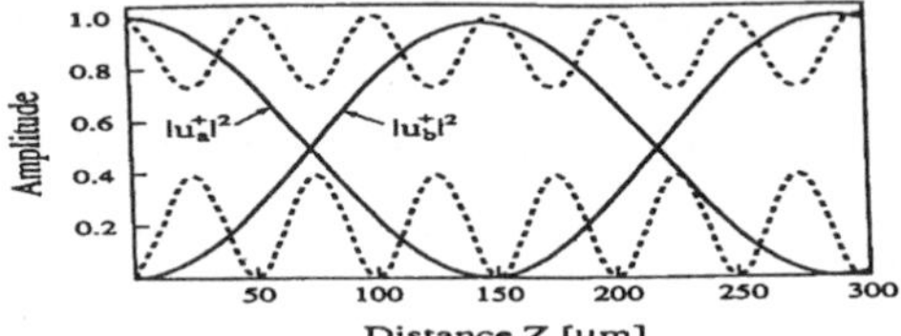

Fig. 2 forward coupling, with and without grating.

On the other hand, in the general case of grating assisted backward coupling, one should not reduce the four coupled equations, due to the simultaneous interaction among all four-waves.

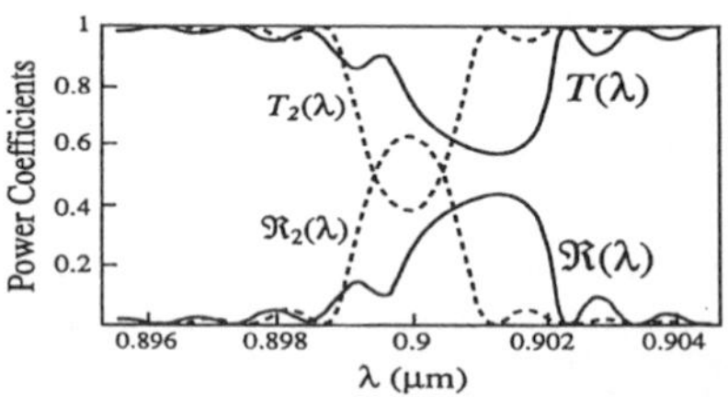

Fig. 3 Grating assisted backward coupling, between two parallel waveguides.

As an example for one such case, consider two slab waveguides with core refractive indices of $n_a = 3.497$ and $n_b = 3.5$, and cladding index of $n_1 = 3.4$. Their thickness is 0.4 μm, and the distance between them is 0.5 μm. The grating period is Λ=130.39 nm, with groove depth of 0.1 μm. Figure 3 depicts the spectral response [reflectivity-$\Re$ (λ) and transmissivity-T(λ)], derived from the four-wave model (in solid lines), with wave injection at z=0 into waveguide-a. In addition, it provides (in dashed lines) the spectral response [$\Re_2$(λ) and T_2(λ)] based on the degenerate unified two-wave model which considers only interaction between $u^+_a(z)$ and $u^-_b(z)$. Prior Two-wave models[7,8] give similar results to the above degenerate two-wave model. One can easily see the extreme differences between the two models. Contrary to the two-wave model, the full unified model yield an asymmetric broadened peak, and the maximum interaction is deviated by 1.5 nm from λ_0=0.9 μm. However, there are cases for which a two-wave interaction may suffice, and some hints to distinguish them from others will be given.

The perturbation shown in Fig. 1 is a single rectangular grating of constant period Λ, and groove depth of Δh. However, the unified formalism is not limited to this case, and we shall discuss also other perturbations such as: chirped, cascaded, and parallel gratings. As a result, the structure may provide many important functions, such as: grating assisted forward and backward coupling, add drop filters, equalizers, and dispersion compensation elements. The TMM is suitable to solve the unified CMT also for the general case of aperiodic gratings (e.g. chirped and parallel gratings).

Increasing the groove depth strengthen backward coupling in both waveguides. The reflectivity is increased especially in waveguide b for which the grating was planned for. Whereas, the transmission through the waveguides is decreased. The location of the grating between the two waveguides is also important. The shorter the distance between the grating and the injected waveguide, the smaller is the required groove depth (Δh) in order to achieve similar effects. The influence of the groove's duty-cycle provides a symmetric backward coupling, where it is maximum at duty cycle of ½. The influence of increasing the difference between the refractive indices within the grooves is similar to increasing the groove depth, except for the result that maximum transmission in waveguide b is obtained for $n_{g1} = n_1$. This is actually a case where a segmented waveguide of refractive index n_{g2} is embedded in the clad between the two waveguides.

Two parallel gratings can be used to obtain similar response to a single grating of double length. Most of the response is similar, however we could not obtain identical response at all wavelengths simultaneously.

The unified CMT can also be used to analyze the case of imperfect grating fabrication or deliberately sinusoidal deformation of the grating envelope periodicity. That is, the grating's grooves length changes sinusoidally over its length. Specifically, we design the grating grooves length according to $\Lambda(i) = \Lambda_0 + \gamma \sin(i2\pi/N)$, where Λ_0 is the average period. The sinusoidal amplitude is γ, the current groove number is counted by i, and the number of grating grooves within each sinusoidal period is denoted by N. It turns out that generally for a more significant grating response we should reduce the parameters N or γ, or both, or increase the total length of the grating. As an example, we compare the spectral response of constant periodic grating ($\gamma = 0 \Rightarrow \Lambda = \Lambda_0$) with the case of: $\gamma = 10^{-10}\mu m$ and N=25 (in both cases $\lambda_0 = 0.9\mu m$, L=300μm, and the grating is in the middle between the two waveguides). The case of sinusoidal envelope periodicity provides a lower backward coupling. In addition, its spectral response is much smoother, there are less side lobes, and the peaks are more isolated, than with the case of constant periodic grating.

References

1. H. Kogelnik and C. V. Shank, "Coupled wave theory of distributed feedback lasers," *J. Appl. Phys.*, **43**, 2327-2335 (1972).
2. D. G. Hall, "Optical waveguide diffraction gratings: coupling between guided modes," *Progress in Optics*, **29**, 1-63 (1991).
3. S. S. Orlov, A. Yariv and S. V. Essen, "Coupled mode analysis of fiber-optic add-drop filters for dense wavelength-division multiplexing," *opt. Lett.* **22**, 688-690 (1997).
4. A. Hardy, "A unified approach to coupled-mode phenomena," *IEEE J. Quant. Electr.,* **34**, 1109-1116 (1998).
5. A. Hardy and W. Streifer, "Coupled mode theory of parallel waveguides," *J. Lightw. Tech.*, **3**, 1135-1146 (1985).
6. T. Makino and J. Glinski, "Transfer matrix analysis of the amplified spontaneous emission of DFB semiconductor laser amplifiers," *IEEE J. Quant. Electr.,* **24**, 1507-1518 (1988).
7. T. L. Koch et al. "Vertically grating-coupled ARROW structures for III-V integrated optics," *IEEE J. Quant. Electr.,* **23**, 889-897 (1987).
8. P. Yeh and H. F. Taylor, "Contradirectional frequency-selective couplers for guided-wave optics," *Appl. Opt.,* **19**, 2848-2855 (1980).

REFRACTIVE BEAM SHAPING ELEMENTS FOR FIBER AND SWITCHING APPLICATIONS

E.-Bernhard Kley, Lars Wittig, Matthias Cumme, and Rolf Goering*

Friedrich Schiller-University Jena, , Max Wien Platz 1, 07743 Jena, Germany
Tel: ++49 3641 657647, FAX: ++49 3641 657680, kley@iap.uni-jena.de
** Piezosystem jena GmbH, Wildenbruchstraße 15, D-07745 Jena, Germany.*

Abstract

The combination of micromachined optical elements with optical fibers and piezo drivers offers new approaches in miniaturized optics. We have designed and tested a setup for switching the shape of a beam into different distributions. A monomode fiber was used as the light source and the switching was done by piezoelectric movement of different refractive beam shaping elements (switching time 2 ms). The conversion efficiency of the beam shaping element is independent on the wavelength and was measured to be >99% in the whole visible range. For the fabrication of the beam shaping elements gray tone lithography based on e-beam written gray tone masks and proportional etching of fused silica has been used.

1. Introduction

For industrial application light sources with special shapes of the outgoing beam are needed frequently. Such intensity distributions are mostly top hat distributions with lateral shapes like circles, rings, rectangles and others. Based on the limitation of the source intensity the shaping of the beam should be done with a high efficiency. Our application was demanding a fast alternation of different shaped beams as well as different wavelengths in the visible range. To fulfill this demand we decided to use the advantages of a micro-optical setup based on the combination of fibers, a piezoelectric fiber switches, a piezo driven translation stage and miniaturized refractive beam shaping elements.

2. The optical setup

Figures 1 shows a sketch of the setup. A commercial piezoelectric driven singlemode fiber switch of the company "piezosystem Jena" [1] was been used for the wavelength selection. For the movement respectively for the selection of the beam shaping elements in front of the singlemode fiber output we have used a piezo driven x/y translation stage [2] of the same company. Both devices are described below.

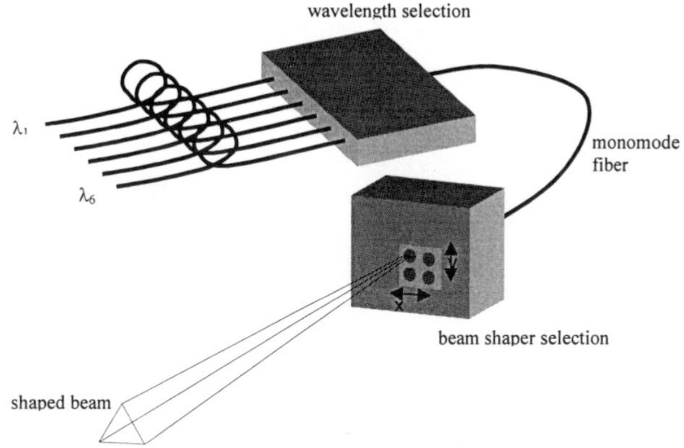

Fig. 1: Scheme of the optical setup.

Piezoelectric beam shaper selection:		***Piezoelectric wavelength selection:***	
2D translation stage PXY 100 SG [1]		1x6 singlemode fiber switch F-SS1 [2]	
x/y motion	100 µm	wavelength	488 nm - 635 nm
accuracy (strain gauge)	50 nm	insertion loss	< 2 dB
resonant frequency		crosstalk	- 45 dB
x-direction	380 Hz	switching time	< 2 ms
y-direction	480 Hz	dimensions	53 x 25 x 30 mm^3

The lateral dimension of the beam shaping elements have to be very small in order to select one of the 4 different beam shaping elements by a maximum x/y motion of 100µm. To be on the save side (an additional alignment area was necessary) the single element size of 80 µm x 80 µm was the maximum.

3. Design and fabrication of the beam shaping elements

For getting a "100%" conversion efficiency and a good top hat intensity distribution the size of the beam shaping elements should be 4 times more than the Gaussian diameter of the beam itself. This means only beam diameters of 20 µm or below can be used in connection with the translation stage described. Due to this 20 μm and the numerical aperture of the fiber of 0.12 we designed the beam shaping element for a fiber to element distance of 100 μm and a spherical wave radius of 106 μm. Especially for the triangular intensity distribution a modified iterative Fourier transform algorithm method described in [5] was used to design the beam shaping element as refractive one. Figure 2 and 3 are showing the surface profiles calculated as well as the measured intensity distributions.

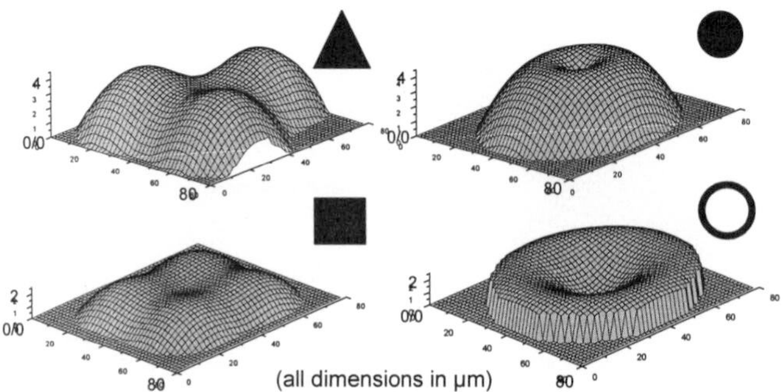

Fig. 2: Surface profile calculated for a BK7 glass material

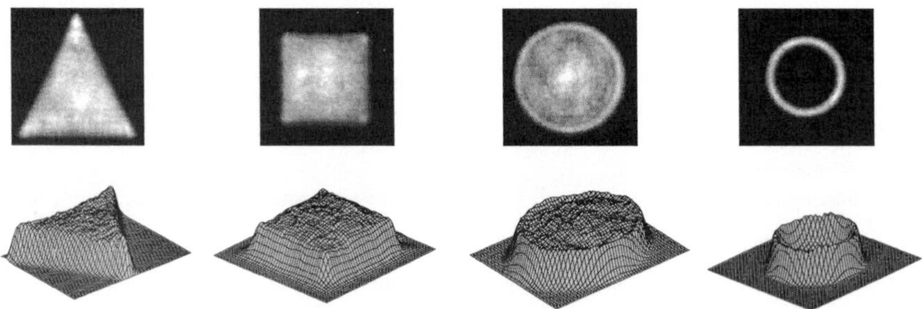

Figure 3: Measured intensity distributions at the far field

Due to the small dimension of the beam shaping elements the design results in a very shallow profile depth. Therefore, the fabrication of the elements was possible by gray tone lithography and proportional transfer into glass described in [4].

4. Results

The optical performance of the fabricated elements turned out to be nearly independent on the wavelength. We measured a conversion efficiency of > 99.5% in the visible range and a rms of the top hat uniformity about 5%. The switching time for the wavelength- as well as the shape-switching was in the range of 2 ms.

Acknowledgements: The authors would like to thank H.-J. Fuchs, D. Schelle, W. Gräf, H. Schmidt and V. Schmeisser for cooperation and investigation of several samples.

References:

[1] piezosystem jena, product catalog, p.4

[2] St. Glöckner et al; "Piezoelectrically driven micro-optic fiber switches", Opt. Eng. 37(4), 1229-1234 (1998)

[3] R. Göring, B. Götz and P. Bücker; "Integration techniques of microoptical components for miniaturized optomechanical switches", Proc. SPIE vol. 3631 "Optoelectronic Integrated Circuits and Packaging III", 148-155 (1999).

[4] E.-B. Kley, L. Wittig, M. Cumme, U. D. Zeitner, P. Dannberg, Fabrication and properties of refractive micro optical beam shaping elements, Micromachining and microfabrication , SPIE Vol. 3879, Santa Clara USA, 1999

[5] M. Cumme, E.-B. Kley, Refractive beam shaping elements for complicated intensity distributions, . this volume

WIGNER ALGEBRA FOR ACHROMATIC OPTICAL FOURIER TRANSFORMATION

Dayong Wang, Avi Pe'er, Asher A. Friesem, and Adolf W. Lohmann*

Department of Complex Systems,
Weizmann Institute of Science, Rehovot 76100, Israel

Abstract

A novel method for designing an achromatic Fourier transform system is proposed. It is based on relatively simple Wigner matrix algebra, where the elements of the Wigner matrix represent specific characteristics that must be modified in order to correct for chromatic dispersion. The basic principles and the design approach for obtaining perfect and approximate achromatic Fourier transformation are presented.

1. Introduction

In general, optical Fourier transformation systems involve monochromatic coherent light sources such as lasers, so they are extremely sensitive to optical noise and misalignment errors of the optical components. The use of polychromatic illumination, either spatially coherent or spatially incoherent, can alleviate these sensitivities and are receiving greater attention lately[1-5]. Unfortunately polychromatic illumination leads to the chromatic dispersion that must be overcome. Some design methods have been proposed to reduce such chromatic dispersion. Most are based on either geometrical optics considerations or paraxial Fresnel diffraction theory, and concentrate on combining holographic diffractive zone plates and achromatic refractive lenses[1,2,6,7]. In this paper, we propose a different design methods that includes Wigner matrix algebra applied to the coordinates of the Wigner distribution function (WDF)[8,9]. This design method is based on relatively simple matrix multiplications, and gives a clear physical insight.

2. Wigner Algebra and Linear Canonical Transforms

Every complex amplitude $u(x,y)$ can be described indirectly but uniquely by a Wigner distribution function (WDF). Using one-dimensional notations, we consider the consequences on the WDF if the input signal $u_{\text{in}}(x_0)$ is transformed into $u_{\text{out}}(x)$ by a first-order optical system. In the space domain, this transformation can be described by a linear canonical transform[10], that is completely specified by its parameters A, B, C and D, which construct a 2×2 unimodular transformation. The canonical transform integral can be inserted into the WDF, to yield

$$W_{\text{out}}(x,v) = W_{\text{in}}(Dx - Bv, -Cx + Av)\,. \tag{1}$$

where the subscripts "in" and "out" denote the input and output planes of the optical system, respectively and v the spatial frequency.

The deformation of $W_{in}(x_0, v_0)$ into $W_{out}(x, v)$ in Eq.(1) can be described as a vector-matrix product, which relates the input variables (x_0, v_0) to the output variables (x, v):

$$\begin{pmatrix} x \\ v \end{pmatrix} = \begin{bmatrix} A & B \\ C & D \end{bmatrix} \begin{pmatrix} x_0 \\ v_0 \end{pmatrix}, \quad AD - BC = 1. \tag{2}$$

We refer to the 2×2 matrix in Eq.(2) as the Wigner matrix.

3. Achromatic Fourier Transformation

The Wigner matrix of an optical system that can perform a perfect Fourier transformation is characterized by $A = D = 0$. When $A = 0$ and $D \neq 0$, the Fourier transformation is imperfect(or inexact), where an additional quadratic phase terms exist in the output distribution. Such an imperfect Fourier transformation is adequate in those applications where only the intensity (power spectrum) is needed. If the scaling factor B is independent of wavelength, at least to the first-order approximation, then the Fourier transformation is achromatized.

3.1 Imperfect Achromatic Fourier transformation

The imperfect achromatic Fourier transformation could be realized with two zone plates and one refractive achromatic lenses[6,7], shown in Fig.1. Here, z_1 is the distance from the input object to the first zone plate ZP_1, z_2 is the distances from ZP_1 to lens L_1, z_3 is the distance from L_1 to the second zone plate ZP_2, and z_4 is the distance from ZP_2 to the output plane. In our notation, assuming the input object is illuminated with a parallel beam of spatially coherent white light, the Wigner matrix of the transformation system can be obtained by cascading seven matrices, as

$$\begin{pmatrix} A(\lambda) & B(\lambda) \\ C(\lambda) & D(\lambda) \end{pmatrix} = \begin{pmatrix} 1 & \lambda z_4 \\ 0 & 1 \end{pmatrix} \begin{pmatrix} 1 & 0 \\ -\dfrac{1}{\lambda_0 f_{02}} & 1 \end{pmatrix} \begin{pmatrix} 1 & \lambda z_3 \\ 0 & 1 \end{pmatrix} \begin{pmatrix} 1 & 0 \\ -\dfrac{1}{\lambda f_1} & 1 \end{pmatrix} \begin{pmatrix} 1 & \lambda z_2 \\ 0 & 1 \end{pmatrix} \begin{pmatrix} 1 & 0 \\ -\dfrac{1}{\lambda_0 f_{01}} & 1 \end{pmatrix} \begin{pmatrix} 1 & \lambda z_1 \\ 0 & 1 \end{pmatrix}, \tag{3}$$

where f_1 is the focal length of the achromatic lens L_1, and f_{02} are the focal lengths of ZP_1 and ZP_2 for the designed wavelength λ_0, respectively. In order to obtain an achromatic Fourier transformation, it is necessary to correct the chromatic dispersion of the scaling factor $B(\lambda)$, while requiring $A(\lambda) = 0$. Usually, it is impractical to satisfy these requirements completely, so we must consider first-order and higher-order approximations, resulting several simple and useful configurations.

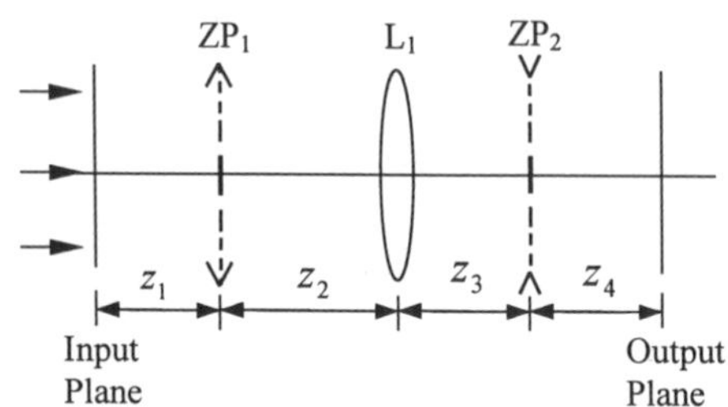

FIG.1 Setup for performing an imperfect achromatic Fourier transform.

3.2 Perfect Achromatic Fourier transformation

Perfect achromatic Fourier transformation requires the removal of the quadratic phase factor at the output. The quadratic phase factor is characterized by the fourth element D in the Wigner matrix. A possible approach for removing the quadratic phase factor is to add some lenses at the output plane of an imperfect achromatic Fourier transformation system. This approach has the advantage that it doesn't change anything else but the quadratic phase factor at the output plane.

Acknowledgments:

This research was supported in part by the Israeli Ministry of Science and the Albert-Einstein-Minerva-Center for Theoretical Physics. Dayong Wang is also with Department of Applied Physics, Beijing Polytechnic University, Beijing 100022, P.R China.

*Permanent address: Lab. Für Nachrichtentechnik, Erlangen-Nürnberg University, 91058 Erlangen, Germany

References:

1. E. Tajahuerce, et.al., Appl. Opt. **37**, 6164-6173 (1998).
2. E. Tajahuerce, et. al., Opt. Comm. **151**, 86-92 (1998).
3. C. L. Adler, et. al., Opt. Comm. **114**, 375-380 (1995).
4. P. Andrés, et.al., Opt.Lett. **24**, 1331-1333(1999).
5. A. Pe'er, Dayong Wang, A. W. Lohmann, and A. A. Friesem, Opt.Lett. **24**, 1469-1471(1999).
6. G. M. Morris, Appl. Opt. **20**, 2017-2025 (1981).
7. P. Andrés, J. Lancis and W. D. Furlan, Appl. Opt. **31**, 4682-4687 (1992).
8. M. J. Bastiaans, J. Opt. Soc. Am. **69**, 1710-1716 (1979).
9. A. W. Lohmann, D. Mendlovic and Z. Zalevsky, *Progress in Optics* **38**, Ed. E.Wolf, 263-342(1998).
10. M. Nazarathy and J. Shamir, J. Opt. Soc. Am. **72**, 356-364 (1982).

OPTICAL AND THERMAL SPECTROSCOPIC INVESTIGATIONS OF PHOTOREFRACTIVE VANADIUM-DOPED CdTe.

Mayer Tapiero

Institut de Physique et Chimie des Matériaux de Strasbourg. Groupe d'Optique Nonlinéaire et d'Optoélectronique 23, rue du Loess, F-67037, Strasbourg Cedex, France

Abstract

By means of thermal spectroscopy and photoconductivity measurements, deep levels in CdTe:V have been characterized. The two charge levels of vanadium, V^{2+} and V^{3+}, have been identified and their roles in the photorefractivity properties determined.

1. Introduction

Semi-insulating CdTe:V has been revealed very interesting for photorefractive (PR) applications in the near IR region, at room temperature. Its intensive investigation is motivated by the compatibility of its operating wavelengths (1-1.55 μm) with semiconductor lasers and with fiber optics communications.
Vanadium was found to be a suitable dopant, making the material semi-insulating and introducing the appropriate deep levels for PR. In the CdTe matrix, V has two electronic configurations : V^{2+} and V^{3+}. These two charge states are responsible for the absorption and for the photogeneration of charge carriers. Knowledge of their thermal and optical parameters and their roles as trapping, absorption and recombination centers is of prime importance, because they affect directly the PR properties.
To this end we have carried out thermal and optical spectroscopic investigations on vanadium doped CdTe with regard to the undoped CdTe. To see the V^{2+} and V^{3+} charge states independently, codoping with Cl and with As transforms the semi-insulating CdTe:V into n-type and p-type crystals respectively. Such crystals were also studied to get more information on the nature of the vanadium ions.

2. Thermal spectroscopy.

We used the PICTS (PhotoInduced Current Transient Spectroscopy) method, suitable for the characterization of trapping levels in the 0.1-1.0 eV energy range in high resistivity materials. The transient current, including both the steady state and the decay of photoconductivity (PC), induced by a square light pulse in a biased sample is recorded every 2 K during a temperature scan from 80 up to about 400 K. Two types of numerical processing of the stored data allow to extract the parameters of the traps. The four-gate data processing is a very sensitive method allowing the detection of the traps independently of their concentrations. Ten levels were observed (Fig.1, curve 2). The corresponding Arrhenius plots (Figure of the right) allow the determination of their signatures (Energy E_t and Capture cross section S_t).

The double-gate spectra exhibit peaks, whose magnitudes are proportional to the concentration, corresponding only to the densely populated levels (Fig.1, curve 1). This gives the concentrations of the traps when the sample is illuminated till saturation.

Comparison of the PICTS spectra of highly and slightly vanadium doped CdTe with those of undoped samples shows that at least two of the detected levels are related to the presence of vanadium. The level, located around 0.95 eV may be due to V^{3+}, while V^{2+} is located around 0.75 eV.

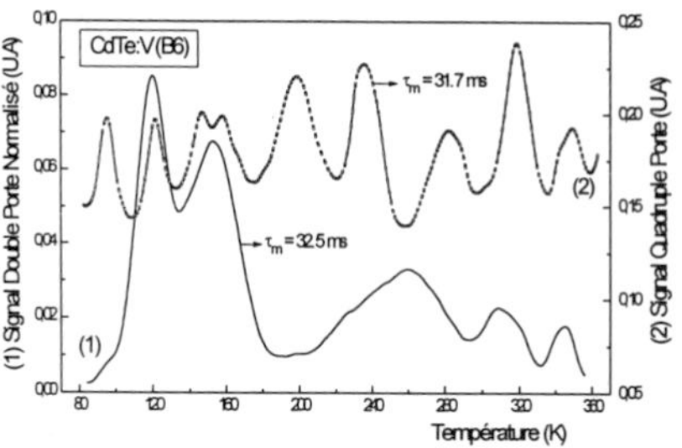

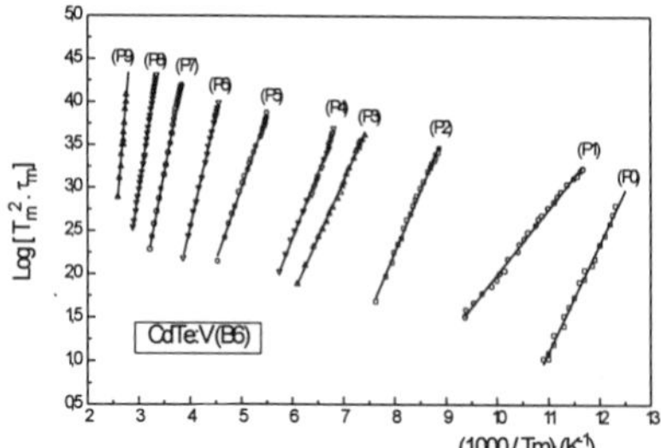

Figure1. PICTS spectra of a CdTe:V. The four-gate spectrum (curve 2) exhibits the peaks corresponding to all levels, independently of their concentrations. The normalized double-gate spectrum (curve 1), from which the concentration is deduced, shows the peaks of the more densely populated levels. Arrhenius plots obtained from the four-gate spectrum (figure of the right) give the thermal ionization energy E_t and the capture cross section S_t.

3. Photoconductivity studies.

3.1 – Spectral Response of steady-state photoconductivity at room temperature.

The steady-state PC spectra were measured at ~ 300 K for samples B23 (CdTe:V), B28 (CdTe:V:As) and B19 (CdTe:V:Cl), in regard of the undoped specimen ND1 (CdTe). The curves presented here were normalized versus the photon flux, which was nearly constant (between 5 10^{12} and 3 10^{13} cm^{-2} s^{-1}).

Figure 2 (left) shows the spectral responses of PC of samples B23 (CdTe:V) and ND1(undoped CdTe). The first noteworthy point is the sharpness of the peak corresponding to intrinsic PC (around 1.5 eV). This is due to the steepness of the fundamental optical absorption edge of CdTe (direct band-gap). The strong decrease on the high-energy side of the spectrum indicates strong surface recombination.

The second point to note is the fact that the undoped crystal exhibits only the sharp peak at ~ 1.5 eV. This tells us that the bands between 0.7 and 1.4 eV observed for doped crystal, are really correlated with the presence of V. This is confirmed by absorption and MCD measurements.

Codoping with chloride provides a donor level which should compensate the V^{3+} ions. Sample B19 (CdTe:V:Cl) should contain essentially V^{2+}. Fig. 2 (right) shows that some peaks disappear or are

strongly reduced. The largest contribution to the PC of B19 is found to occur above 1.0 eV and to extend to as high as 1.4 eV. The deduced assignments to V^{2+} are in good agreement with MCD conclusions.
Codoping with arsenic provides an acceptor level which should compensate the V^{2+} ions. Crystal B28 (CdTe:V:As) should contain only V^{3+}. It is, in fact, observed that some peaks are reduced and others are strengthen in regard to the reference spectrum of CdTe:V. One can deduce the assignment to V^{3+}.
Finally, P1 (1.17 eV) and P2 (0.82 eV) are attributed to V^{2+}. P3 (0.86 eV), P4 (0.92 eV), P5 (1.02 eV) and P6 or P7 (0.96-0.98 eV) to V^{3+}. The broad band, located at 1.25-1.35 eV, may be due to both V^{2+} and V^{3+}.

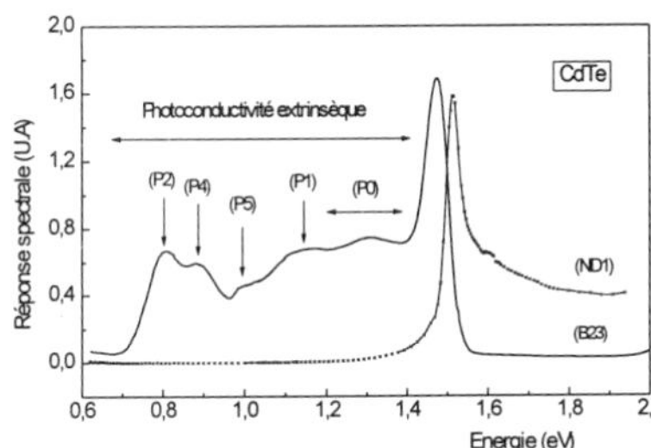

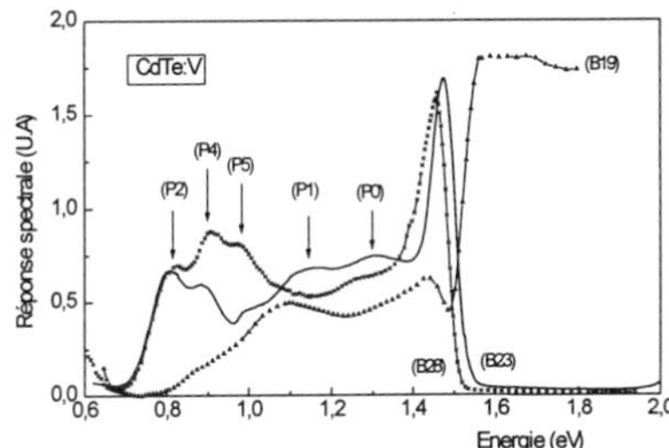

Figure 3. Spectral response of PC for crystal B23 (CdTe:V), compared to that of undoped CdTe (ND1), shows the presence of band structures related to the presence of V. On the right, spectra of codoped samples with Cl (B19) and with As (B28) in regard to that of B23. The intensive peaks around 1.5 eV correspond to the intrinsic PC.

3.2 – *Spectral response of steady-state Photoconductivity at variable temperature*

Detailed steady-state PC measurements were carried out for a series of temperatures on the semi-insulating CdTe:V crystals. For sake of clarity only three spectra of sample B23 are shown on Fig. 3, left.
The modification of the intrinsic spectrum results from the temperature dependence of the band gap and of the PC gain (curve labeled "Gap" in figure of the right). We find a linear shift of its maximum of 0.46 meV/K, in good agreement with that of the literature.
The detailed study of the extrinsic region has been revealed very interesting from two points of view : the thermal dependence of the energy of each structure and the thermal dependence of its magnitude.
Thermal variation of the position of all transitions shows that, besides peak P3, all the energies are quite constant, at least up to room temperature. This suggests that these transitions are internal transitions or transitions from a vanadium level to one of the bands to which they are tied. Peak P3, whose energy is 0.87 eV at room temperature, varies with a slope of 0.24 meV/K, half of that observed for the bandgap. It

is known that, when temperature is lowered, the width of the bandgap increases, the conduction and

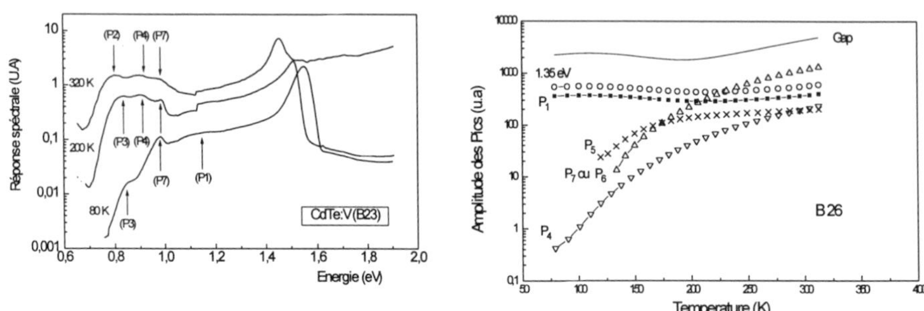

valence bands move away, one from the other, symmetrically. This suggests that P3 is fix wen T varies.

Figure 3. Spectral responses of PC obtained for crystal B23 (CdTe:V) at temperatures 80, 200, and 320 K. On the right, thermal variation of the magnitudes of the extrinsic transitions and of the width of the band gap.

The figure on the right shows the thermal dependence of the magnitude of all transitions.

The curve labeled "Gap" has a variation which reflects that of the PC gain or transport parameter $\mu\ \tau$, the product of the mobility by the lifetime of the photocarriers. Peaks of "high energy", labeled 1.35 eV and P1 (1.17 eV) have the same thermal behavior as the band-to-band transition (Gap) suggesting that they are due to direct transitions from V^{2+} centers to the conduction band. The released photocarriers, electrons in these cases, are governed by the same transport parameter $\mu\ \tau$. This is coherent with measurements of the photorefractive gain which show clearly that electrons are the majority carriers when CdTe samples are illuminated with 1.06 μm (1.17 eV) radiation, the V^{3+} concentration being increased simultaneously. All the other peaks (energies ≤ 1 eV) have also the same behavior for T > 230 K. Below 230 K the change is drastic. At 80 K, peak P5 has a value 10 times lower; peak P6 decreases by 2 decades; the magnitude of peak P4 decreases by more than 3 decades, while band P3 almost vanishes at liquid-nitrogen temperature. It is note worthy that the lower is the energy of the transition, the higher is the decrease of its magnitude with decreasing temperature. At this stage, it is possible to qualitatively interpret the change in PC spectra as the temperature decreases, by assuming a two-step photo-thermal ionization process. This means that the internal $d \rightarrow d^*$ transitions are presumably followed by a subsequent thermally stimulated $d^* \rightarrow CB$ transfer.

PART 12

FIBER OPTICS AND SENSORS

Chairperson: *M. Tur, Israel*

Annals of the Israel Physical Society, v. 14

ELECTRO-OPTICS and MICROELECTRONICS
Eds: Raphael LAVI and Ehud AZULAY

APPLYING HIGH-ORDER-MODE FIBER TECHNOLOGY TO MANAGE CHROMATIC DISPERSION OF SMF AND NZDSF ACROSS THE TRANSPORT BAND

Yochay Danziger, Eran Herman, Doug Askegard

Chief Technical Officer, Research Physicist, Product Planner
C/o LaserComm Inc. Kiryat Atidim, Building # 4
P.O. Box 58147, Tel Aviv 61580 Israel

Abstract

High-order-mode fiber provides the correct chromatic dispersion compensation across the transport band thereby enabling utilization of the full bandwidth potential of fiber at 10 Gb/s and higher bit-rates. Dispersion management devices have been fabricated verifying the practical application of the technology.

1. Introduction

The many types of optical fibers available today are evidence of the need for greater bandwidth. The standard “C” band of Erbium doped fiber amplifiers (~1530-1565nm), positioned along long-haul systems, contains up to 40 channels with 100 GHz spacing, each operating at a rate of up to 10 Gb/s each. In order to accommodate these many channels, dispersion management devices must provide the correct dispersion compensation across the entire band. Dispersion management becomes even more of a challenge with the expansion in to the “L” band portion of the spectrum (1570-161nm), and the increase of transmission rates to 40Gb/s and beyond. These developments call for innovative dispersion management devices with greater accuracy and superior performance.

New technology has been developed recently for facilitating migration to 10Gb/s in deployed WDM system and enabling next-general 40 Gb/s DWDM long-haul transmission. This technology is based on the use of high-order-modes (HOMs) that may be configured to offer successive compensation for many values of chromatic dispersion and dispersion slope (across the entire transmission band).

2. Dispersion management

Dispersion management is the process of balancing positive and negative dispersion over the length of the transmission fiber so that as the signals that travel through the fiber, they will always experience some chromatic dispersion but when they reach the receiver, the total dispersion would be near zero, or within an acceptable limit. The maximum allowable accumulated chromatic dispersion for a 1 dB penalty is a function of the bit rate transmitted and is determined by: DL(ps/nm)<$10^5/R^2$, where R is the bit rate in Gb/s. Table I presents the total allowable dispersion for various data rates.

Table 1: Allowable dispersion in fiber

Data Rate	Total Allowable Dispersion
2.5 Gb/s	12,000 to 16,000 ps/nm*
10 Gb/s	800 to 1000 ps/nm*
40 Gb/s	60 to 100 ps/nm*

*Actual limits depend on transmitter laser spectral bandwidth and on receiver design/sensitivity.

3. Dispersion in Transport Fibers

Chromatic dispersion is well characterized, known, and controlled in fibers of recent manufacture (Table II). The conventional single mode fiber (cSMF), which has a zero dispersion wavelength in the 1330nm region, was widely installed in the 1980's and early 1990's. The various types of the non-zero dispersion shifted fibers (NZDSF) yield lower levels of dispersion per unit length, however, because of significant difference in dispersion over the 1530nm to 1565nm range, dispersion slope compensation becomes critical to achieving maximum spacing between generators.

Table 2: Typical Dispersion and Slope

Type of Fiber	Typical Dispersion @1550nm	Typical Slope @ 1550nm	Dispersion range over 1530 to 1565nm
Conventional SMF	17 ps/nm-km	0.057 ps/nm^2-km	15.9 to 17.8 ps/nm-km
NZDSF (early) –type 1 -type 2	2.6 ps/nm-km 3.5 ps/nm-km	0.067 ps/nm^2 0.067 ps/nm^2-km	1.3 to 3.6 ps/nm-km 2.2 to 4.6 ps/nm-km
NZDSF (large effective area)	3.8 ps/nm-km	0.1 ps/nm^2-km	1.8 to 5.3 ps/nm-km
NZDSF (reduced slope)	4.4 ps/nm-km	0.45 ps/nm^2-km	3.5 to 5.1 ps/nm-km
NZDSF (new large eff. Area	2.7 ps/nm-km	0.07 ps/nm2-km	1.3 to 3.7 ps/nm-km
NZDSF (new light fiber)	8.0 ps/nm-km	0.057 ps/nm^2-km	6.8 tp 8.9 ps/nm-km

Modes can be thought of as guided optical waves, propagating along the optical fiber. Mathematically, modes are solutions of Maxwell Equations in the optical fiber, subject to appropriate boundary conditions. Depending on its dimensions and refractive index profile, an optical fiber at a given optical frequency can support either: (a) many modes – such as in a multi-mode fiber; (b) a single mode (LP_{01})- conventional single-mode fiber or NZDSF fiber; or (c) few modes – a few-mode (or HOM) fiber. Only the fundamental mode (LP_{01}) exists for all wavelengths. Each of the other HOMs has a cutoff wavelength, above which it can no longer propagate. Modes differ from one another in: (a) the spatial distribution of their amplitude and intensity; (b) their phase and group velocities; and (c) their dispersion properties. Figure 1 shows typical dispersion characteristics of different modes. Dispersion and dispersion slope of HOM fibers can be controlled by their design and manufacturing process.

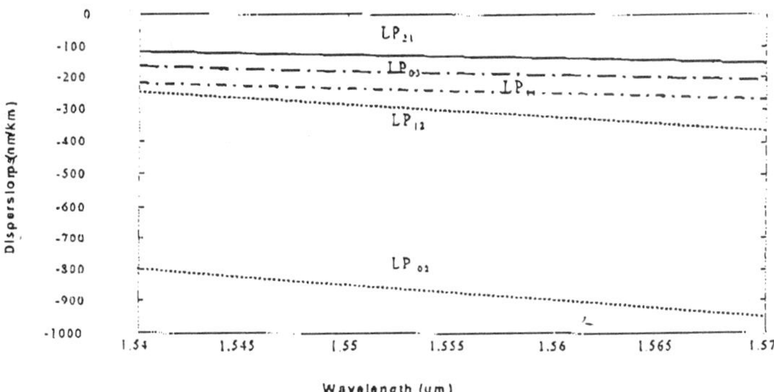

Figure 1: Typical dispersion properties of HOM fibers

5. A Dispersion Management Device Utilizing HOM Fiber

Chromatic dispersion can be effectively managed by HOM fiber, however, it requires that signal energy from basic LP_{01} mode must be transformed to a desired HOM and back to basic mode (the conversion process is proprietary and not discussed in this paper). Figure 2 presents a schematic illustration of a HOM dispersion management device.

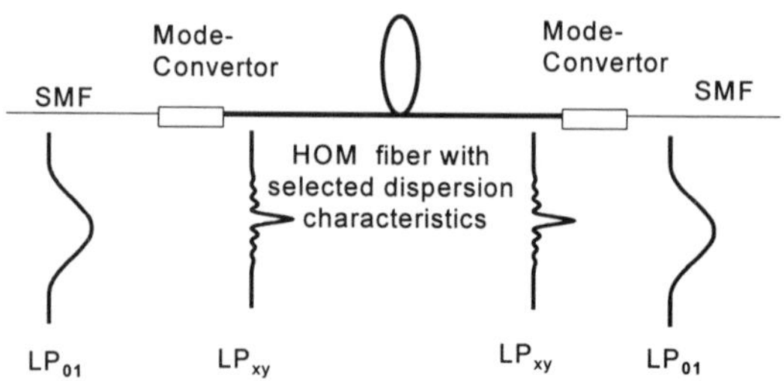

Figure 2. Dispersion management device, using HOM fiber

Figure 3 presents dispersion measurements at various wavelengths, achieved by HOM dispersion management device (HOM DMD), characterized by a nominal dispersion of –270 ps/nm @1550 nm with a slope over the "C" band of approximately –6.6 ps/nm^2. All measurements were taken at the basic LP_{01} mode at the device terminations.

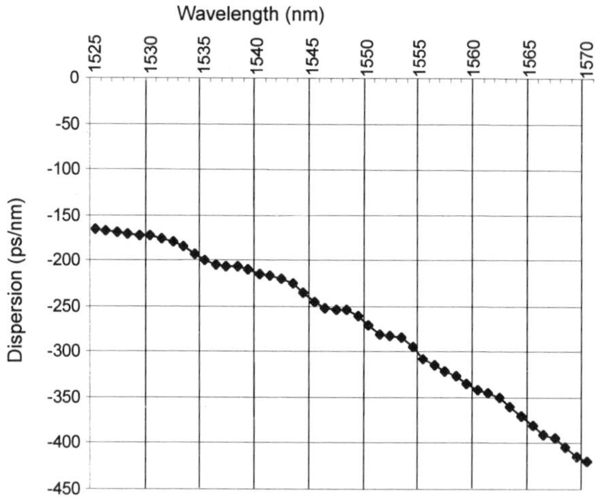

Figure 3. Performance of HOM DMD

References:

[1] C. Poole, et al., "Helical-Grating Two-Mode Fiber Spatial-Mode Coupler", Journal of Lightwave Technology, Vol. 9, No. 5, May, 1991.

[2] I. Kaminow, T. Koch, "Optical Fiber Telecommunications IIIA", 1997.

[3] Y. Liu, et al., "Advanced Fiber Design for WDM Optical Networks: An Economical & Future Proofing Solution", Technical Digest, AMTC'98, Volume 1, 1998.

[4] J. Lively, "Dealing With the Critical Problem of Chromatic Dispersion", Lightwave, September, 1998.

Optical Fiber Heater

Haim Lotem, Jacob Kagan* and Dan Sagie*

NRCN, P.O.Box 9001, 84190 Beer-Sheva, Israel
*Rotem Industries, ROTEM Ind. Park, P.O. Box 9046, Beer-Sheva

Abstract

An optical heater that is based on an absorbing optical fiber is proposed and demonstrated. It is composed of a laser source coupled to a high absorption fiber through a high transmission fiber. Homogeneous heat dissipation, independent of location along the heater may be achieved by z dependence absorption coefficient $\alpha(z) = -\ln(1-\beta \cdot z)/z$, where β is a constant absorption rate. The fiber heater advantages are that it is immune to electric and magnetic fields without producing electromagnetic perturbations. Additionally it is safe for use in hazardous chemical environments. A fiber heater was demonstrated using glass fiber and a 2.1 μm laser.

Since the original introduction of optical fibers there was an enormous advance in enhancing the transmission coefficient of fibers. With the present technology, near IR commercial fibers with better than 0.2 dB/km loss are commonly used in long distance communication networks. In such systems, optical amplifiers separated by fiber sections of up to 80 km long are common. A completely different fiber technology trend is the development of fibers with controlled spectral absorption, especially designed for applications such as optical sensors and laser amplifiers.

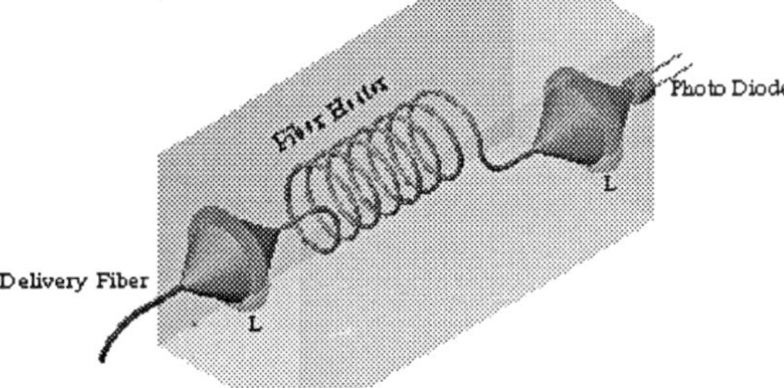

Figure 1: Fiber heater

In the present work we discuss a possible use of absorbing fibers in a device called optical fiber heater. In such a heater, the transmitted light absorbed along a fiber is used as a heat source that controls the temperature of its surrounding, see Figure 1. The generated heat in the fiber may be transferred to its destination by direct heat coupling or by convection. The fiber heater system is composed of a laser source coupled to a high absorption fiber section through an intermediate high transmission delivery fiber, see Figure 2. The control of the total power dissipated along the fiber may be achieved by varying the absorbed power in the fiber. This may be done by varying the optical power coupled into the fiber or by varying the laser wavelength in the vicinity of absorption spectral band of the fiber. Even more important is the capability to control the power dissipation along the fiber either by controlling the doping, or by an appropriate coating of the fiber cladding.

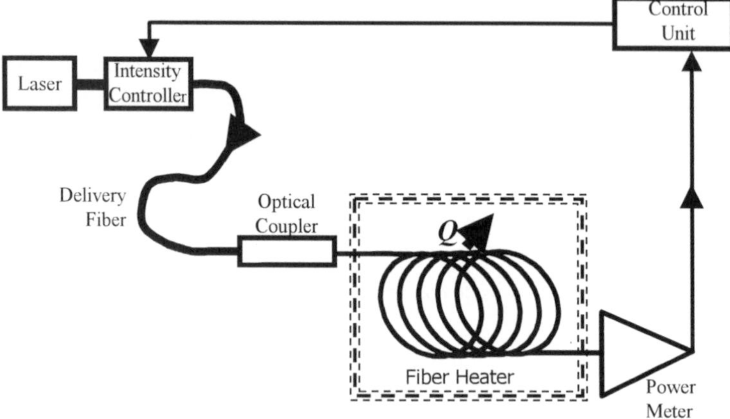

Figure 2: Fiber heater system

A fiber heater is characterized by several unique advantages. In contrast to a common electrical heater, the fiber heater exhibits an extremely low electrical conductivity, and it radiates no electromagnetic fields below the optical region. This makes the fiber heater very useful in high voltage or high magnetic field environments, as well as in systems that are extremely sensitive to electromagnetic perturbations, such as mechanical gyros. Since an electrical voltage does not derive the system, the heater is totally immune to any electrical discharge, and the need for electrical isolation of the heated medium is eliminated. Also, since the fiber material composition is not chemically active, a fiber heater is a safe device for use in most hazardous chemical, especially inflammable environments.

The light intensity *I(z)* along a standard commercial optical fiber is monotonically decreased due to absorption and scattering effects. With a constant loss fraction per unit fiber length,

$$\frac{dI(z)}{dz} = -\alpha \cdot I(z) \qquad (1)$$

the intensity is exponentially decreased, $I(z) = e^{-\alpha \cdot z}$, as shown in Figure 3A. Due to the exponential intensity decay along a fiber axis, *z,* the expected dissipated energy per unit length of the fiber strongly depends on location*, z*, as shown in figure 3B. The analyzed example is a fiber sample in which 90% of the input energy is internally dissipated.

In general, from a practical point of view, it is advantageous to have a heater wire with constant heat dissipation per unit length, i.e. with,

$$\frac{dI(z)}{dz} = -C \qquad (2).$$

Since the light intensity along a fiber is decreased with *z*, in order to obtain constant heat dissipation per unit length the absorption coefficient α should increase with *z* as:

$$\alpha(z) = -\ln(1 - \beta \cdot z)/z \qquad (3),$$

(where β is a constant that controls the absorption rate). The light intensity is then linearly decreased along the fiber, and thus, constant dissipated heat per unit length condition is achieved. This behavior is demonstrated in Figure 3 in which the absorption coefficient $\alpha(z)$, the laser light intensity *I(z)* and the dissipated heat per unit length are shown as a function of location, *z*. (See curves marked H in the figure). With an absorption coefficient β=98.2 m^{-1}, Eq. (1) implies that 90% of the input light energy is converted into heat in a 10 m long fiber section.

The absorption coefficient of a fiber may be controlled, in general, by doping the fiber core as well as the fiber cladding. A doping procedure may start first by coloring the fiber external surface using an evaporated coating on, or by immersing the fiber in a coloring solution. The dopant may then be diffused into the fiber cladding and core by a high temperature treatment. The dopant concentration and thus, the absorption coefficient are controllable by selecting the duration of the coloring stage, and \ or the diffusion process. This way, in principle, any *z* dependence of an absorption coefficient, $\alpha(z)$ may be tailored.

A demonstration of a basic fiber heater was performed using a 600 μm core diameter glass fiber of several meters long. The fiber was directly coupled to a 30 W continuous wave beam of a 2.1 μm Holmium laser at. A short focal length IR-quartz lens performed the coupling. Due to OH ion absorption at 2.1 μm, more than 50% of the laser energy was converted into heat in the fiber. The generated heat dissipated through the fiber envelope is used as a controlled heat source.

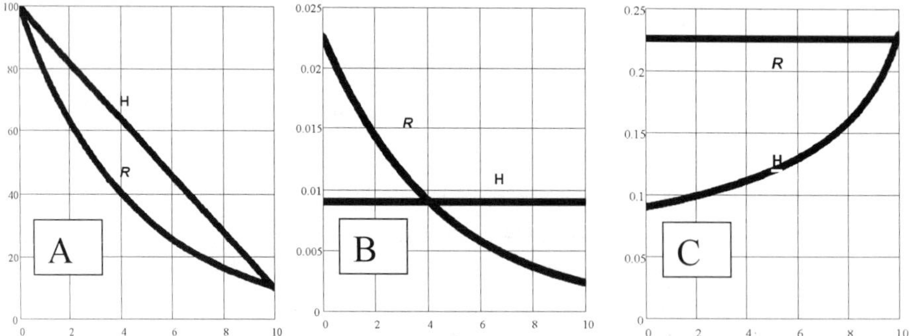

Figure 3: The calculated light intensity (A), The dissipated energy per unit length (B), and the absorption coefficient (C), are given in arbitrary units as a function of location ***z*** along the fiber.

Regular fiber case is marked "R", while a tailored absorption fiber is marked by "H".

In conclusion, an optical heater based on an absorbing fiber and an optical light source is proposed. This device may be useful in unique environments since it is chemically inert and it does not generate electric or magnetic fields. Homogeneous heating may be obtained by having constant heat dissipation per unit length of the fiber. This specification may be achieved using a particular dependence of the absorption coefficient on location, *z,* given here.

MINIATURE FIBER-OPTIC NON-CONTACT DISTANCE GAUGE

Ehud Shafir and Garry Berkovic

SOREQ NRC, Yavne 81800, Israel

Abstract

We suggest a fiber-optic device by which distances in the range of a few millimeters to hundreds of millimeters can be measured at hard-to-reach and adverse locations. The measurements are done opposite a tip of a fiber, are absolute in nature and take milliseconds to few seconds – depending on the resolution required. An FM chirped tiny diode-laser serves as the light source while the distance information is obtained from the beat frequency between the light reflected from the target and a local oscillator. In addition, for a moving target, the velocity can be simultaneously measured by the Doppler contribution to the beat frequency. The basic concept and the prototype performance characteristics are presented.

1. Introduction

Distance measurement by optical means has long been used where fast, non-contact and accurate results are needed. The most prevalent optical method involves measuring the time taken for a light pulse to travel to a target and back. This method is mainly suitable for distances longer than tens of meters. At shorter distances, interferometric techniques are more common. In this manuscript we demonstrate an optical fiber system, employing readily available components, for short-range distance measurements. According to this approach, a fiber optic tip is positioned facing the target, and the absolute distance between them is optically measured. This configuration is particularly attractive when distances are to be measured in hard-to-reach or adverse locations.

2. Principle of Operation

Consider a continuous-wave laser beam launched onto, and reflected from, a target to be measured, where the frequency of the beam is linearly modulated in time (chirped). When the reflected beam returns to the system, its optical frequency differs from that of the beam then being emitted from the system, due to the chirp. This frequency difference is proportional to the time elapsed during the roundtrip of the beam, and thus may be used to measure the unknown distance (X) between the system and the target, for a known refractive index of the medium between the system and the target (n). When the returning and the outgoing (local oscillator) beams interfere their frequency difference manifests itself as a readily measurable beat frequency, f_{Beat} [1-2]. It is straightforward to show that

$$X = \frac{c}{2n \, {}^{d\nu}\!/_{dt}} \cdot f_{Beat} \tag{1}$$

where c is the speed of light, and ${}^{d\nu}\!/_{dt}$ is the chirp rate. The saw-tooth command voltage V_{PTP} producing this chirp at a modulation frequency f_{Mod} enables us to re-write equation (1) as:

$$f_{Beat} = \frac{2X}{c} \cdot \frac{d\nu}{dV} \cdot f_{Mod} \cdot V_{PTP} \qquad \text{since} \qquad \frac{d\nu}{dt} = \frac{d\nu}{dV} \cdot f_{Mod} \cdot V_{PTP} \tag{2}$$

$\frac{d\nu}{dV}$ can be found once for each diode laser by a calibration measurement at a known distance.

Note that if the target is moving at velocity v during the measurement, the measured beat frequency will also include a Doppler component, $f_{Doppler} = \frac{2v}{\lambda_0}$, which is independent of the modulation frequency and voltage. For instance, for a velocity of 0.1 mm/sec and 1.5 μm laser (λ_0), the Doppler shift will be 130 Hz. The static (distance) and Doppler (velocity) components can be individually determined by two consecutive measurements of the beat frequency under different conditions of the $f_{Mod} \cdot V_{PTP}$ product.

3. Fiber Optic Embodiment and Experimental Results

Previous embodiments of the above principle were demonstrated with bulk-optic components, requiring delicate alignment procedures. We propose a simpler embodiment utilizing optical fibers eliminating the need for alignment and self-producing the local oscillator term. Figure 1 depicts this approach, showing the entire system as well as a magnified view of the fiber-tip and the target. We use a 50μm fiber coupler, and a 60mm focal length lens collimates the outgoing beam. {Note that a GRIN lens could replace the bulk lens, enabling measurement of distances down to the mm range. In this case the gap between the fiber tip and the GRIN lens should be filled with an index-matched material, and the system will determine the distance between the far end of the lens and the target}. The laser (New Focus 6328) is frequency modulated by application of a saw-tooth voltage to a PZT modulating the laser external cavity. Other modulation techniques like direct current modulation may be considered.

As shown in Figure 2, a beat signal is clearly observed when the laser is modulated. The distance calculated from the beat frequency matches the actual distance (X) to within 2%. To further verify equation 2, we measured the beat frequency for several values of the chirp rate while keeping the target fixed. As shown in Figure 3, for a wide range of V_{PTP} and f_{Mod} a highly linear relation between the beat frequency and the $f_{Mod} \cdot V_{PTP}$ product is found.

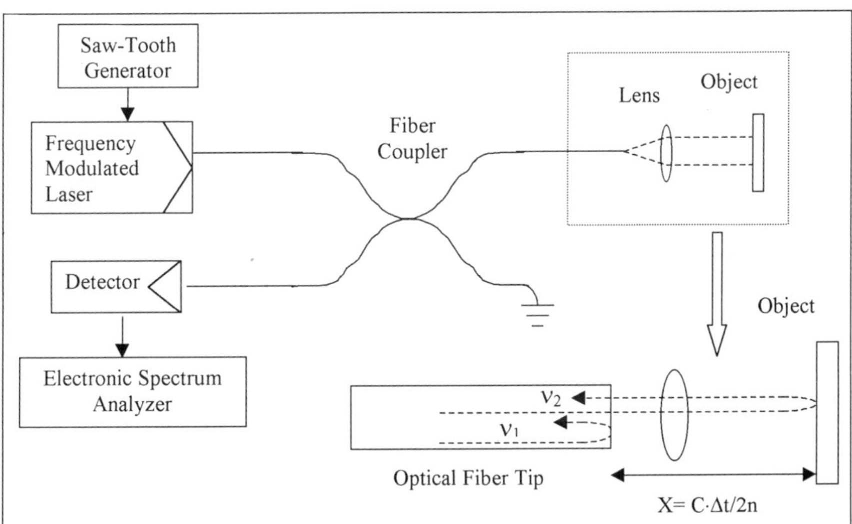

Figure 1. Schematic of fiber optic system for distance measurement. The lower right hand corner shows a magnified picture of the fiber tip and the object.

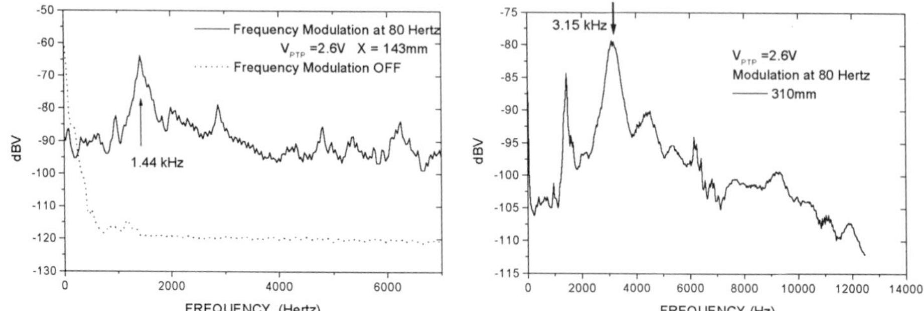

Figure 2 : Comparison of Spectrum Analyzer Outputs with and without laser modulation. A clear peak is seen at 1.44 kHz, and this peak moves to 3.15 kHz as the target distance is changed from 143mm to 310mm. At a target distance of 460mm a beat frequency of 4.69 kHz is obtained (not shown). Traces are the averages of 256 scans at 32 scans/sec.

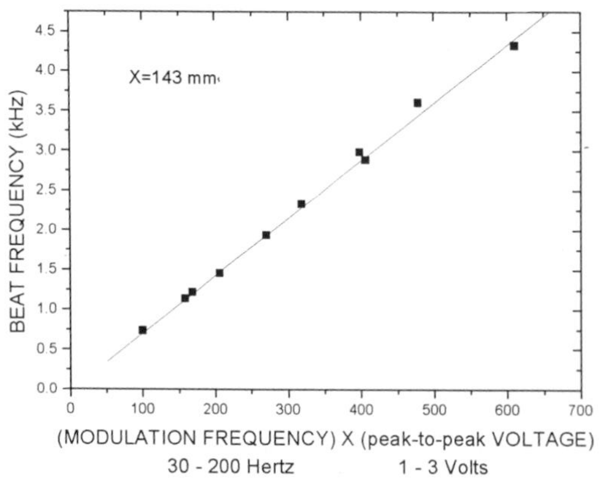

Figure 3. Measurement of f_{Beat} for various combinations of saw-tooth modulation voltage and frequency, for a fixed target distance.

References:

[1] G. Beheim and K. Fritsch, *Elect. Lett.* 21 (1985) 93.

[2] U. Minoni, L. Rovati and F. Docchio, *Rev. Sci. Instrum.* 69 (1998) 3992.

FIBER OPTIC DEVICE FOR RESPIRATION MEASUREMENT

Alexander Suhov, Anatoly Babchenko, and Meir Nitzan

Jerusalem College of Technology
21 Havaad Haleumi St., POB 16031, Jerusalem 91160, Israel
Tel: 972-2-6751212, Fax: 972-2-6751200, e-mail: babchenk@mail.jct.ac.il

Abstract.
A fiber optic sensor for the measurement of the respiratory depth has been developed. The sensor is composed of a bent optic fiber, which is connected to an elastic section of a chest belt, so that its radius of curvature changes during respiration. The signal is transmitted from the sensor to a processing part. This method provides the capability to measure the influence of physical activity on breathing parameters. The method was tested on two groups of subjects: before and after physical activity. The test showed high sensitivity and precision of the sensor.

1. Introduction

Optical measuring devices take an important place in the modern medical diagnosis and are very suitable for many medical applications, mainly due to their reliability. Light is not influenced by electro-magnetic fields, so fiber optic sensors are used, for example, in magnetic resonance imaging (MRI) environment etc. In the current study an optical sensor based on a bent optical fiber (Fig. 1) was used for respiratory measurement applications.

The respiratory monitoring is very important in cases of different diseases, especially in improper functioning of respiratory system and for medical research. Such quantitative monitoring could be done either by direct measuring, which requires nonleaking connection to the patient's airways, or of body dimension changes. These techniques are used for clinical tests of pulmonary function and supply different characteristics of respiratory waveform. In this study the correlation between respiration and arterial blood pressure was examined.

2. Materials and Methods

The device for respiration measurements utilizes a quartz-polymer optical fiber of 400 μm core diameter bent into the shape of an ellipse. The bent fiber is connected to a chest belt. The radius of the bend of the sensing part of the fiber in expiratory state is 5 cm. An infrared light emitting diode emitting 845 nm wavelength light is disposed adjacent to the input end of the optical fiber to feed the light through it. The intensity of returned light measured by P-I-N silicon photodiode and translated into chest dimension change. The light level transmitted in the fiber depends on a curvature radius of the fiber. The photodetector's output was amplified, digitized and sampled (250 samples per second) by an analog-to-digital (A/D) converter. Measurements were performed on five healthy untrained subjects aged

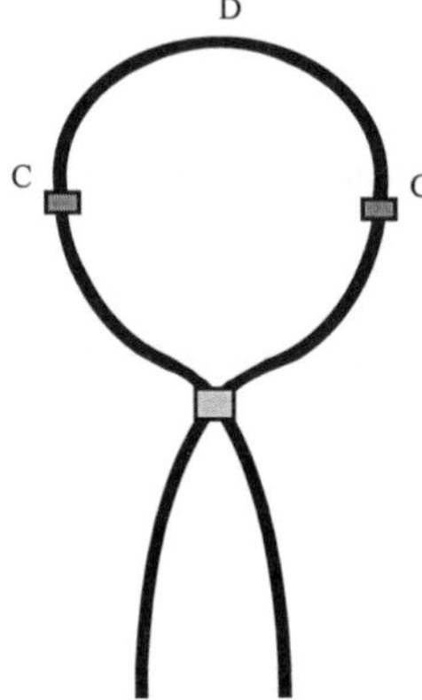

Fig. 1 Bent optical fiber of breathing depth sensor. C indicates the points of connection to the nonelastic part of the belt. D is the sensing region.

21 - 46 years. The first examination was performed in the sitting position, after a 10 min rest. The chest belt was placed on subject. The subjects were stripped to the waist in order to avoid an incidental delay in waveform. Each examination included a 5 min simultaneous measurement of respiratory depth and arterial blood pressure, measured by means of Ohmeda 2300 Finapress. The Finapres monitor measures arterial blood pressure in the finger using a PPG signal kept constant. The cuff was attached to the right-hand index fingertip of a subject, and the hand was held at heart level. The Finapres output was sampled by the same A/D converter used for the breathing depth measurement. After the first examination every subject was asked to run on a running path during 5 min. The second examination was performed immediately after the physical activity. Respiratory depth and arterial blood pressure were measured again.

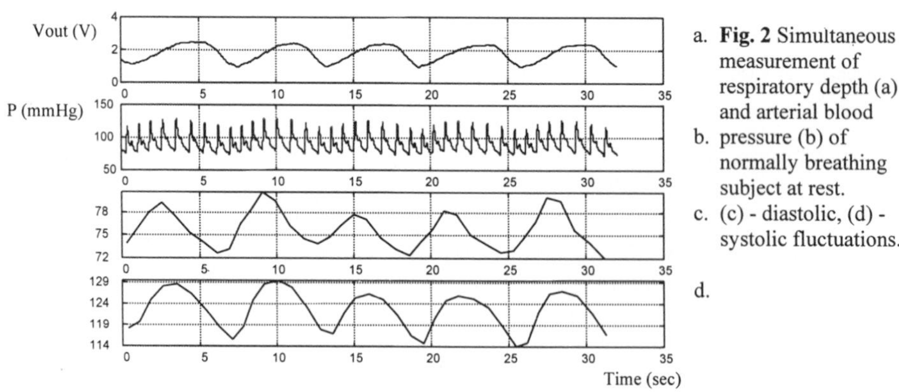

Fig. 2 Simultaneous measurement of respiratory depth (a) and arterial blood pressure (b) of normally breathing subject at rest. (c) - diastolic, (d) - systolic fluctuations.

3. Results and Discussion

Fig. 2 shows the simultaneous record of the arterial blood pressure (b) and breathing curve (a) of a typical subject before a physical activity. The breathing period is 6.5 sec and amplitude is 2.3 volt. The heart beat

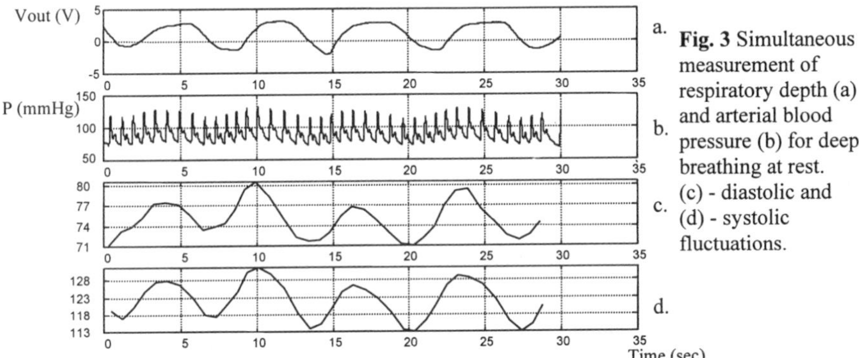

Fig. 3 Simultaneous measurement of respiratory depth (a) and arterial blood pressure (b) for deep breathing at rest. (c) - diastolic and (d) - systolic fluctuations.

period is 0.83 sec. After the examination, the pressure signal was analyzed and the minimum and maximum of each pulse were digitally determined in order to obtain the systolic (d) and diastolic (c) curves. The systolic pressure varies between 110 to 130 mmHg. The diastolic pressure varies between 70 to 80 mmHg. The systolic and diastolic curves are positively correlated with the breathing curve. The correlation coefficient found is about 0.92. Similar results were obtained in most of the other examinations.

Fig. 3 shows the typical results of an examination preformed before physical activity with deep breathing subject. The breathing period is 6.6 sec the amplitude is 5.5 volt. The heart beat period is 0.65 sec. The arterial blood pressure characteristics almost didn't change. The correlation coefficient in this case is 0.85. The correlation coefficient found for the examinations after physical activity were very low - below 0.50.

4. Conclusions

In the current study, a fiber optic device was used for measurement of breathing depth. The device is sensitive, simple, and easy to use and can therefore be used for monitoring respiration rate and depth. A correlation was found between the breathing depth and arterial blood pressure measured by Finapres device. The fluctuations in systolic and diastolic pressure are due to the regulatory mechanism of the autonomic nervous system, and a method for a quantitative evaluation of this effect is important both for physiological cardiovascular studies and for clinical neuropathic examinations.

5. Reference:

[1] A. Babchenko, B. Khanokh, Y. Somer, and M. Nitzan, "Fiber optic sensor for the measurement of respiratory chest circumference changes," J. Biomed. Opt. **4(2)**, 224-229 (1999)

PART 13

OPTICAL AND ELECTRONIC DEVICES AND APPLICATIONS

Chairperson: *S. Sarusi, Israel*

Annals of the Israel Physical Society, v. 14

ELECTRO-OPTICS and MICROELECTRONICS
Eds: Raphael LAVI and Ehud AZULAY

UNCOOLED MICROMACHINED CMOS COMPATIBLE IR SENSING MICROSYSTEMS

Eran Socher, Ofir Bochobza-Degani and Yael Nemirovsky

MEMS group, Kidron Microelectronics Research Center, Department of Electrical Engineering, Technion-Israel Institute of Technology, Haifa 32000, Israel.

Abstract

The paper presents design, fabrication, characterization and analysis of uncooled micromachined CMOS compatible IR sensing microsystems. The microsystems under study are thermoelectric sensors for thermal uncooled IR detection integrated monolithically on a CMOS chip containing their readout circuits. In the design, a novel optimization scheme was employed and the dependence of performance upon pixel area and aspect ratio was modeled and investigated. Two methods of fabrication were developed, backside wet bulk micromachining and front-side dry RIE micromachining.

1. Introduction

Thermal imaging is expected to be employed in many applications, especially in security and surveillance, as its price decreases. Usage of uncooled thermal sensors lowers the cost of imagers while micromachining allows for integration of thermally isolated elements in integrated circuits. Other uncooled sensing methods, such as bolometers [1] and pyroelectric sensors [2] require the use of materials unconventional to CMOS IC processing as well as temperature stabilization and uniformity to achieve constant responsivity. Thermoelectric sensing, on the other hand, can make use of standard CMOS materials [4], such as polysilicon, doped silicon and aluminum. Integration of sensors with CMOS readout lowers the environment interference and the projected cost. Thermoelectric sensors are less affected by ambient temperature as they measure the rise of temperature with respect to the ambient and not absolute temperature. The absence of cooling and bias requirements makes thermoelectric sensors good candidates for low power thermal imagers in autonomous systems, such as satellites.

CMOS compatible thermoelectric sensors were realized by several groups [5-6]. The important parameters in thermal sensing pixels are the response time (τ) and the Noise Equivalent Power (NEP) or the Noise Equivalent Temperature Difference (NETD). The authors have recently published an analysis and optimization method for the internal configuration of cantilever type thermoelectric sensors with fixed pixel dimensions [7]. Using this optimization method, sensors can be designed for best detectivity for a given response time and vice versa, showing a tradeoff between these two figures of merit. In this paper we show the effects of pixel area and aspect ratio upon its performance and the tradeoff between detectivities and response time. Realization approaches are also presented.

2. Design Schemes and Optimization

For a given pixel dimensions, the internal structure of the sensor determines its properties and among them, its Noise Equivalent Power (NEP) and the response time τ. Analytic and numerical models show that it is best to take the metal leads as narrow as possible. The design of the pixel is therefore subject to the number of thermocouples and the width of the polysilicon leads. Calculations show that for every response time the best NEP is achieved when the design uses as much as possible of the pixel width for the polysilicon leads. Such designs are shown in Fig. 1 and illustrate the tradeoff between the sensor detectivity and response time.

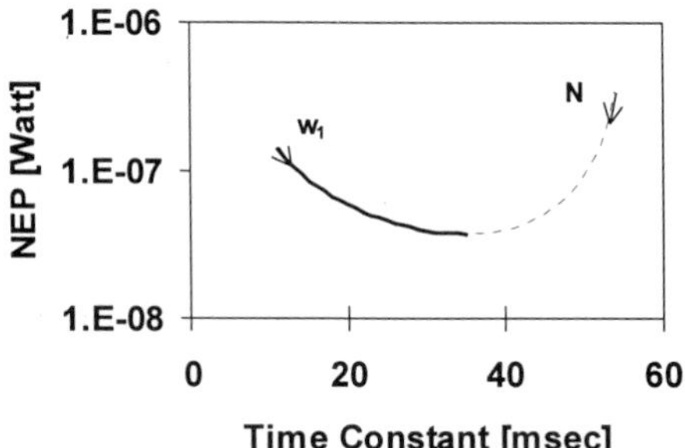

FIG. 1 The optimal design line showing the tradeoff between sensor NEP and response time by controlling the ratio between the number (N) and width (w_1) of the polysilicon leads.

The minimum point of the NEP can be found approximately using:

$$N_{opt} \cong \sqrt[3]{\frac{v_{n,readout-in}^2}{4k_B T_0 B R_{s1}} \frac{K_d W t_d + K_1 W t_1}{K_2 t_2 w_2} \frac{W}{L}}$$

$$R_{opt} \cong \left(\frac{v_{n,readout-in}^2 \sqrt{R_{s1}}}{4k_B T_0 B} \frac{K_d t_d + K_1 t_1}{K_2 t_2 w_2} \right)^{2/3} \sqrt[3]{A}$$

The tradeoff in performance is enhanced when the pixel aspect ratio and area can be controlled. In the case of aspect ratio dependent cantilevers and bridges, the optimal performance with respect to NEP and response time follows approximately an analytical rule:

$$NEP\sqrt{\tau} \cong \frac{4\sqrt{4k_B T_0 B R_{opt} + v^2_{n,readout-in}}\sqrt{K_2 t_2 w_2 t_d \rho_d c_d}}{\pi \varepsilon_{eff}(\alpha_2 - \alpha_1)}\sqrt[4]{\frac{R_{s1} A}{R_{opt}}}$$

This rule does not depend on the aspect ratio itself, only on the area. When keeping a constant aspect ratio (e.g. 2) and changing the area, a slightly different relation is achieved for cantilevers:

$$NEP\tau^{3/8} \cong \frac{(v_{n,readout-in} K_2 w_2 t_2)^{3/4} 2^{9/4} R_{s1}^{1/8} (t_d \rho_d c_d)^{3/8}}{\varepsilon_{eff}(\alpha_{2.} - \alpha_1)\pi^{3/4}}\left(\frac{k_B T_0 B}{K_1 t_1 + K_d t_d}\right)$$

For bridges performance in this case is inferior and the product is bigger with a factor of $2^{1/4}$.

3. Fabrication and Results

Two methods of fabrication were pursued, both being compatible with CMOS technology, so that processed CMOS chips were subsequently micromachined to complete the microsystems. The first approach was to define a masking layer (LTO) on the backside of the chip and then etch the whole thickness from the backside using EDP as an anisotropic bulk silicon etching material. Release of the suspended structure is achieved after reactive ion etching (RIE) of another LTO layer that was deposited previously for protection on the front side. A post-processed CMOS chip can be seen in Fig. 2.

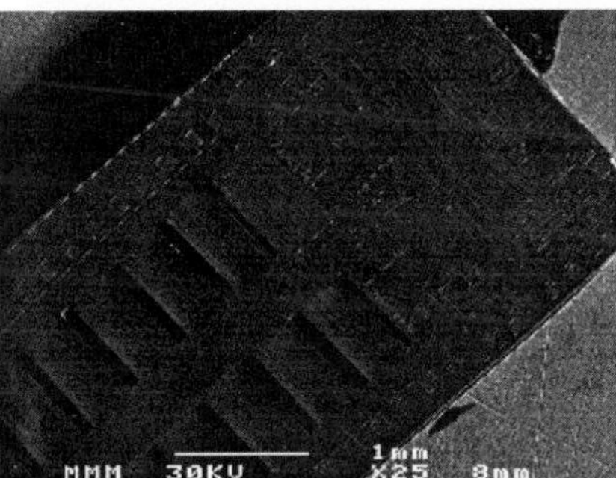

FIG. 2 Post-processed IR imager using backside wet bulk micromachining.

The second approach utilizes only RIE for front side etch and release. The CMOS passivation layer serves as a masking layer for the dry etching of the silicon near and under the suspended structure. Underetching is possible due to the isotropic nature of the etching. This novel approach allows integration of 2D arrays of sensor pixels, and uses a novel spiral design of the pixel structure to allow easy underetching and increase sensitivity. Such post-processed pixel in a CMOS chip can be seen in Fig. 3.

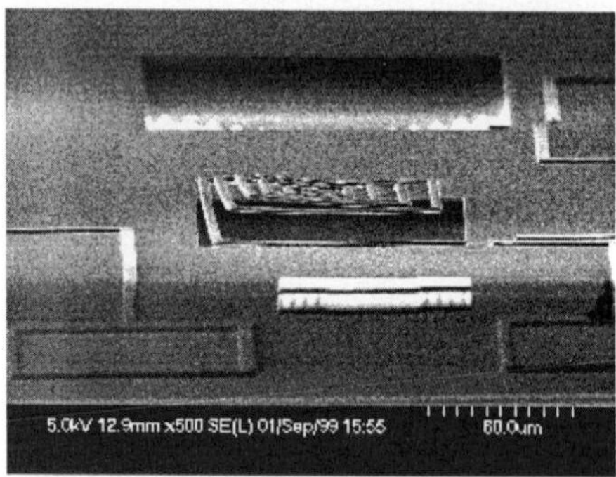

FIG. 3 Post-processed IR sensor pixel using front side dry bulk micromachining.

Measurements done on the fabricated pixels showed performance that agrees with both the modeling and the optimization schemes and relations.

References:

[1] B. E. Cole, R. E. Higashi and R. A. Wood, "Monolithic Two-Dimensional Arrays of Micromachined Microstructures for Infrared Application", Proc. IEEE, August 1998, pp. 1679-1686.

[2] N. Fujitsuka et. al., "Monolithic pyroelectric infrared image sensor using PVDF thin film", Dig. Tech. Papers Transducers '97 Chicago IL June 1997, pp. 1237-1240.

[4] H. Baltes, O. Paul and O. Brand, "Micromachined Thermally Based CMOS Microsensors", Proc. IEEE, August 1998, pp. 1660-1678.

[5] R. Lenggenhager and H. Baltes, "Improved thermoelectric infrared sensor using double poly CMOS technology", in Dig. Tech. Papers Transducers '93, Yokohama, Japan, 1993, pp. 1008-1011.

[6] A. W. van Herwaarden and P. M. Sarro, "Thermal sensors based on Seebeck effect", Sensors and Actuators, 10 (1986) 321.

[7] E. Socher, O. Degani and Y. Nemirovsky, "Optimal Design and Design Considerations of CMOS Compatible Thermoelectric Sensors", Sensors and Actuators A 71 (1998) 107-115.

CHARACTERIZATION OF CMOS PHOTODIODES FOR IMAGE SENSORS

Igor Brouk, Avner Ezion, and Yael Nemirovsky*
Department of Electrical Engineering
Technion – Israel Institute of Technology, Haifa, 32000, Israel
*- *Tel.: ++972-4-8294688 Fax.: ++972-4-8323041 Email: nemirov@ee.technion.ac.il*

Abstract
The results of electro-optical characterization of CMOS photodiodes with different optical window dimensions are presented in this paper.

1. Introduction

The growing interest, which recently has been shown in the photodiodes, is caused by the fact that lately they have been widely adopted as photodetectors in design of CMOS image sensors and microelectromechanical systems (MEMS). The CMOS technology, in particular, is specifically attractive for this application since it allows a wide range of analog and digital signal processing circuits to reside on the same substrate with the photodetector obtaining better performance.

2. Design of CMOS photodiodes

There are three types of front-side illuminated photodiode structures which under study here: implemented by n^+(source implantation) / P-substrate, N-well / P-substrate, and p^+(source implantation) / N-well / P-substrate. All of them were fabricated, tested and compared. In this paper only the characterization results of the first photodiode type are presented.

The structure of this photodiode is shown in Fig.1. It is made by implantation of *n*-type impurity into an initially *p*-type semiconductor so that a thin n^+-type surface layer is formed, as shown in Fig.1, the contact p^+ to the substrate is shown too. The optical window is defined by value L_{opt}. The exact dimensions are scaled according to the applied 2 μm CMOS technology provided by MOSIS [4].

3. The simulated results and comparison with measurements

3.1 Dark Current

Simulation and measurement of the dark current extracted from the current–voltage characteristics at forward bias, i.e. the diffusion component of the dark current, show that the diffusion component of the dark current density is of ~ $3\cdot10^{-11}$-$2\cdot10^{-10}$ A/cm^2, which is determined mainly by the impurity concentration in p-substrate region and is valid for the two other photodiode types as well. In addition, it was found from simulation and measurement results that diffusion component of the dark current density decreases as photodiode area increases. This is caused by lateral current contribution to the total dark current, which is more significant for photodiodes with small dimensions.

3.2 Quantum Efficiency

The measured quantum efficiency is presented in Fig. 2. The simulated quantum efficiency, shown in Fig. 3, was calculated by means of two-dimensional model based on the numerical solution of the two-dimensional continuity equations. On the basis of these results, a number of interesting observations can be made:

The simulation and measurement results show that maximum quantum efficiency reaches approximately 60 %. The quantum efficiency decreases as the photodiode dimensions are reduced. This due to the fact that the diffusion of the charge carriers in the neutral regions takes place in all directions and, therefore, some part of the generated charge carriers, which does not diffuse in the junction direction, will not contribute to the photo-current.

All graphs of the measured quantum efficiency are modulated with certain frequency that is caused by wave interference in the transparent passivation layer, covering the chip from the top side.

3.3 Noise

In accordance to basic noise mechanisms, the only important noise source, which takes place in the semiconductor photodiode, is the shot noise, whose power spectral density in the case of photodiode is given by: $S_{SN} = 2q\,(I_S + I_\lambda)$, where I_λ is the photo-current, I_S is the inverse saturation current, and q is the electron charge. Fig.4 shows the dependence of noise power spectral density taken at frequency of 1kHz on the photodiode current, while the whole chip was illuminated uniformly and the photo-current was generated in each photodiode in accordance with its type and dimensions. The data points are fitted with Eq.(2) (see dashed line of Fig.4), indicating that the shot noise is indeed the dominant noise.

4. Conclusions

It was found that there is strong dependence of the quantum efficiency and the dark current density on the photodiode form and dimensions, which is explained by the lateral diffusion. The shot noise, is the main mechanism, which is responsible for noise in all photodiode types tested here. The simulation results are in the good agreement with the measurements. When comparing the above-mentioned diode structures it was found that n^+ implantation - P-substrate diode structure is the prefered one from point of quantum efficiency and the dark current.

References:

[1] C. C. Wang, I. L.Fujimori, C. G. Sodini, "Characterization of CMOS Photodiodes for Imager Application", 1999 IEEE Workshop on Charge-Coupled Devices and Advanced Image Sensors, pp.76-79 June, 1999.

[2] E.R.Fossum, "CMOS Image Sensor: Electronic Camera-On-A-Chip," IEEE Tran. On Elec. Dev., pp.1689-1698, Vol.44, No.10, Oct 1997.

[3] I. Brouk, M.Sc. Thesis, "CMOS Low Noise Analog Readout for Visible Photon Detection," supervised by Y. Nemirovsky, to be published in February, 2000.

[4] MOSIS (Metal Oxide Semiconductor Implementation Service). A multiproject fabrication service run by ARPA (The Advanced Research Projects Agency).

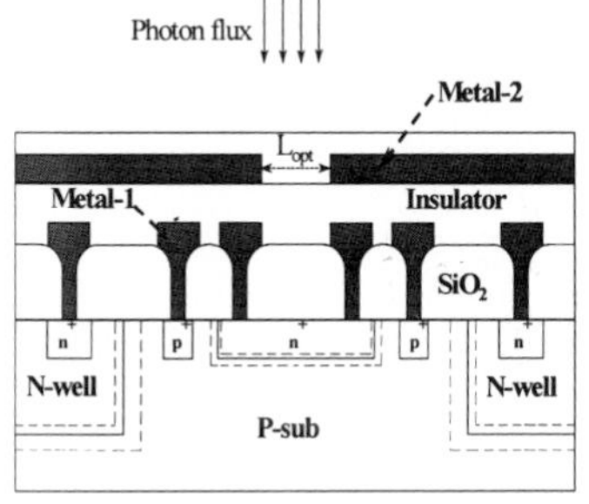

FIG.1. The structure of the vertical front-side illuminated photodiode implemented by n^+ - implantation / P-substrate.

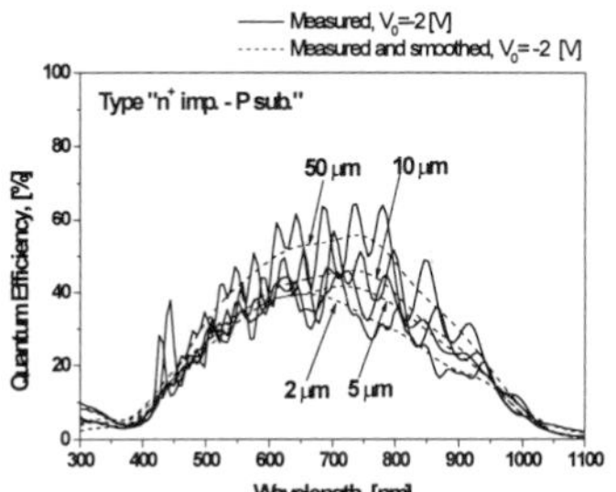

FIG.2 Quantum efficiency measured with uniform illumination for the photodiodes implemented by n^+ implantation / P substrate.

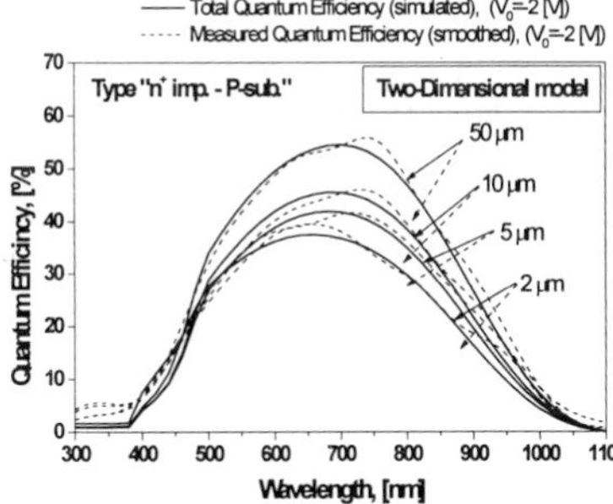

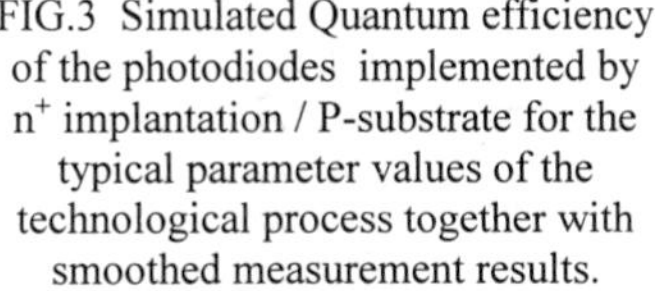

FIG.3 Simulated Quantum efficiency of the photodiodes implemented by n^+ implantation / P-substrate for the typical parameter values of the technological process together with smoothed measurement results.

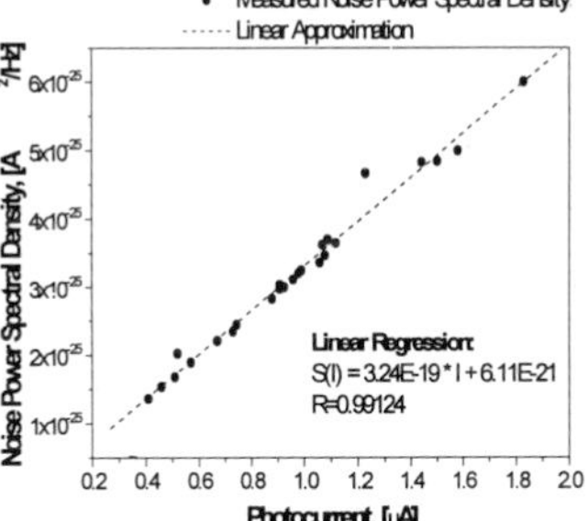

FIG. 4 Dependence of the photodiode noise power spectral density taken at frequency of 1kHz on the photodiode current.

INTEGRATED SILICON OPTOELECTRONIC CIRCUITS (OEICs) IN CMOS AND BICMOS TECHNOLOGY FOR DVD SYSTEMS

A. Ghazi, T. Heide, K. Kieschnick, H. Zimmermann and P. Seegebrecht

University of Kiel, Chair for Semiconductor Electronics,
Kaiserstraße 2, D-24143 Kiel, Germany

Abstract

The innovative integration of photodetectors and amplifiers on the same substrate allows new kinds of large volume consumer products. CMOS and BiCMOS optoelectronic integrated circuits (OEICs) with enhanced data rates for applications in CD-ROM or Digital-Versatile-Disk (DVD) systems are introduced.

1. Introduction

By integrating photodetectors and amplifiers on the same substrate smaller die sizes, better immunity against electromagnetic interference as well as faster systems can be achieved. In addition higher reliability and lower system costs can also be obtained. The integration of photoreceivers in CMOS and BiCMOS technology offers a wide range of large-volume low-cost products, especially for the red spectral range (635-650 nm). At these wavelengths one of the most important commercial applications at the present time are low-offset receivers for optical storage systems like CD-ROM or Digital-Versatile-Disk. This paper gives an overview of optoelectronic integrated circuits for DVD systems in CMOS and BiCMOS technology.

2. PIN CMOS Integration

In contrast to the additional process complexity for the integration of PIN photodiodes in bipolar technologies as reported in [1], the PIN CMOS integration was suggested to result in much less additional process complexity compared to the original process [2]. The integration of the PIN photodiode in the standard twin-well CMOS process requires only two additional steps: (i) one mask in order to block out the originally unmasked threshold implantation in the PIN photodiode area and (ii) a back-contact metallization [3]. The PIN CMOS OEIC is integrated on wafers with an epitaxial layer. The doping concentration in the epitaxial layer must be low (e.g. 5×10^{13} cm^{-3}) in order to achieve fast PIN photodiodes. The MOSFETs are placed in wells, and therefore a reduction of the doping concentration in the epitaxial layer does not deteriorate the electrical performance of the MOSFETs. The p^+-source/drain island can serve as the anode, the epitaxial layer as the intrinsic region and the highly doped n-substrate as the cathode of the fast PIN photodiode. A special antireflection coating layer is considered here in order to optimize the sensitivity of the PIN photodiodes in the red spectral range.

3. CMOS DVD OEIC

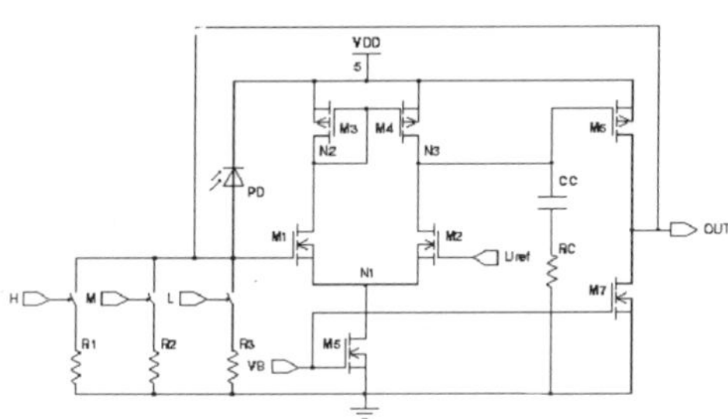

FIG. 1 Voltage follower of the CMOS DVD OEIC

We investigated an 8-channel OEIC for universal focusing and tracking methods of optical storage systems like DVD. This OEIC consists of 8 channels with PIN photodiodes and voltage followers. The gain of the amplifiers is switchable (high, medium, and low), so that three levels of photocurrent can be detected and amplified. We have integrated this OEIC on n-substrate wafers with an epitaxial layer. The used two-stage CMOS opamp is compensated to ground (see Figure 1).

We also have a gain switch circuit and a compensation circuit for the resistors on chip. The compensation circuit consists of a polysilicon reference resistor and an opamp. Process deviations concerning the MOS resistors are largely compensated with this circuit. The supply voltages of the OEIC are 5 V and 2.5 V. The laser light with a wavelength of 638 nm is coupled into the photodiodes of the OEIC on the wafer prober via a single-mode fiber. We have used a network analyzer for the modulation of the laser and for the frequency response measurements of the OEIC. The total power consumption is approximately 70 mW. The –3 dB bandwidth was measured with an active probe head, whereby the amplifiers were loaded with C_L = 11.2 pF and R_L =1 kΩ. The –3 dB bandwidth is 33 MHz at high gain, 50 MHz at medium gain, and 54 MHz at low gain. The offset voltage is smaller than 10 mV and the sensitivity is 3.3 mV/μW without ARC. Test structures showed that the sensitivity can be enhanced to 5.6 mV/μW with a special antireflection coating layer. The noise level at 10 MHz with a resolution bandwidth (RBW) of 30 kHz is –89 dBm.

4. BiCMOS Double Photodiode

For advanced optical storage systems, standard PN photodiodes with bandwidths of 10-20 MHz are not fast enough due to the large penetration depth of the light and the resulting slow diffusion of photogenerated carriers. The approach in [1] uses three additional masks in order to integrate fast PIN photodiodes. In contrast, no process modifications are required to implement a so-called double photodiode (DPD) [4]. The slow carrier diffusion is avoided by the use of two vertically arranged PN junctions. Both anodes are connected to ground. A –3 dB bandwidth of 229 MHz has been achieved for the DPD. By an anti-reflection coating for a wavelength of 638 nm a quantum efficiency of up to 95 % can be realized.

5. BiCMOS DVD OEIC

An 8-channel DVD OEIC consisting of four fast and four sensitive channels for tracking control was fabricated in a self-adjusting twin-well 0.8 μm BiCMOS process.

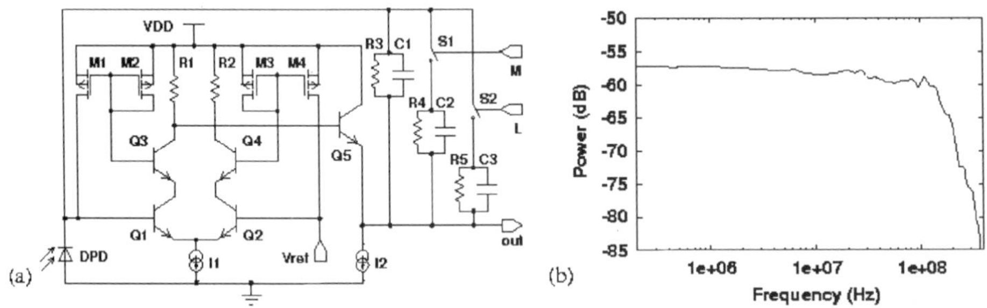

FIG. 2 (a) fast channel (A-D), (b) sensitive channel (E-H) of BiCMOS OEIC

In Figure 2(a) the circuit diagram of a fast channel (A-D) of the OEIC for optical storage systems is shown. An integrated DPD is connected to a transimpedance amplifier in order to obtain a low output offset voltage compared to a reference voltage of 2.5 V as is required for applications in optical storage systems. For read, erase and write access, the gain is switchable by MOS transfer gates between high, medium and low with a ratio of approximately 1/3 each. Poly-poly-capacitors are used for frequency compensation with C1-C2. Only NPN transistors are applied in the signal path of the operational amplifier in order to achieve high –3 dB bandwidths. The NPN transistors Q3 and Q4 are implemented in order to sense the base currents of Q1 and Q2, respectively. These currents are mirrored by the PMOS current mirrors to the bases of the input transistors Q1 and Q2, respectively. This biasing of the input transistors Q1 and Q2 reduces the systematic output offset of approximately 110 mV, which would result from the base current of Q1 across the resistor R3 (20 kΩ), by more than one order of magnitude. For the fast channels the following bandwidths were achieved [4]: High gain: 91.0 MHz, medium gain: 94.7 MHz, low gain: 94.8 MHz. The bandwidth of more than 92 MHz exceeds that of common commercial OEICs by a factor of about two. An even higher operating speed can be achieved by a transimpedance amplifier without gain switching and with active PMOS load elements. The bandwidth of this OEIC channel is 118.6 MHz (Figure 2(b)). At the four sensitive channels E-H with a ten times larger sensitivity for tracking control NMOS source followers are added in front of the bipolar difference amplifier Q1 and Q2. In such a way the input currents and the resulting offset voltages across the feedback resistors of about 200 kΩ are reduced and a high sensitivity of 88 mV/μW in combination with a low offset voltage can be realized. For this amplifier bandwidths of more than 5.3 MHz are measured.

6. Conclusions

The performance of the PIN CMOS OEICs is enhanced when they are integrated on substrates with a low-doped epitaxial layer. This results in a faster frequency response of the photodiode. With the achieved results, double speed DVD systems with PIN CMOS OEICs are possible. A BiCMOS-OEIC with bandwidths in excess of 90 MHz was realized in full custom design. The results demonstrate that it is possible to avoid the slow carrier diffusion problem by implementing a double-photodiode. This can be employed in high-speed CMOS and BiCMOS OEICs.

Acknowledgements: The authors would like to thank R. Popp and R. Buchner from the Fraunhofer IFT in Munich as well as H. Pleß and Thesys Mikroelektronik GmbH in Erfurt, Germany, for chip processing. Parts of this work were carried out for projects, which are funded by the German `Bundesministerium für Bildung, Wissenschaft, Forschung und Technologie' under references 01BS604/5 and 01BS607/8.

References:

[1] M. Yamamoto, M. Kubo, and K. Nakao, "Si-OEIC with a Built-in Pin-Photodiode", Transactions on Electron Devices, vol. 42, no. 1, 1995, pp. 58-63.

[2] H. Zimmermann, "Monolithic Bipolar-, CMOS-, and BiCMOS-Receiver OEICs", CAS'96 Proceedings, Int. Semicond. Conf. 1996, IEEE catalog number: 96TH8170, pp. 31-40

[3] H. Zimmermann, U. Müller, R. Buchner, and P. Seegebrecht, "Optoelectronic Receiver Circuits in CMOS-Technology", Mikroelektronik'97, GMM-Fachbericht 17, München, March 1997, pp. 195-202.

[4] H. Zimmermann, K. Kieschnick, M. Heise, and H. Pless, "High-Bandwidth BiCMOS OEIC for Optical Storage Systems", 1999 IEEE ISSCC, Digest of Technical Papers, pp 384-385.

POROUS SILICON LIGHT EMISSION ENHANCEMENT BY NF3/UV PHOTO-THERMAL SURFACE TREATMENT

S. Stolyarova, A. El-Bahar and Y. Nemirovsky

Electrical Engineering Department, Technion-Israel Institute of Technology, Haifa 32000, Israel

Abstract

A novel NF_3/UV photo-chemical surface treatment of porous silicon is presented. This treatment strongly enhances the photoluminescence of porous silicon. The increase of PL intensity is about 1 – 2 orders of magnitude at treatment temperatures in the range of 300 - 400°C. The overall effect of photoluminescence enhancement is suggested to be due to the NF_3/UV photothermal etching of the as-formed native oxide, as well as, to the cleaning and passivation of the porous silicon surface with fluorine, followed by a rapid growth of a more stoichiometric oxide SiO_x ($x \approx 2$) layer in air.

1. Introduction

Porous silicon attracts continuous interest due to its visible photoluminescence (PL). The intensity of PL from porous silicon (PS) is found to be determined by different factors such as porosity, thickness of the porous layers, surface morphology, oxidation, etc [1]. The issue of PL intensity enhancement is extremely important for possible applications of porous silicon in various optoelectronic devices.

In this letter, we report a strong effect of porous silicon S-type PL enhancement related to NF_3 photo-thermal reactions. The NF_3 is known as an etchant gas capable at certain conditions to remove native oxide from silicon surface and/or to etch the silicon [2]. In the presence of UV photons, provided in this study by an excimer lamp, the activity of the reactions at the surface is significantly enhanced [3].

2. Experimental

The p-type PS samples were prepared by anodisation of p-type silicon wafers in HF/ethanol (1:1) solution, rinsing in ethanol and drying by pentane. The 3 μ thick layer with 60 % porosity was prepared for 2 min with current density of 30 mA/cm^2. The treatments of PS were carried out in an MOCVD system with a horizontal quartz reactor. The samples were placed onto graphite susceptor, heated by infrared lamps. The PS treatment was performed by introducing of a mixture NF_3/N_2 (1:1) at 50 – 400°C, 60 Torr total pressure and 500 sccm total flow rate, with and without UV excitation. An excimer lamp situated above the samples provided the UV irradiation at 222 nm.

Photoluminescence (PL), atomic force microscopy (AFM) and Auger spectrometry measurements were used to characterize the porous silicon. The PL was measured at room temperature by Dylor spectrometer using Ar laser irradiation of 514.5 nm.

3. Results and Discussion

Fig. 1 shows the effect of NF_3/N_2 treatment on the photoluminescence of 3 μ porous silicon performed at 300°C for 10 min, with and without UV excitation. The sample not subjected to NF_3/N_2 treatment served as a reference. The drastic increase of PL maximum intensity from I_{ref} = 150 to I = 3000 (I/I_{ref} = 20) was achieved by the NF_3/N_2 thermal treatment. The incorporation of UV photons in the treatment provided an additional enhancement of PL (I/I_{ref} = 28).

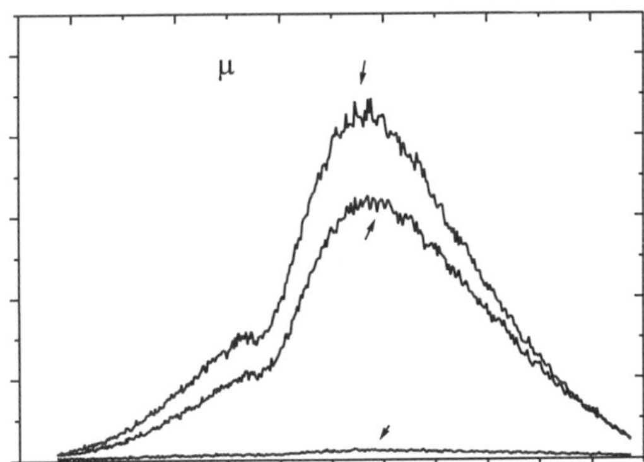

FIG. 1. Photoluminescence spectra of 3 μ porous silicon treated by NF_3/N_2 at 300°C for 10 minutes with and without excimer UV activation. Reference sample corresponds to the as prepared porous silicon without photothermal treatment.

To reveal the role of NF_3 in the PL enhancement effect, the photothermal treatments of porous silicon were carried out separately in N_2, H_2 and in vacuum (residual gases) environments. In the absence of NF_3, the treatments did not lead to an increase of the PL intensity. Thus the observed effect of PL enhancement is proved to be due to reactions induced by NF_3 on the surface of porous silicon.

To study the effect of temperature, the treatments were conducted at 50, 150, 300, 400°C. No effect was observed at 50 and 150°C indicating the thermo-activated nature of the process. The effect was found to be stronger at 300°C than at 400°C. This phenomenon can be related to the etching of PS by NF_3 and the modification of the layer thickness.

The AFM observation of PS surface after the treatments showed the significant change of surface morphology. A layer of pyramidal crystallites, sometimes as high as 1.5 microns, was formed on PS surface. The PL increase correlated with the coverage of PS surface by the crystallites: fully covered surface resulted in much higher PL intensity.

The Auger spectrometry was applied to determine the nature of the chemical components on the PS surface. Table 1 shows the atomic composition of surface layer, calculated from Auger spectrum, as well as, the corresponding PL intensities. For the samples, which did not exhibit PL increase and for which the growth of new phases was not found, the relationship between Si and O signals, together with the shape of the Si peak in the Auger spectra, indicated the presence of a native oxide. The O/Si concentration ratio, calculated from the Auger spectrum, was in the range of 0.5 - 1. In contrast, for the

Table 1. Atomic concentrations (%) of elements at the PS surface calculated from Auger spectra.

SAMPLE	Si	O	C	F	N	O/Si	PL, a. u
Reference, 3	59	28	13	-	-	0.5	150
N2/300°C /3	30	30	40	-	0.1	1	85
H2/300°C/3	45	50	5	-	-	1	80
NF_3/150°C/3	52	30	16	0.1	2	0.6	170
NF_3/300°C/3	37	56	7	0.1	0.1	1.5	4,500
NF_3/400°C/3	36	61	3	0.1	-	1.7	3,000
NF_3/300°C/10	32	64	2	2	-	2	10,000

samples, which showed strong PL increase and exhibited a continuous layers growth, the ratio of O and Si concentrations was in the range of 1.5 - 2. These results indicate that the treatment proposed here, removes the as-formed oxide and enhances the growth of an oxide SiO_x layer with higher oxygen content (up to stoichiometric SiO_2) on the PS surface. In addition, the incorporation of up to 2% of fluorine , as well as, a decrease of carbon content in the treated samples, can be seen from Auger data.

To show, that the observed effect of the PL increase was not caused just by the cleaning of the PS surface, we exposed porous silicon to H_2/UV photo-thermal treatment at 300°C, which was supposed to remove carbon contaminants. Indeed, we observed the decrease of carbon content by the H_2/UV annealing, but without any increase of the PL intensity.

Furthermore, no emission could be detected either in the chamber or in the moment the samples were taken out of the reactor. Only after about 10 minutes in air it could be seen. Thus, it can be deduced that the PL increase is a result of interaction of NF_3 treated PS surface with the air ambient.

Thus, the AFM observation and Auger analysis indicate that the enhancement of PL signal from porous silicon can be correlated with PS surface etching and the growth of oxygen enriched oxide layer with traces of fluorine.

The luminescence of porous silicon and its intensity is known to be sensitive to the interface properties between the porous silicon skeleton and the covering oxide [4]. The removal of the highly defective native oxide, following the electrochemical process, the cleaning of the surface from contaminants and its passivation with fluorine in the NF_3/UV process, stimulate the growth of a more stoichiometric oxide layer. An improved interface is formed, resulting in the observed significant increase of porous silicon photoluminescence.

Acknowledgments: The research was supported by the 0.25 μ Consortium, funded by the Ministry of Industry and Commerce of Israel. We thank Dr. R. Brenner for Auger measurements and useful discussions. The technical assistance of A. Shay and Y. Leibowitz is highly appreciated. The contribution of the Gilady Program of Ministry of Immigrant Absorption of Israel is gratefully acknowledged.

References

[1] Properties of Porous Silicon, L. Canham Ed., EMIS Datareviews Series No.18, 1997, 406 p.

[2] M. Hirose, S. Yokogama and Y. Yamakage, J. Vac. Sci. Technol., B3(5), 1998, p.1445.

[3] I. W. Boyd and J. Y. Zhang, Nuclear Instruments and Methods in Physics Research, B 121, 1997, 349.

[4] J. C. Vial, A. Bsiesy, F. Gaspard, R. Herino, M. Ligeon, F. Muller and R. Romestain, Phys. Rev. B, 45, 1992, p. 14 171.

PART 14

MACHINE VISION AND IMAGE PROCESSING

Chairpersons: *I. Wilf, Israel; N.S. Kopeika, Israel*

Annals of the Israel Physical Society, v. 14
ELECTRO-OPTICS and MICROELECTRONICS
Eds: Raphael LAVI and Ehud AZULAY

APPLICATION OF THE NONLINEAR CELLULAR NEURAL FILTERS FOR NOISE REDUCTION, EXTRACTION OF IMAGE DETAILS, EDGE DETECTION AND SOLUTION OF THE SUPER-RESOLUTION PROBLEM

Igor Aizenberg, Naum Aizenberg, Taras Bregin, Konstantin Butakov, Elya Farberov

Company Neural Networks Technologies Ltd. 3, Hashmonaim str., Bnei-Brak, 51264, Israel
Tel: (972) 3 619-9711, Fax: (972) 3 619-9170
e-mail: {nnt, igora}@netvision.net.il

Abstract

Nonlinear cellular neural filters (NCNF) are based on the non-linearity of the activation functions of universal binary neurons (UBN) and multi-valued neurons (MVN). NCNF, which include the multi-valued non-linear filters (MVF) and cellular Boolean filters, are extremely powerful mean for solution of the various image processing problems. The keyword idea of the NCNF is use of the complex non-linearity. The main applications of NCNF are the following: noise reduction, extraction of image details, precise edge detection, correction of the super-resoluted images, etc.

1. Introduction

NCNF have been introduced for the first time in [1]. Their two components are multi-valued filters (MVF) [2] and cellular neural Boolean filters (CNBF) [3]. These filters are based on the complex non-linearities of activation functions of the multi-valued and universal binary neurons, respectively.

The mentioned non-linearities define a specific nonlinear averaging and therefore they may be applied to design of the new nonlinear filters. The NCNF family of spatial domain filters is described by the following equation:

$$\hat{B}_{ij} = F(w_0 + \sum_{\substack{i-n\le k\le i+n \\ j-m\le l\le j+m}} w_{kl} Y_{kl}). \qquad (1)$$

F in (1) is defined in one of the following similar, but alternative ways:

$$F(z)=j, \ if \ 2\pi(j+1)/k > \arg(z) \ge 2\pi j/k, \qquad (2)$$

$$F(z)=(-1)^j, \ if \ 2\pi(j+1)/m > \arg(z) \ge 2\pi j/m \qquad (3)$$

The equation (2) defines the activation function of multi-valued neuron [2], and the equation (2) defines the activation function of universal binary neuron [4]. Thus (1)-(2) defines multi-valued filter [2], and (1)-(3) defines cellular neural Boolean filter [3] that together have been united to the family of nonlinear cellular neural filters in [1]. To clarify the notions in (1) it is necessary to note that an additional non-linearity is used in NCNF. Let $0 \le B \le 2^s - 1 = k - 1$ is a dynamic range of a 2-D signal. We can put the root of unity

$$e^B = \exp(i2\pi B/k)=Y \tag{4}$$

to the correspondence with the value real-valued value B. When Y is defined by (4) then (1) is becoming the MVF. When Y_{kl} in (1) is a value of the kl-th pixel in t-th binary plane ($t = 0, 1, ..., 2^s - 1$) of the image then (1) is becoming the CNBF. In the last case all the binary planes of the image have to be obtained directly, without thresholding, then they have to be processed separately and reintegrated into the resulting gray-scale image. Generalization of the NCNF to color image processing is natural: each channel of a color image has to be processed separately as a gray-scale image.

2. Applications of the NCNF

Let us consider the most effective applications of the NCNF on several examples.

1) *Noise reduction.* MVF-component of the NCNF is the most effective for this application. It should be successfully applied to noise reduction with a fine preservation of the image boundaries itself, but it is also very interesting to connect the MVF with other nonlinear filter, e.g., rank-order one. Such a connection makes possible to reduce so complex noise as a speckle one with the maximal preservation of the image boundaries (Fig. 1).

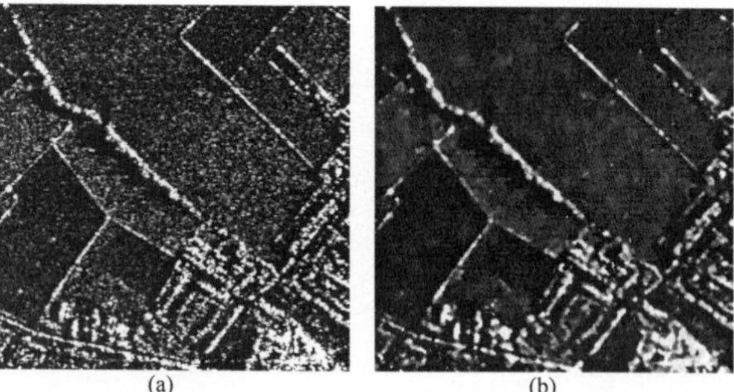

(a) (b)

FIG.1 Speckle noise reduction using rank-order MVF – (a) –noisy image; (b) – the results of the processing

2) *Edge detection.* CNBF-component of the NCNF is using for solution of the precise edge detection problem including the edge detection by narrow direction. The precise edge detection algorithm is much more effective than classical algorithms. It also may be used for edged segmentation of the image. An example of precise edge detection is shown in Fig. 2.

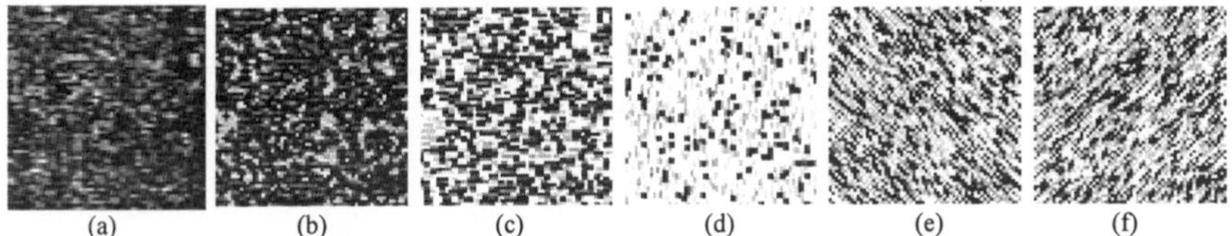

(a) (b) (c) (d) (e) (f)

(a) - original image of the sea surface; (b) - global edges; (c) – edges in the direction West↔East; (d) – edges in the direction North ↔ South; (e) –edges in the direction North-West ↔ South-East; (f) – edges in the direction South-West ↔ North-East

FIG. 2 Edge detection by narrow direction (palette of the edged images is inverted)

3) *Extraction of image details.* To extract the smallest image details or details of the given sizes, both components of the NCNF should be used. The way to obtain the optimal results in this direction has been based in [4]. Here instead of linear filters for frequency correction we are using the multi-valued ones that are much more effective. An example of extraction of the image details is shown in Fig. 3.

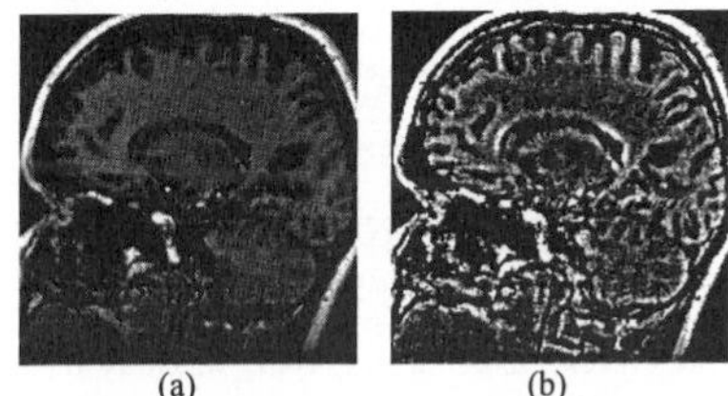

(a) (b)

(a) – original tomogram; (b) – the results of the processing

FIG.3 Extraction of the image details using NCNF-MVF.

4) *Solution of the super-resolution problem.* To increase the image resolution an iterative procedure of the highest frequencies spectral coefficients approximation is used [2]. To obtain the final approximation of the super-resoluted image MVF are very effective. Using MVF it is possible to remove the noisy artifacts from the resulting image and to sharp it. An example is shown in Fig.4.

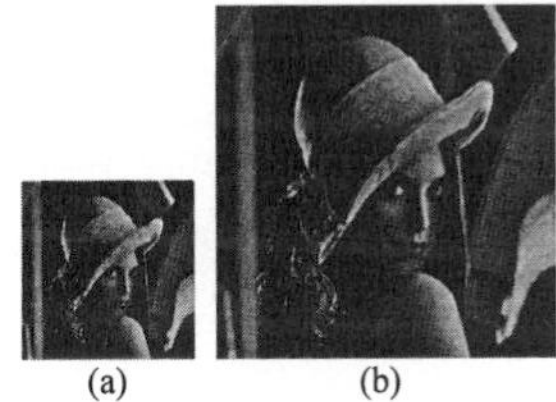

(a) (b)

FIG.4 Super-resolution: (a) - input 256x256 image; (b) – resulting 512x512 image

References

[1] Aizenberg I.N., Aizenberg N.N., Agaian S., Astola J., Egiazarian K. "Nonlinear cellular neural filtering for noise reduction and extraction of image details", *SPIE Proceedings*, Vol. 3646, 1999, pp.100-111.

[2] Aizenberg I.N., Aizenberg N.N. "Application of the neural networks based on multi-valued neurons in image processing and recognition", *SPIE Proceedings*, Vol. 3307, 1998, pp. 88-97.

[3] Aizenberg I.N., Aizenberg N.N., Astola J, Egiazarian K. "Cellular Neural filters based on universal binary neurons", *Proceedings of IEEE Nordic Signal Processing Symposium* (NORSIG-98), Vigso (Denmark), pp. 141-144, June, 1998.

[4] Aizenberg I.N. "Processing of noisy and small-detailed gray-scale images using cellular neural networks", *Journal of Electronic Imaging*, vol. 6, No 3, July, 1997, pp. 272-285.

EVALUATING THE AGREEMENT BETWEEN VARIOUS IMAGE METRICS AND HUMAN DETECTION PERFORMANCE OF TARGETS EMBEDDED IN NATURAL AND ENHANCED INFRARED IMAGES

G. Aviram, S. R. Rotman

Ben-Gurion University of the Negev, Department of Electrical and Computer Engineering.
P.O. Box 653, Beer-Sheva, 84105, Israel.

Abstract

Three image metrics: the DOYLE local contrast metric, the POE global clutter metric and the ICOM global/local texture metric, are described and used in this paper to evaluate human target detection performance and image quality judgments, as obtained from psychophysical experiments. High correlation was found between the DOYLE metric and the data obtained from the qualitative image comparison experiment, and between the POE metric and the data obtained from the quantitative target detection performance experiment. The ICOM texture metric was found in good agreement with the results of both experiments. These results were used to define classification rules and decision thresholds.

1 Introduction

This paper briefly summarizes a major part of a research work [1]-[3], done in order to evaluate the effects of image enhancement on human detection performance of targets embedded in infrared images. The research work uses the results obtained from two (qualitative and quantitative) specially designed psychophysical experiments. The enhancement effects are analyzed in various profiles, and according to the agreement with several image metrics [4]-[9], which try to imitate different perceptual cues such as clutter, contrast and texture. Classification rules are defined on the basis of these metrics and used to predict whether a natural or enhanced infrared image is classified as a "High" or a "Low" performance image, in the context of the ease to detect the embedded target.

2 Image Metrics

2.1 *ICOM Textural metric*

The ICOM textural metric is an improved version of the COM metric described in [4]. The metric is based on the Markov co-occurrence matrix, which contains information about both the intensity values distribution and the possible transitions among neighbor pixels in the examined image area. For the ICOM metric, the co-occurrence matrix is used twice. First, the textures of image areas in the size of the target are examined, looking for those areas that contain a target-like texture (equation 1). Second, the textures of image areas in the size of the target and its local background are examined, looking for those areas that differ from the target size area texture (equation 2). An intelligent combination of the two examinations can specify a target-like area, which is distinct from its local background (equation 3). Such an area is probably either the real target, or an area very similar to it attracting the human eye, and causing the human observer to classify it as a real target.
Figure 1 illustrates the metric principles.

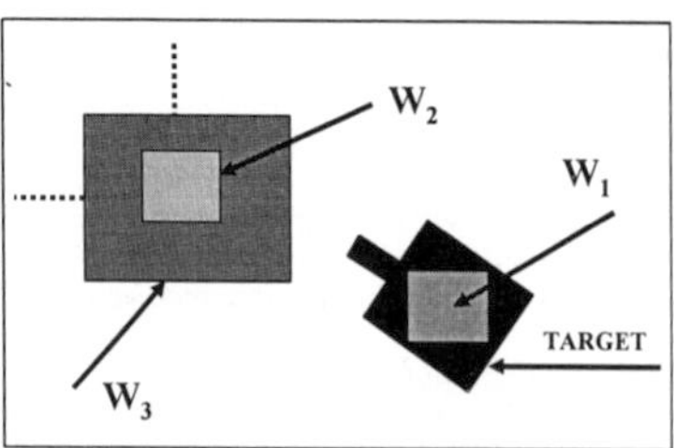

Fig.1 Image Areas for ICOM Illustration

$$x = \sum (CM_{W1} - CM_{W2})^2 \qquad (1)$$

$$y = \sum (CM_{W3} - CM_{W2})^2 \qquad (2)$$

$$z = (1-x)^4 * y^{\frac{1}{4}} \qquad (3)$$

2.2 *POE - Global Edge Based Clutter Metric*

The POE metric is designed to imitate the human visual system, which is assumed to function as a HPF (High Pass Filter), causing the human eye to fixate on image edges. In order to calculate the image POE one should first enhance the image edges, and normalize the resulting image to 0-255 gray scale values, than calculate the number of points that exceed a predefined threshold ***T*** in each edge enhanced image block ($POE_{i,T}$), and calculate the RMS for the whole image according to the following equation

$$POE = \sqrt{\frac{1}{N}\sum_{i=1}^{N} POE_{i,T}^2} \qquad (4)$$

where N is the number of image blocks whose size are about twice the apparent target size.
A more detailed description of the algorithm and its properties including several examples can be found in references [2],[6]-[8].

2.3 *DOYLE - Local Contrast Metric*

The DOYLE local contrast metric measures the target and its local background mean and variance pixel values to assess the target distinctness from the background. The metric is calculated according to the following equation

$$DOYLE = \sqrt{(\mu_t - \mu_b)^2 + k*(\sigma_t - \sigma_b)^2} \qquad (5)$$

where,
μ_t , μ_b - target and local background means.
σ_t , σ_b - target and local background variances.
k - weighting coefficient.

A more detailed description on the metric and its properties followed by several examples can be found in references [2],[9].

3 The Experimental Process

3.1 *Infrared Image Database*

The image database for this work contains 12 natural 3-5 microns infrared images, taken from a staring thermal sensor attached to an air vehicle flying at about 1500 feet above the surface. The images were taken in the spring season in a desert climate area, at four different periods of the day: morning (6-8 am), noon (12-2pm), evening (6-8pm) and night (0-2am) , 3 images per day period.
The digitized image size is 512*512 pixels. The dynamic range of the displayed "black and white" image is 256 gray levels or 8 bits per pixel.
The targets embedded in the images are tanks, Armed Personnel Carriers (APC) or trucks. The thermal condition of the targets is warm (engine working) or cold (engine off). Several images from the database can be found in references [1]-[3].

3.2 *Infrared Image Enhancement*

Infrared image enhancement techniques are designed to deal with the special characteristics of the natural infrared image, which is dominated by the background radiation at the average scene temperature. In general, infrared image enhancement methods can be spatial, point or transform operations, which are applied to the natural image. For this work we chose to implement the well-established point operation algorithms (HE-Histogram Equalization, HP-Histogram Projection and WH-Weighted Hybrid Mapping) and a spatial operation algorithm (MD-median filtering). Detailed description and examples of these techniques can be found in references [1]-[2], while descriptions of several other widespread infrared enhancement methods including the above mentioned techniques can be found in reference [10].

3.2.1 Histogram Equalization - HE

Histogram equalization converts the original histogram to an approximately uniform histogram by moving sparsely occupied adjacent gray levels closer together while reserving greater dynamic range for high pixel count gray levels. Due to the fact that typical infrared images contain relatively high noise level and a small target, applying histogram equalization yields wide dynamic range allocation for the background and noise variations, and narrow dynamic range for the target. After applying histogram equalization, the resulting image loses its natural view and inside details of a small target, but edges and boundaries are enhanced.

3.2.2 Histogram Projection – HP

Histogram projection converts the original histogram in a way whereby the total dynamic display range (8 bits per pixel in our case) is assigned equally to each occupied gray level, regardless of how many pixels occupy this level. Histogram projection is a nonlinear transform, which keeps the natural view of the image, and improves the target details resolution.

3.2.3 Weighted Hybrid Mapping – WH

Weighted hybrid mapping combines the advantages of both, the HE and HP algorithms. More specifically, each original pixel is set to the weighted sum of the HE and HP transformations according to equation (6).

$$WH = \omega \cdot HP + (1 - \omega) \cdot HE \qquad (6)$$

where,
WH, HP, HE – pixel gray level after applying weighted hybrid mapping, histogram projection and histogram equalization respectively.
ω – weight coefficient.
The natural view and the enhancement level of the edges is determined solely by the weighting coefficient, which was empirically set in our work to 0.75.

3.2.4 Median Filtering – MD

Median filtering is a nonlinear spatial algorithm that replaces each pixel with the median of its neighbors in a predefined window. The algorithm is used to smooth sharp edges and reduce image noise, hopefully without affecting the image spatial resolution. The level of filtering is determined by the window size, which usually varies from 3x3 to 7x7 pixels.

3.3 Experimental Methodology.

The psychophysical experiment was assembled from two independent experiments of different type:

3.3.1 Image Quality Comparison Experiment

This experiment is a qualitative one, whereby each of 20 participating observers is exposed to 120 different possible image couples (each one of the 12 natural scenes is enhanced by 4 different enhancement methods to produce 10 possible different image couples for each scene, and a total of 120 different image couples for the whole database), and is asked to decide which of the two images shown simultaneously is preferable, from the point of view of ease to detect the embedded target. This kind of experiment is defined as a comparative judgment experiment and is interpreted according to the LCJ - Law of Comparative Judgment [11]. A similar experimental procedure was used in references [5] and [9].

3.3.2 Target Detection Rate and Probability Performance Experiment

This experiment is a quantitative one, whereby each one of the participating observers is asked to view an image, to decide on the presence of one or more targets, and to mark their positions as quickly as possible. Each observer was exposed one half of the image database that was built of 60 images containing targets (12 different scenes times 4 enhancement methods plus 12 natural scenes) and 10 clutter images. The detection rates and probabilities are extracted from this experiment. This experiment was interpreted using the ANOVA statistical tool [12].

4 Correlation Analysis

In order to compare the levels of agreement between the enhancement efficiency and different image metrics, we calculated the correlation between target detection experimental data and different image metrics outputs.
The relevant experimental data contain, for each tested scene and for each enhancement method, the following:

1. The image quality value on the psychological continuum (LCJ).
2. The average detection rate calculated for each image from the "time to detect" measures.
3. The probability of detection (Pd) ratio calculated for every image as the ratio between the number of true targets detected by ***n*** observers, and the maximum possible number of true targets that can be detected by ***n*** observers.

The image metrics we compare in this analysis are the ICOM textural metric, the POE clutter metric and the *DOYLE* local contrast metric. These metrics are calculated for each tested scene and for each enhancement method.
Averaging the resulting correlation values over the tested scenes yielded the results shown in figures 2.a, 2.b and 2.c for LCJ, Detection Rate and Pd respectively.

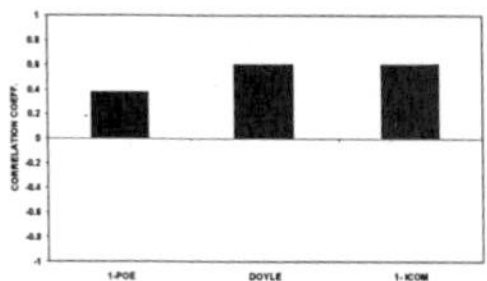

Fig 2.a Average Correlation with LCJ

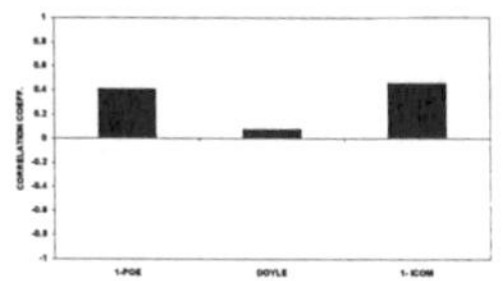

Fig. 2.b Average Correlation with Detection Rate

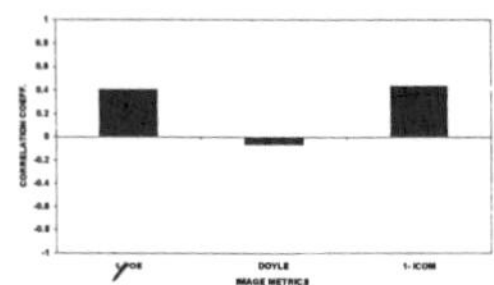

Fig 2.c Average Correlation with Pd

Thsee figures show that relatively high correlation values were obtained between the ICOM textural metric and both the LCJ scale values and the detection performance (Pd and Detection Rate) values. on the other hand relatively high correlation values were obtained between the DOYLE local contrast metric and the LCJ scale values only, and between the POE global clutter metric and the detection performance (Pd and Detection Rate) values only.

These results are attributed to the following:

The observers participating in the image comparison experiment were cued only to the local area of the target, therefore the DOYLE local contrast metric was found to be the most adequate to describe the LCJ scale values dependency on the enhancement methods.

The observers participating in the time detection experiment had no preliminary knowledge of the target position, therefore the POE global edge based metric was found most appropriate for describing the detection rate and the Pd values dependency on the enhancement methods.

As for the new ICOM metric, the good agreement found with the results of both experiments is related to its unique structure that incorporates both local and global perception cues. With regard to the target-to-background distinctness perception cue, the part of the metric measuring the differences between the textures of a suspected target area and its local background, basically does an equivalent job to the DOYLE metric. Apparently, the human observer, even if he knows the location of the target, is bothered by other objects resembling the target, therefore producing high correlation values with the LCJ scale values. On the other hand, mere edges do not bother him. As to the HPF model of the human visual system, the part of the metric giving rise to the textural matches between the real target and other image areas, locates the areas that attract the human eye. This function is similar to the function of the edge based POE metric, and therefore produces high correlation values with the detection performance (Pd and Detection Rate) values.

5 Classification Analysis

Image classifiers, nearly independent of the enhancement methods, and based on the image metrics, were used to distinguish between "High" and "Low" performance images. The classifiers efficiency (Th_{eff}) is calculated, for each enhancement method, using the "High" and "Low" performance definitions (table 1), and equations (5) and (6).

Table 1. "High" and "Low" performance values.

	High Performance	Low Performance
Detection Time (Seconds)	< 8	> 8
Detection Rate (Seconds^{-1})	> 0.125	< 0.125
Pd	> 0.6	< 0.6
LCJ (Normalized)	> 0.6	< 0.6

$$TH = \frac{\frac{1}{K_H}\sum_{i=1}^{K_H} M_H(i) + \frac{1}{K_L}\sum_{i=1}^{K_L} M_L(i)}{2} \qquad (5)$$

$$TH_{eff} = \frac{\sum(I_H \rangle TH) + \sum(I_L \langle TH)}{I_H + I_L} \qquad (6)$$

where

K_H, K_L - number of "high" and "low" performance scenes.

M_H, M_L - values of image metrics of "high" and "low" performance scenes.

I_H, I_L – total number of "high" and "low" performance scenes.

$\sum(I_H \rangle TH)$ - total number of correctly classified "high" performance scenes.

$\sum(I_L \langle TH)$ - total number of correctly classified "low" performance scenes.

For example, figure 3 shows the efficiency of the ICOM classifier, when applied to classify the natural scenes to "High" and "Low" performance regions, according to the associated experimental Pd measures. In this case 90% classification efficiency was achieved. Averaging the efficiency measures obtained for each classifier (ICOM, POE_{HE} and $DOYLE_{MD}$) over the enhancement methods yields the results illustrated in figures 4.a, 4.b and 4.c for LCJ, Detection Rate and Pd respectively.

These figures strongly support the results obtained in the previous section. The appropriateness of the ICOM based classifier to predict target detection rates, as well as probabilities of detection and LCJ values, of both natural and enhanced infrared images was confirmed. The POE based classifier proved itself adequate mostly for target detection rates and probability of detection, while the DOYLE based classifier had the advantage in predicting the LCJ values.

In order to specify the dependence on the enhancement methods, we plot for each enhancement method the classification efficiency results obtained from each classifier. The results are compared with the target detection experimental data, and shown in figures 5.a and 5.b for (Pd vs. ICOM) and (LCJ vs. ICOM) as examples.

Note: ICOM and LCJ values are normalized to the 0-1 range.

The excellent match obtained in both examples confirms the excellent agreement between the ICOM metric and the experimental data.

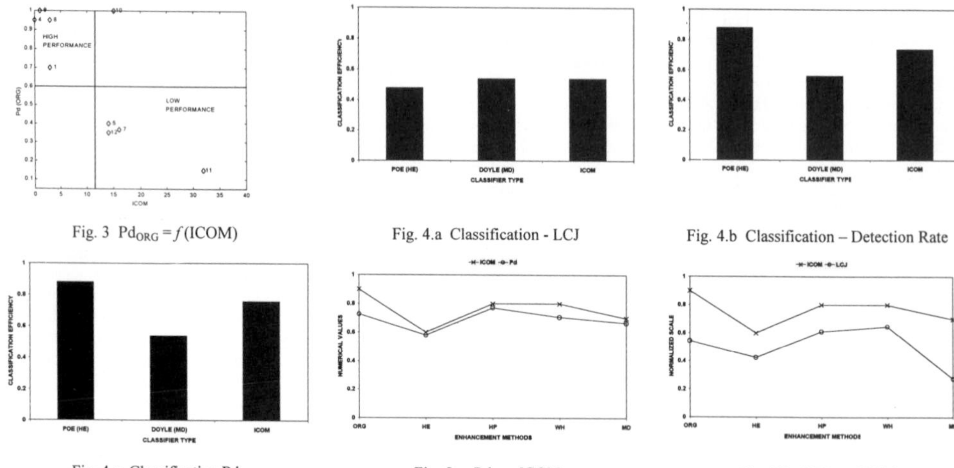

Fig. 3 $Pd_{ORG} = f(\text{ICOM})$

Fig. 4.a Classification - LCJ

Fig. 4.b Classification – Detection Rate

Fig. 4.c Classification Pd

Fig. 5.a *Pd* vs. *ICOM*

Fig. 5.b *LCJ* vs. *ICOM*

6 Summary and Conclusions

The DOYLE local contrast metric, the POE global clutter metric and the ICOM global/local texture metric were introduced and used in this paper to evaluate the agreement between human target detection performance as obtained from psychophysical experiments and the metric products. High correlation was found between the DOYLE metric and the data obtained from the qualitative image comparison experiment, and between the POE metric and the data obtained from the quantitative target detection performance experiment. The ICOM texture metric was found in good agreement with the results of both experiments. According to these results robust classifiers were defined and used to classify natural and enhanced images to "High" and "Low" performance regions. Again best classification efficiencies regarding the ability to predict target detection rates, as well as probabilities of detection and LCJ values, of both natural and enhanced infrared images was achieved using the ICOM textural metric. It is believed that the unique combination of global and local features of the ICOM metric best imitates the human target perception process, and therefore produces the best agreement with the experimental data.

7 References

[1] G.Aviram, S.R.Rotman, "Human Detection Performance of Targets Embedded in Infrared Images – The Effect of Image Enhancement", Optical Engineering, Vol. 38, No. 8, pp. 1433-1440, (Aug. 1999).

[2] G.Aviram, S.R.Rotman, "Evaluation of Human Detection Performance of Targets Embedded in Infrared Images Using Image Metrics", Accepted for publication in Optical Engineering.

[3] G.Aviram, S.R.Rotman, "Evaluating Human Detection Performance Of Targets And False Alarms, Using An Improved Textural Image Metric, Submitted for publication in Optical Engineering.

[4] S.R.Rotman, D.Hsu, A.Cohen, D.Shamay, M.L.Kowalczyk, "Textural Metrics for Clutter Affecting Human Target Acquisition", Infrared Physics & Technology, Vol. 37, pp. 667-674, (1996).

[5] A.C.Copeland, M.M.Trivedi, "Signature Strength Metric for Camouflaged Targets Corresponding to Human Perceptual Cues", Optical Engineering, Vol. 37, No. 2, pp. 582-591, (Feb. 1998).

[6] S.R.Rotman, G.Tidhar, M.L.Kowalczyk, "Clutter Metrics for Target Detection Systems", IEEE Trans. Aerospace Electron. Syst., Vol. 30, No. 1, pp. 81-91, (Jan. 1994).

[7] S.R.Rotman, M.L.Kowalczyk, V.George, "Modeling Human Search and Target Acquisition Performance: fixation-point analysis", Optical Engineering, Vol. 33, No. 11, pp. 3803 - 3809, (Nov. 1994).

[8] G.Tidhar, G.Reiter, Z.Avital, Y.Hadar, S.R.Rotman, V.George, M.L.Kowalczyk, "Modeling Human Search and Target Acq. Performance: IV. Detection Probability in The Cluttered Environment", Opt. Eng., Vol. 33, No. 3, pp. 801 - 808, (Mar. 1994).

[9] A.C.Copeland, M.M.Trivedi, J.R.McManamey, "Evaluation of Image Metrics for Target Discrimination Using Psychophysical Experiments", Optical Engineering, Vol. 35, No. 6, pp. 1714 - 1722, (Jun. 1996)..

[10] J.Silverman, V.E.Vickers, "Display and Enhancement of Infrared Images", Chapter 15 in Electro-Optical Displays, M.A.Karim, Marcel-Dekker Inc. (1992).

[11] W.S.Torgerson, "Theory and Methods of Scaling", Chapter 9, John Wiley & Sons Inc. (1967).

[12] G.Keppel, "Design and Analysis", Chapters 9-12, Prentice-Hall, (1973).

EVALUATING THE EFFECT OF ATMOSPHERIC BLUR, NOISE DISTORTION AND WIENER FILTER RESTORATION ON HUMAN TARGET DETECTION PERFORMANCE IN NFRARED IMAGES

G.Aviram, S.R.Rotman

Ben-Gurion University of the Negev, Department of Electrical and Computer Engineering. P.O. Box 653, Beer-Sheva, 84105, Israel.

Abstract

The effects of atmospheric blur, gaussian noise distortion and Wiener filter restoration on human target detection performance is addressed in this paper. A specially designed psychophysical experiment shows significant degradation in detection performance caused by the contrast and noise limiting effects of the atmospheric blur and noise distortion. The experiment also shows that although Wiener filter restoration improves the distorted image sharpness and contrast, it amplifies the noise and does not compensate for the performance degradation caused by distortion. The paper presents a spectral analysis approach, which uses the system's MTF and the spectral characteristics of the original, distorted and restored images to measure quantitatively the efficiency of the restoration and to define the nature (i.e contrast or noise limited) of images. The analysis results together with the Johnson chart are then used to predict human detection performance, and to evaluate the level of agreement between the predicted and experimental detection probabilities. The method was applied to the experimental infrared image database and showed good agreement with the related experimental probabilities of detection.

1 Introduction

This paper briefly summarizes a research work [1]-[5] done in order to evaluate the effects of image enhancement, distortion and restoration on human detection performance of targets embedded in infrared images. The part of the research described here refers to [5] and uses the results obtained from a specially designed psychophysical experiment and a spectral analysis approach to face and answer the following questions:

1. How can one measure quantitatively the efficiency of restoration and how is this measure related to the spatial spectrums of contrast and noise limited images ?
2. How can the degradation in the observers' probability of detection performance regarding distorted and restored natural infrared imagery, be evaluated and predicted ?

2 Experimental Database, Procedure and Results

2.1 Experimental Database

The image database for this experiment contained total number of 36 images, including 12 natural images, 12 distorted images and 12 restored images. The natural images are 3-5 microns infrared images containing military targets (tanks, Armed Personnel Carriers (APC) or trucks). The digitized image size is 256*256 pixels. The dynamic range of the displayed "black and white" image is 256 gray levels or 8 bits per pixel. The original images intensity distribution functions (image histograms) were linearly mapped to exploit the full available image dynamic range. Figure 1 presents one of the original linearly mapped images of the database, taken in the morning day period. The embedded target (marked by a white rectangle) is a cold tank. Each of the 12 natural images was distorted according to the same distortion procedure to produce 12 distorted images.
The distortion procedure contained the effect of relatively severe atmospheric conditions [6] and relatively low noise level. The atmospheric conditions were modeled by the atmospheric MTF according to equations (1)-(3). The resulting MTF is shown in figure 2. The values of the MTF parameters and noise level are presented in table 1. According to these conditions, the spectrum of each original, linearly mapped image was multiplied by the atmospheric MTF to produce blurred images. Noise was then added according to the desired SNR. Applying this procedure to the image presented in figure 1 yields a distorted image as shown in figure 3. One can easily notice the resolution degradation caused by distortion.

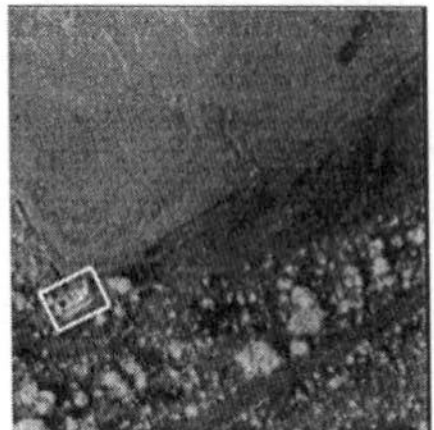

Fig. 1 Original (Linear Mapped) Image

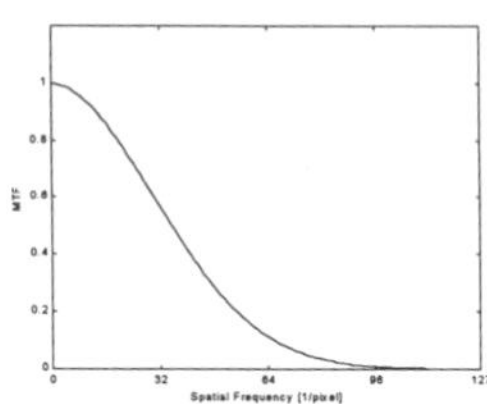

Fig. 2 Atmospheric MTF

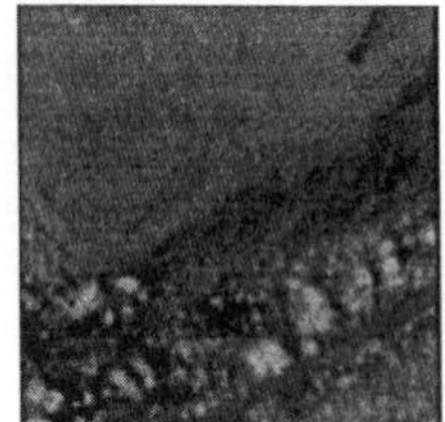

Fig. 3 Distorted Image

$$MTF_{ATMOSPHERE} = MTF_{TURBULENCE} \cdot MTF_{AEROSOL} \quad (1)$$

$$MTF(\nu)_{TURBULENCE} = \exp(-21.57\nu^{\frac{5}{3}}C_n^2\lambda^{-\frac{1}{3}}R) \quad (2)$$

$$MTF(\nu)_{AEROSOL} = \begin{cases} \exp\left[-A_a R - S_a R(\nu/\nu_c)^2\right] & \nu \le \nu_c \\ \exp\left[-A_a R - S_a R\right] & \nu > \nu_c \end{cases} \quad (3)$$

Parameter	Value
λ - wavelength	0.5 x 10^{-6} [m]
C_n^2 - turbulence strength	1 x 10^{-13} [$m^{-2/3}$]
R - range	10000* [m]
A_a - absorption coefficient	0.01 [1/Km]
S_a - scattering coefficient	5 [1/Km]
ν_c - aerosol cutoff frequency	10000 [cyc/rad]
SNR - signal to noise ratio	25 [dB.]

Table 1. Atmospheric and Noise Conditions

Restoration was implemented using the well-established Wiener filter. The filter's transfer function is given in equation (4). The constant variable Γ, is proportional to 1/SNR and describes the ratio between the power spectrums of the noise and the blurred images. The filter transfer function is shown in figure 4 for severe atmospheric conditions and $\Gamma = 1/[25dB.]$.

$$MTF_{WIENER} = \frac{MTF_{ATMOSPHERE}}{MTF^2_{ATMOSPHERE} + \Gamma} \quad (4)$$

Restoration is achieved when the distorted image spectrum is multiplied by the Wiener filter MTF. The image presented in figure 5 is obtained after applying the restoration procedure.

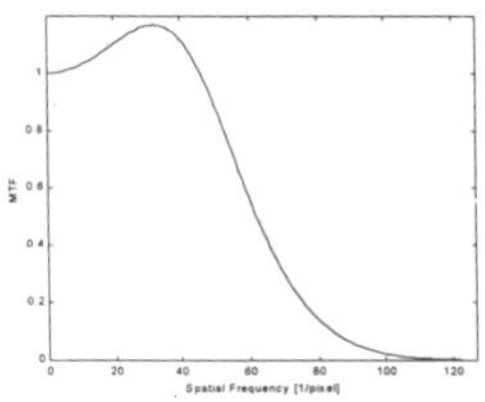

Fig. 4 Wiener Filter MTF

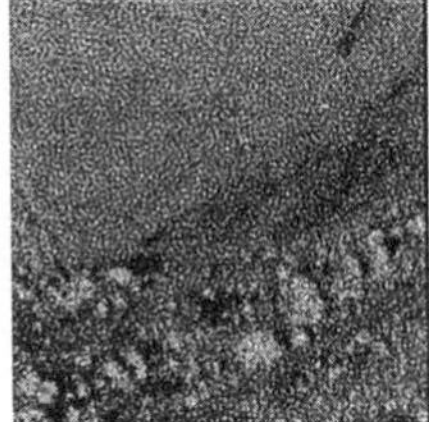

Fig. 5 Restored Image

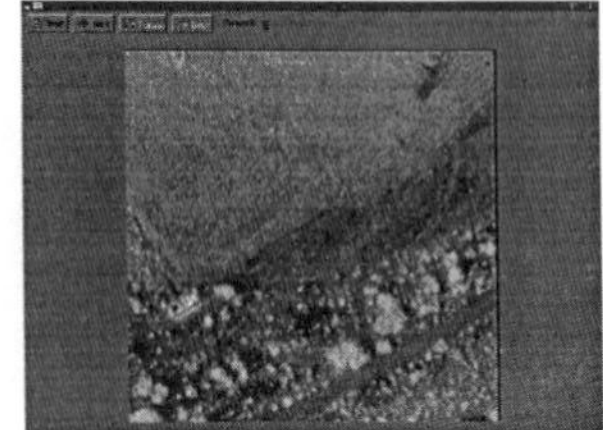

Fig. 6 Experimental Operational Interface

2.2 *Experimental Procedure*

20 observers viewed each of the 36 images. Each observer was exposed no more than twice to the same scene. If that situation occurred, the order of the images was such that the poorer image was viewed prior to the better one. The observers were asked to view each image, to decide on the presence of the targets, and to mark their positions as quickly as possible. The observers were not told whether the image contains one, several or no targets at all, and they were limited to a period of 60 seconds for each image. Image coordinates marked by the observers and the time from image appearance to detection were recorded. The 36 images were shown to each observer in the middle of a 15", 640*480 pixels computer screen, using a simple, special

designed, operating interface which includes a "start" button, which when pressed, brings the first image on screen, and a "next" button for showing the next image. The operating interface is shown in figure 6.

Preliminary explanations as well as few examples were introduced to each observer before starting the experiment. All the participating observers were adult, inexperienced with this kind of experiment and imagery, educated and healthy. The experiment was conducted in a dim lighted room with an office-like background noise (air conditioning, casual chat, and so forth). Observers sat in a comfortable office chair, viewing the screen from a distance of about 40 cm.

2.3 Experimental Results

The experimental results contain two major measures:

1. Pd (probability of detection), defined as the ratio between the number of true targets detected by the 20 observers, to the maximum possible number of true targets that can be detected by the 20 observers.
2. Drate (detection rate), defined as 1/(detection time) which is the time passed from the image appearance to correct detection.

For each image these two measures were calculated and averaged (in the case of Drate) over the 20 observers. Figures 7 and 8 show Pd and Drate results (averaged over the 12 scenes) as obtained for the original linear mapped, the distorted and the restored images.

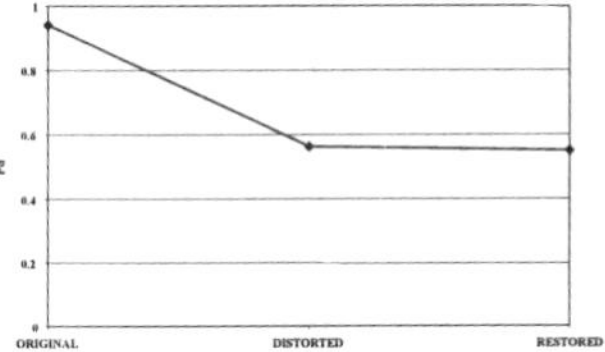

Fig. 7 Pd – Probability of Detection

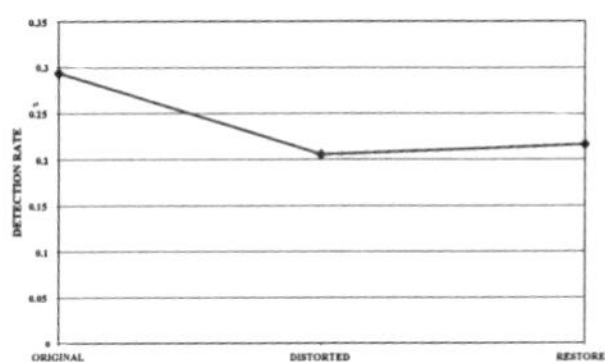

Fig. 8 DRate – Detection Rate

These figures illustrate, as expected, that distortion degrades both the detection probability and rate. The restoration, although expected to improve the degraded performance, fails to do so.

3 Restoration Efficiency

In order to understand how restoration efficiency is determined we first must briefly address two main factors of the imaging system that influence and limit the observer ability to resolve an embedded target. These factors are resolution and noise.

3.1 Resolution (Contrast) Limited Imagery

The imaging system MTF determines which spatial frequencies will reach the human visual system and at what level. Nevertheless, because our visual system has a limited ability to resolve high spatial frequencies of an image spectrum, not all the available image spectrum will reach the visual system. Although, in practice, the limit varies from person to person, there is a representative model [7] that describes the typical eye spatial frequency transfer function, which determines the minimal contrast needed at each spatial frequency for target perception. This model is given in equation (5) as follows

$$MTF_{EYE}(\nu) = \frac{\nu_e}{\exp(-c_1\nu) - \exp(-c_2\nu)} \qquad (5)$$

ν - angular spatial frequency (cycles/degree).
ν_e=0.001033, c_1=0.1138°, c_2=0.3250°

The intersection between the eye threshold contrast MTF curve and the system MTF curve, determines the maximum resolved spatial frequency (fmax) for imagery produced from that system, or in other words, determines the system effective bandwidth. Figure 9 shows a typical system MTF and an eye threshold contrast MTF curves and the intersection between them.

3.2 Noise Limited Imagery

In noise limited imagery the system noise level limits the observer's maximum resolvable spatial frequency and as a consequence his ability to resolve a target. As the Signal to Noise Ratio (SNR) threshold increases, the lower is the maximum resolvable spatial frequency. Moreover, if this frequency is smaller than the frequency determined by the resolution analysis, the image is classified as a noise limited image. Figure 10 shows the case of noise limited imagery.

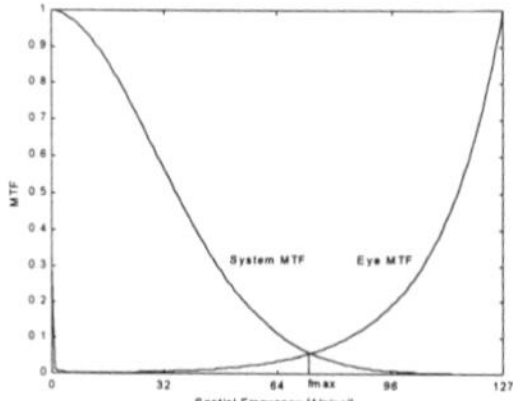

Fig. 9 Bandwidth of Contrast limited Image

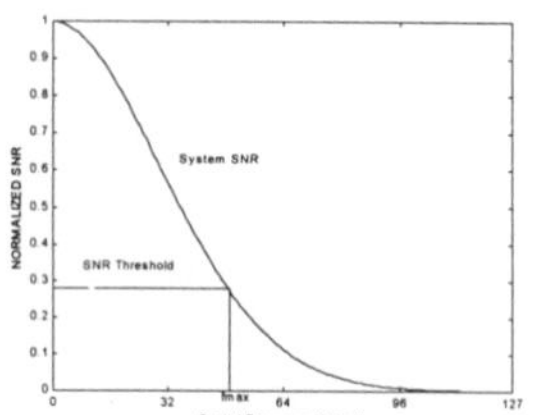

Fig. 10. Bandwidth of Noise limited Imagery

3.3 *Spectral Analysis*

By understanding the system constraints that limit human perception and the relations between them, we can now address the problem of restoration efficiency as mentioned in the paper introduction. Working in the spatial frequency domain, we analyzed the original, distorted and restored images' spatial frequency functions. The analysis method was to calculate a circular integral for the two-dimensional spatial frequency function characterizing each image. This integral sums up and normalizes the energy in each angular spatial frequency. We called this integral the PSD – the Power Spectral Density function. The integral is calculated according to equation (6). I (r,θ) is the image spectrum in polar coordinates.

$$PSD(r) = \oint I(r,\theta)d(\theta) \qquad (6)$$

Figure 11 shows the PSD's of the original linearly mapped image spectrum, the atmospheric blurred image spectrum (distorted), the restored image spectrum (without noise) and the restored image spectrum (with noise).

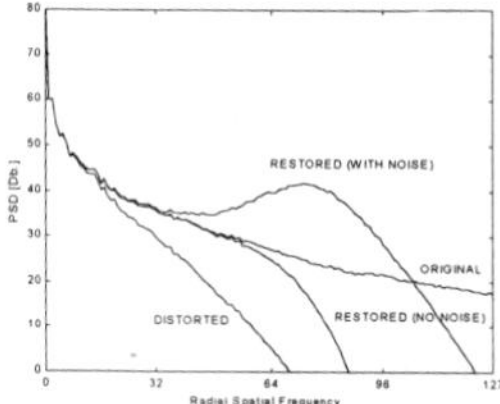

Fig. 11. Original, Distorted and Restored (constant Γ) Image PSD's

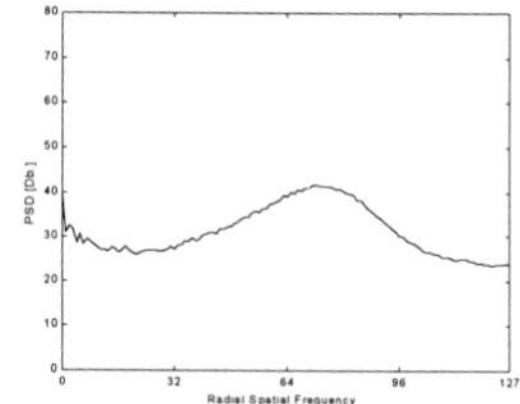

Fig. 12. PSD of the Original and Restored Difference Image Spectrum (constant Γ)

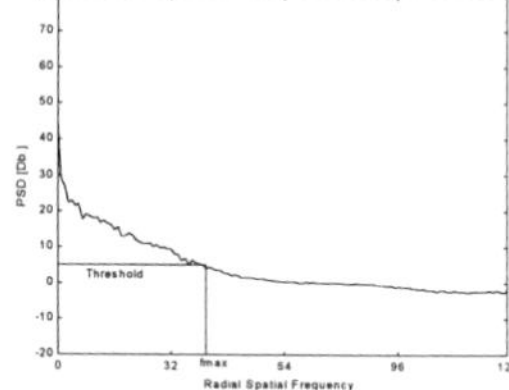

Fig. 13. Cutoff Frequency Threshold (constant Γ)

From figure 11 one can learn the following:

1. The distorted image bandwidth is, as expected, much narrower than the original image bandwidth.
2. The restored image bandwidth (without noise) is, as expected, wider than the distorted image bandwidth, but still the high frequencies are attenuated as a result of the Γ variable in the Wiener filter MTF.
3. The restored image bandwidth (with noise) is much wider than the distorted image bandwidth, but it is followed by a significant energy increment in the medium range of the spatial frequencies, which modifies the image spectrum.

The last statement explains why restoration, although characterized by a relatively wide bandwidth, when applied to the noise added distorted images, was not able to raise the detection performance back to the level obtained for the original images. Furthermore, we suspect that the spatial frequency, from which the original image PSD and the restored image PSD (with noise) begin to separate, is a cutoff frequency that limits target perception.

To check these arguments we analyzed the spectrum of the difference between the original (O) and the restored (R) images, defined in equation (7) as follows

$$DIFF(u,v) = FFT(O(x,y) - R(x,y)) \qquad (7)$$

The PSD of this spectrum is shown in figure 12, which clearly shows the added spectral components in the mid-range of spatial frequencies.

From this result, we calculated the ratio between the original image PSD and the subtracted images PSD, and set an empirical threshold, which determines the spatial frequency from which this ratio goes below a certain level. Using the evaluated spatial cutoff frequency as an input to the target acquisition model for the evaluation of the probability of detection (see section 4), forces the use of an adaptive threshold. The reason for that is that the target acquisition model [7] and the Johnson criterion [8] are not optimized for images with clutter. The use of a clutter dependent adaptive threshold enables to obtain higher cutoff frequencies for low clutter images and lower cutoff frequencies in the case of high clutter images, and therefore to weigh the clutter effect onto the maximal resolvable spatial frequency. The adaptive threshold was set according to the value of the global probability of edge (POE) clutter metric [2]. Three levels of clutter were defined – high, moderate and low. For high clutter images the threshold level was the highest (typical threshold level of 10 [dB.]). For low clutter images, the threshold level was the lowest (typical threshold level of 4 [dB.]). The threshold for moderate clutter level was 7 [dB.]. Figure 13 shows the PSD of the ratio between the original image PSD and the subtracted images PSD, and the cutoff spatial frequency obtained by setting the threshold to a level of 4 [dB.]. The evaluated frequency, which equals approximately 44 radial spatial frequency units, is assumed to be the maximum resolvable spatial frequency, and therefore, the one that limits the observer's detection performance in this particular image.

4 Detection Performance Prediction

The familiar target acquisition model [7] and Johnson chart [8], although not necessarily optimized for this kind of imagery, were both used to relate the resulting spatial cutoff frequencies with the experimental probabilities of detection. Assuming a linear relation between the spatial frequency and the effective number of line pairs on the target [9], and having a way to estimate the effective number of line pairs on the original undistorted target, leads to a way to calculate the effective number of line pairs on the restored target as follows:

$$n_r = n_o \frac{f_{\max}}{128} \qquad (8)$$

n_o , n_r – effective number of line pairs on the original and restored targets.
$f_{\max}$ – spatial cutoff frequency.

The number of effective line pairs on the original target (n_o) is estimated from the expression relating target acquisition probability, given infinite observation time, with the effective number of line pairs on the target as follows:

$$Pd_\infty = \frac{(n/n_{50})^E}{1+(n/n_{50})^E} \qquad (9)$$

$E=2.7+0.7(n/n_{50})$
Pd_∞ - probability of detection, given infinite observation time.
n - effective number of line pairs on the target.
n_{50} - effective number of line pairs on the target needed to obtain a probability of detection of 0.5 (taken from the Johnson chart).

In practice we set Pd_∞ equal to the experimental probability of detection and obtained the effective number of line pairs on the original target. The value of n_{50} was taken as one line pair. The use of equation (9) for this application is not ideal because the observation time was limited during the experiment. Nevertheless, it is sufficient for illustrating the main ideas of this research. After evaluating from equation (8) the number of line pairs on the restored target, we used equation (9) again, only at this time the known variable was n (the value of n_r) and the estimated value was the probability of detection associated to the restored images. Figure 14 show the relation obtained between the experimental Pd and the estimated Pd. The solid line in the figure refers to an ideal fit. The agreement level found in the analysis can be quantified according to the mean and the standard deviation of the distance of the measures from the ideal fit curve. The lower are the mean and the standard deviation, the higher will be the agreement level. Accordingly, we formulated the Agreement Level Measure (ALM) as follows

$$ALM = \left|\mu_D \cdot \sigma_D^2\right| \qquad (10)$$

where μ_D, σ_D – mean and standard deviation of the distances between the ideal and the experimental curves.

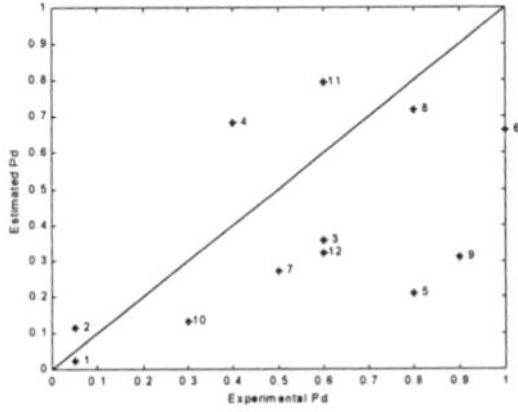

Fig. 14. Experimental and Estimated Pd

ALM value of 4.41 was obtained for the data presented in figure 14.

5 Summary and Conclusions

The performance degrading effects of atmospheric blur, noise distortion and image restoration are illustrated in this paper via the results of a specially designed psychophysical experiment. A spectral analysis approach, which was applied to the image database, enabled a quantitative definition of the restoration efficiency by means of a spatial cutoff frequency.
The results obtained from the analysis was used to address two issues:

1. Defining whether contrast or noise limits target perception in a given image.
2. Predicting human detection performance of targets embedded in restored images.

The first issue was handled by comparing the spatial frequency obtained from the analysis with that obtained from the intersection of the system MTF curve with the eye threshold MTF curve. The lower spatial frequency determines the nature of the image. In most analyzed cases, restoration caused the image to change from being contrast-limited to noise-limited. This is the leading conclusion regarding the inability of Wiener filter restoration to compensate for performance degradation caused by distortion. The second issue was handled by using the target acquisition model and the Johnson chart to convert the effective number of line pairs on the restored target to estimated Pd values. Applying this technique to the experimental data yielded good agreement between the experimental and the estimated Pd values.

6 References

[1] G.Aviram, S.R.Rotman, "Human Detection Performance of Targets Embedded in Infrared Images – The Effect of Image Enhancement", Optical Engineering, Vol. 38, No. 8, pp. 1433-1440, (Aug. 1999).

[2] G.Aviram, S.R.Rotman, "Evaluation of Human Detection Performance of Targets Embedded in Infrared Images Using Image Metrics". Accepted for publication in Optical Engineering.

[3] G.Aviram, S.R.Rotman, "Evaluating Human Detection Performance of Targets and False Alarms, Using an Improved Textural Image Metric". Submitted for publication in Optical Engineering.

[4] G.Aviram, S.R.Rotman, "Analyzing the improving effect of modeled histogram enhancement, on human target detection performance of infrared images". Accepted for publication in Infrared Physics & Technology.

[5] G.Aviram, S.R.Rotman, R.Succary, "Evaluating and Predicting Human Detection Performance of Targets Embedded in Distorted and Restored Infrared Images". Submitted for publication in Optical Engineering.

[6] D.Sadot, N.S.Kopeika, S.R.Rotman, "Target Acquisition Modeling for Contrast-Limited Imaging: Effects of Atmospheric Blur and Image Restoration", Optical Engineering, Vol. 12, No. 11, pp. 2401-2414, (Nov. 1995).

[7] S.R.Rotman, E.S.Gordon, M.L.Kowalczyk, "Modeling Human Search and Target Acquisition Performance: First Detection Probability in a Realistic Multitarget Scenario", Optical Engineering, Vol. 28, pp. 1216-1222, (Nov. 1989).

[8] J.Johnson, "Analysis of Image Forming Systems", paper presented at Image Intensifier Symposium, Ft. Belvoir,VA, AD 220160s, pp. 244-273, (Oct. 1958).

[9] N.S.Kopeika, "A System Engineering Approach to Imaging", Chapters 10-11, SPIE Optical Engineering Press, (1998).

OPTICAL-ELECTRONIC SYSTEM FOR FINGER-PRINT IDENTIFICATION

Soifer V.A., Kotlyar V.V., Skidanov R.V.

Image Processing Systems Institute, the Russian Academy of Sciences,

151, Molodogvardejskaya st., Samara, 443001 Russia

e-mail: soifer@sgau.volgacom.samara.su

Farberov E.

Neural Network Technologie, Ltd Bnei-Brak 51264, Israel

e-mail: nnt@netvision.net.il

Abstract

The developed optical-digital system is intended for real-time finger-print identification.

1. Introduction

Person identification via finger-prints arouses steady interest and is of great importance for security purposes. These problems are tackled using hybrid optical-digital devices in which contour images are recognized through the optical implementation of the Radon transform [1,2] or using a mutual correlation of two images by photo-refractive mirrors [3]. In some available systems finger-prints are entered into the computer in real time by a prism operating in the full internal reflection mode [3,4]. Fast optical input of a finger-print into the computer memory and optical image preprocessing essentially contribute to the speed of identification process, thus providing the advantage of hybrid systems over digital ones. However, the available systems for finger-print identification possess a number of disadvantages. It takes computer some time to analyze and classify finger-print fringes structure using an optical neuron network dedicated for extracting straight-line segments in the image with the aid of the Radon transform [1] or through calculating the Halton number (the number of bifurcations and finger-print fringe centers) [2]. This is a high spatial frequency in the finger-print image which complicates image processing. An optical method for constructing the directions field of structurally redundant images is proposed in Ref. [5]. This method enables the information contained in the images with spatial carrier frequency to be effectively compressed. With this method, an image segment is replaced by a single number equal to the average fringe inclination angle in this segment. This paper presents the results of testing an experimental optical-digital setup for finger-print identification, which uses the constructed finger-print's directions field for identification.

2. Directions field

The directions field of an image is a function of point coordinates equal to the tangent to the line of the image intensity level. Assuming $I(x,y)$ to be the light intensity in the image, the directions field $\varphi(x,y)$ is defined by

$$\operatorname{tg} \varphi(x,y) = -\frac{\partial I(x,y) / \partial x}{\partial I(x,y) / \partial y}, \quad 0 \le \varphi(x,y) < \pi. \tag{1}$$

Obviously, the angle $\varphi(x,y)$ specifies a direction perpendicular to the gradient vector

$$\left(\partial I(x,y) / \partial x, \ \partial I(x,y) / \partial y\right).$$

It is impossible to realize Eq. (1) using optical means.

In an experimental optical-digital setup for finger-print identification, we find in the Fourier-correlator's frequency plane (see Fig. 1, lenses S13 and S14) a spatial filter SF in the form of an amplitude (or phase) transparency composed of binary diffraction gratings that divide the angle of 180 degrees into N sectors, while the other part of the filter transmits no light (see Fig. 1b). A set of "partial" images is found at the output. The finger-print is entered using a prism P and then reduced using a telescopic configuration (lenses S11 and S12). Here, M1 - M3 are turning mirrors.

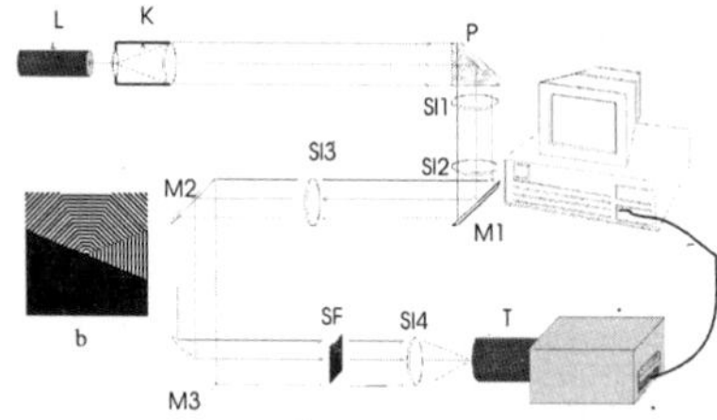

FIG. 1. (a) An experimental optical-digital setup for finger-print identification, (b) spatial filter.

Although the grating sectors that generate the "partial" images do not intersect (Fig. 1b), every point of the original image contributes to every point in the "partial' image. Structurally redundant images (contour images of finger-print type) are composed of a great number of lines with characteristic period or of segments of diffraction gratings, with their individual spatial spectrum contributing only to one of the filter's angular sectors. Therefore, the "partial" images should be expected to be nearly orthogonal, that is the "overlapping integral" for two "partial" images is much less than the total energy of each of them:

$$\int_{-\infty}^{\infty}\int I_n(x,y) I_m(x,y) dx\, dy << \int_{-\infty}^{\infty}\int I_n^2(x,y) dx\, dy. \tag{2}$$

If the inequality (2) is not valid, one may conduct the orthogonalization operation of the "partial" images by the rule:

$$\widetilde{I}_k(x,y)=\begin{cases}1, & I_k(x,y)=\max\limits_n\{I_n(x,y)\}\\ 0, & I_k(x,y)\neq\max\limits_n\{I_n(x,y)\}\end{cases},\qquad (3)$$

where $\max\limits_n\{\ldots\}$ is the maximum point-value of the function among the functions with the numbers $n=\overline{1,N}$.

It can be seen from Eq. (3) that the "partial" images $\widetilde{I}_k(x,y)$ are represented by an orthogonal set of binary functions. These, instead of Eq. (1), allow the direction field to be derived the formula

$$\varphi(x,y)=\sum_{n=1}^{N}\varphi_n\widetilde{I}_n(x,y),\qquad \varphi_n=\frac{\pi n}{N}.\qquad (4)$$

Finger-prints are identified through seeking a minimal Euclidean distance between the feature vectors of the analyzed finger-print and those of a set of standard base finger-prints. The Euclidean distance is derived from

$$d_{mn}=\left[\sum_{k=1}^{K}\left[b_m^{(k)}-a_n^{(k)}\right]^2\right]^{\frac{1}{2}},\qquad (5)$$

where $b_m^{(k)}$ and $a_n^{(k)}$ are the k-th coefficients of m-th base and n-th analyzed feature vectors and K is the number of coefficients of the Hadamard expansion of the directions fields.

3. Practical results

The experimental setup shown in Fig. 2 allows the real-time finger-print identification to be conducted.

FIG. 2. The outside view of the experimental setup for finger-print identification.

To control the setup operation we developed a program in the C++ language, with the interface window shown in Fig. 3. The program allows one to determine whether a given finger-print is found in data base. If a finger-print is not found in data base, the program allows them to be entered in the data base during several learning cycles. In the course of experiment, 50 different finger-prints were subject to identification using a data base of 1060 finger-prints. For every finger, 6 finger-prints were entered in the learning mode and 4 finger-prints for identification. The identification reliability was 0.965 at a 0.9 certainty of the results, i.e. out of 200 finger-prints, 7 finger-prints were identified incorrectly. The identification time is about 1 sec. . The rate of data base exhaustion was 109400 vectors per second on a PC with Pentium 200 processor.

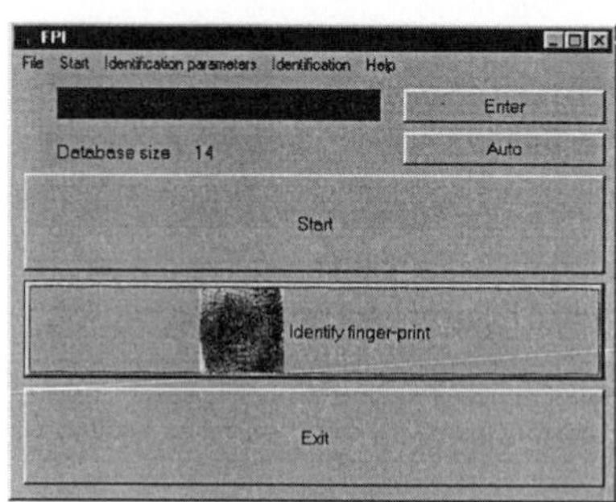

FIG. 3. The control program in operation.

Conclusions

We have developed an experimental optical-digital setup for finger-print identification, which allows real-time finger-print identification with a 0.96 reliability and a 0.9 certainty of the results, the rate of data base exhaustion being 109400 vectors per second.

References

[1] M. Seth, A.K. Datta "Optical implementation of a connectionist model of Hough transform" //Opt. Eng., 1996,v. 35, no 6, p. 1779-1794.

[2] H. Huh, J.K. Pan "Optical digital invariant recognition of two-dimensional patterns with straight links." //Opt. Eng., 1996, v. 35, no 4, p. 997-1002.

[3] J. Rodolfo, H. Rajbenbach, J.P. Haignard "Performance of a photorefractive joint transform correlator for fingerprint identification" //Opt. Eng., 1995, v. 34, no 4, p. 1166-1171.

[4] M. Dawagoe, A. Tojo "Fingerprint pattern classification" Pattern Recognition, 1984, v. 17, p. 295-303.

[5] V. A. Soifer, V. V. Kotlyar, S. N. Khonina, An optical method for constructing the directions field", the Journal of Avtometriya, 1996, issue 1, pp. 31-36 (in Russian).

בס"ד

MULTI-OPTICAL SYSTEMS FOR PANORAMIC VIEWING

Steve Bloomberg, Izhack Zubalsky, Moshe Meidan

Tamam, Israel Aircraft Industries,
P.O. Box 75, Yehud, 56000, Israel

Abstract

Various solutions to achieve a wide, panoramic field of view are discussed. The multi-optics solution using various types of optical switch is recommended to fulfill resolution requirements with an instantaneous panoramic field of view.

1. Introduction

There are many uses for panoramic optics. Security systems and warning systems for various types of vehicles are two examples for which it is necessary to provide the maximum angular coverage.

Today, in the visual spectrum, CCD cameras with wide-field optics are commonly available at low prices. Therefore a 360° field can be achieved at a reasonable cost by combining a number of separate CCD optical systems. However in the infra-red, to provide a cost-effective solution with reasonable resolution, it is necessary to use the minimum number of detectors possible

This paper presents a solution to achieve a panoramic field of view with the necessary resolution using multi-optics combined with an optical switch.

2. Concepts for Panoramic Viewing

Three concepts considered to achieve a full 360° field of view are:

1. Gimbaled system – by using relative narrow optics to provide the necessary resolution, the optics can be rotated by the gimbals through 360°. This method provides the required resolution but does not provide an instantaneous or even near instantaneous panoramic field.
2. Single lens, i.e. a wide field-of-view lens with a single camera. Such a system will have limited resolution, as the camera pixel resolution will be the total field of view (very wide) divided by the number of pixels. However the wide field of view is instantaneous.

An example of a very large field of view system is the Omnicam[1] which has been developed by Dr Shree Nayar of Columbia University. Two concave parabolic mirrors back to back provide a 360° x 70° FOV with a simple lens-camera combination placed beyond the parabola focal point looking at the intermediate image. There are serious problems of distortion and aberrations which are corrected using interpolation algorithms in the software, but the resolution will always be limited. Fish-eye type lenses can also be used for this type of solution but with similar drawbacks.

3. Multi-optics –using several different optics to cover the full 360°. Using this method, the required resolution can be achieved for an instantaneous wide field of view. A multi-optics solution can be achieved either by using a large number of optics-camera units which is bulky and expensive or by using several optical objectives combined with a smaller number of detectors and detector optics. In this paper is discussed the latter solution. An effective way of implementing this solution is to use switching methods between the different optics objectives.

3. System Considerations

The first constraint in the system is usually the type of detector to be used. The pixel size and number of pixels will together with the optics focal length determine the field of view.

The first consideration in determining the optics to be used is the required system resolution or IFOV (i.e. the angular resolution of a single pixel).

Angular pixel resolution = pixel size / focal length

This determines the nominal minimum acceptable focal length of the optics.

It should be noted that there are methods of increasing the resolution capabilities of the system beyond the Nyquist limit, e.g. using microscan dithering[2].

When the minimum focal length to provide the necessary resolution has been determined, the field of view for each individual optics is known, so the number of optics necessary to provide a full 360° FOV can be calculated.

To cover a full 360° range, each optics-detector combination should cover more than its calculated share of the field of view, in order to allow for overlap which will be necessary when joining together the different views, and to account for distortion effects.

After determining the number of optics needed, using the aforementioned considerations, the challenge is to use as few as possible detectors by combining a number of different optics with a smaller number of detectors.

4. Switching Methods

To combine several optics with a smaller number of detectors, the principle of time-sharing must be used. A scan mirror (or alternative method) rotates between the different objective lenses bringing the different parts of the wide FOV to a common detector unit.

Time sharing solutions are possible as the camera integration time is usually of the order of 4 msecs or less. Assuming a refresh rate of 60 Hz (16.7 msec) for the camera, for 4 msec out of each 16.7 msec, the scan mirror must be positioned so that the required optical field is fixed to be seen by the FPA[3]. During the remaining time, the mirror moves so that the next frame will see a view from a different optics. An apparent instantaneous wide FOV can then be built up from the pictures from the different optical objectives. An example of such a solution is shown in fig. 1.

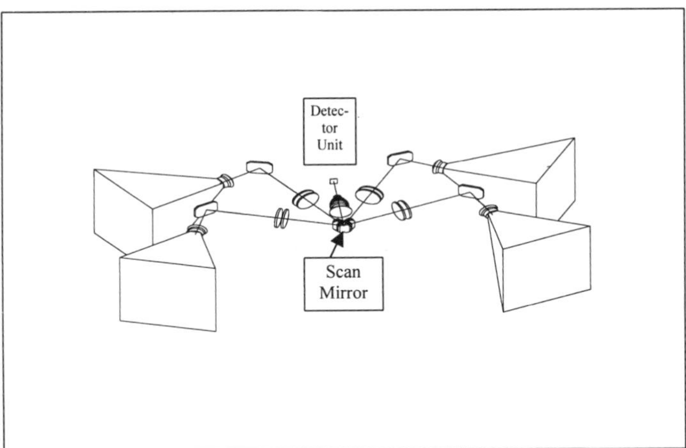

Fig. 1 Time-sharing solution with scan mirror.
160° FOV using one detector unit.

Another type of solution (fig. 2) is a radiation sharing solution. In this solution, two optics are combined with one detector using a 50-50 beamsplitter. A chopper is used to block the view to one of the optics when the second optics is used and vice versa. The advantage of this solution is that the chopper revolves at a constant rate. There is no need for a scan mirror which has to stop during the camera integration time. However there are two problems with this solution:

1. The energy is reduced by a factor of 2 by the beamsplitter. The aperture diameter has to be increased by at least 40% to reach the nominal energy level without the beamsplitter.
2. The chopper itself radiates energy and may have to be cooled to avoid noise from the "blocked" channel disturbing the "open" channel.

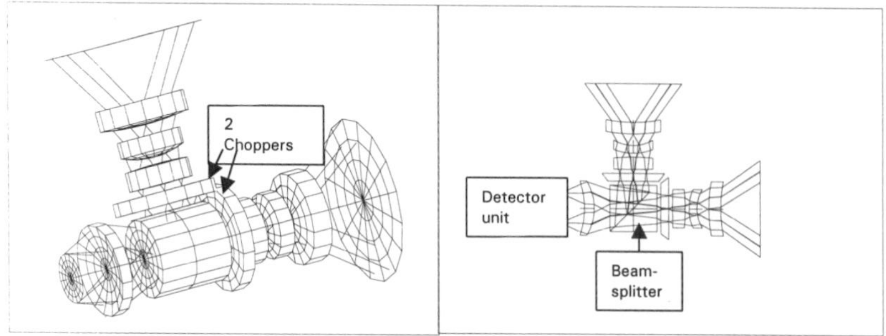

Fig. 2 Energy-sharing solution – 50% beamsplitter

5. Conclusion

To provide a panoramic field of view, a solution of multi-optics using optical switching is advantageous over the single wide optics solution due to the increased resolution and simpler aberration treatment. This solution can be considered instantaneous which is the advantage over conventional gimbaled systems.

6. References

1. S.K. Nayar "Catadioptric Omnidirectional Camera". Proc. Of IEEE Conference on Computer Vision and Pattern Recognition, June 1997
2. "Evaluation of the microscan process". J. Fortin, P.Chevrette, R.Plante SPIE San Diego, 1994
3. Patent Pending

A NEW FOCAL SENSOR FOR LASER SCANNING WAFER INSPECTION SYSTEM

Haim Feldman

Wafer Inspection Product Line, Applied –Materials PDC P.O.Box 601, Yavne Israel.

Oded Arnon

Optical Engineering, 78a Sheinkin St. Tel-Aviv 65223 Israel.

Abstract

A new focal-sensor is developed for laser wafer inspection system. Based on sensing the wavefront curvature of the back-reflected light, the sensor is highly sensitivity, it has a wide capture range and it is highly immuned to wafer microstructure.

1. Introduction

The detection of defects in wafers during the production process is of prime importance in the solid-state industry. One way of doing the detection is by scanning the wafer with a focused laser spot and collecting the reflected-scattered light in different directions. Defects are then detected by comparing the pattern of reflection from different dies on the wafer. While doing so, it is necessary to maintain the laser spot in-focus during the entire scanning process.

Sensing focal errors on wafers poses several difficulties. It is desired to control the focal error to a better than a quarter Depth-Of-Field (DOF = $\lambda/2NA^2$) over the entire scanning area of the wafer. The difficulty derives from the fact that the surface is highly irregular on the microscopic scale, and its reflectivity varies dramatically from close the 100% of the metal lines, to only few percents of the dense memory area.

This article describes [1] a new focal sensor that is developed to overcome these difficulties.

It uses the back-reflected light from the wafer and forms a focal error signal, which is used in combination with a servo system. The optical system is maintained in focus during the entire scanning process.

2. Principle of operation

This focal sensor belongs to the family of Hartmann wavefront sensor [2]. The principle of operation is shown schematically in Figure 1. A laser-scanning beam is focused to a scan-line by a lens L onto the wafer. Light is reflected back from this line, passes through L again in the opposite direction, and reaches the focal sensor, after being reflected by beam-splitter, BS. In principle, the focal sensor is assembled of two slits and a CCD camera. As shown below, in the in-focus position, the reflected light emerges collimated, with plane parallel wavefronts. After passing through the slits, the light propagates straight ahead and forms two spots at the CCD plane, at points A and B. On the other hand, when the wafer is in an out-of-focus position, spherical wavefronts emerge from lens L, to shift the spots from A and B to points A' and B', correspondingly. It is the difference: $FE = \Delta 1 - \Delta 1$, a measurable parameter, that is used to represents the focal error.

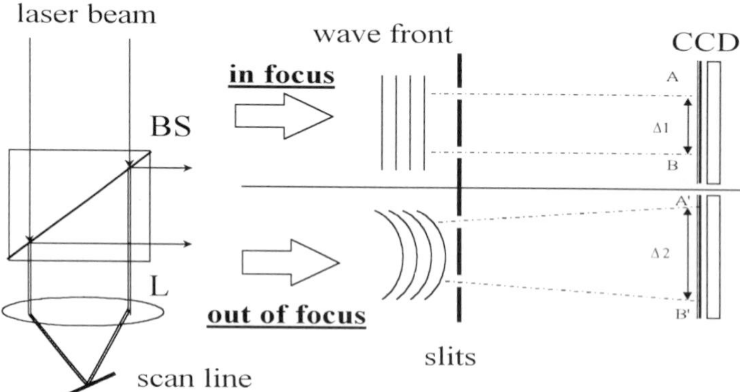

Figure 1: Focal sensor; principle of operation.

3 Practical Implementation

A practical implementation of this simple idea is shown in Figure 2. Lenses (a) and (b) are used to project the plane of the objective stop onto the plane of the slits, so that the reflected light will never miss the slits in all the different scanning positions along the scan-line. Lens (c) is mounted near by the slits. It forms a Fourier transform of the slits aperture on the CCD plane, to keep the width (along the Z direction) of the two spots as small as possible. The two cylindrical lenses operate along the X direction. They project further on the plane of the slits onto the CCD, in a similar way as lenses (a) and (b) did before. This layout, combined with the asymmetrical structure of the slits, produces the images of the spots that are shown in Figure 3.

When the wafer moves in and out-of-focus, one spot goes up, the other goes down, and vise versa. The CCD image is fed to a frame-grabber and a DSP is used to estimate the "center of gravity" of each

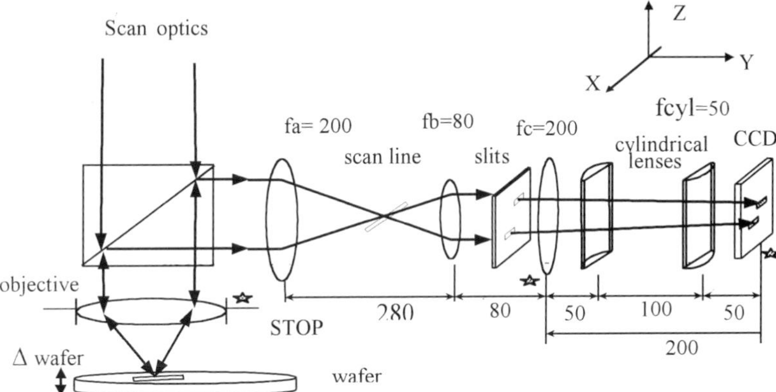

Figure 2: The practical implementation of the focal sensor

spot, to compute the error signal, FE. This particular arrangement has several advantages over the basic concept that is shown in Figure 1: it doubles the dynamic-range of the system and eliminates an unfortunate ambiguity that occurs when the two spots cross the CCD center. Moreover, in this way the spots are formed as a result of integration along the entire scan-line, so that the process averages over the micro-structure irregularity of the wafer, to provide eventually a cleaner signal.

Equation 3.1 gives the out-of-focus sensitivity of this scheme:

3.1 $$\frac{\Delta_{CCD}}{\Delta_{wafer}}\left[\frac{pixel}{DOF}\right] \propto \frac{f_a \cdot f_c}{f_b}$$

Δ_{CCD} is measured in pixels and is the relative displacement of the spots on the CCD. Δ_{wafer} is the wafer out-of-focus displacement, as given in units of DOF. Figure 3 shows - from top to bottom- three CCD images as obtained from three out-of-focus positions: -14 DOF, -0.4 DOF and 13.6 DOF, respectively. A summary of the sensor performance is shown in Table 1, where *resolution* is the minimal measurable focal error, and *contrast-ratio* is the ratio between the minimal and the maximal operational levels of illumination of that sensor.

Table 1: Focus sensor performance.

SENSITIVITY	RESOLUTION	CONTRAST RATIO	CAPTURE RANGE
6 pixel/DOF	0.1 DOF	1:200	± 50 DOF

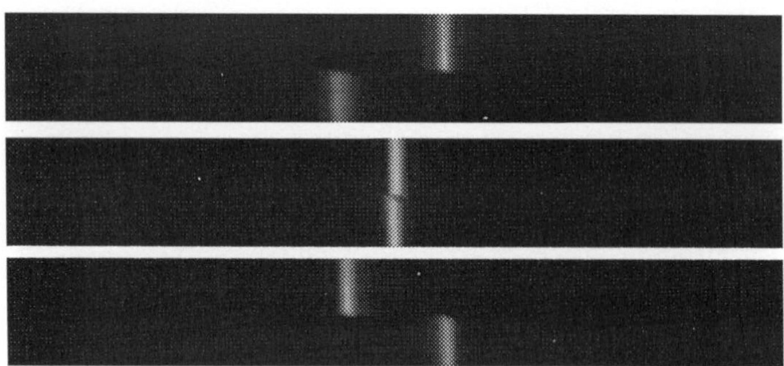

Figure 2: Spot pattern on the CCD at three out-of-focus positions.

3. Conclusion

A new focal sensor is designed and built to optimize the process of wafer inspection in systems that use laser scanning. It offers superior performance over other commercially available focal sensors, in this specific application.

References:

[1] US Patent Application No. 09/220,340
[2] J.M.Geary, "Introduction To.Wavefront Sensors", SPIE Optical Engineering Press 1995.

AUTOMATED TARGET DETECTION USING INFRARED MULTISPECTRAL ANALYSIS

Arik Kershenbaum and Glen Guttman

Israel Aircraft Industries, TAMAM Division.
PO Box 75, Yehud 56100, Israel

Abstract

A system is described for the automatic detection and recognition of objects using thermal infrared imaging. By dividing the infrared spectrum into a number of sub-bands, sufficient information is gathered to solve the radiation equations, thereby deriving values for the true temperature and emissivity for each pixel of the image. A statistical classification process assigns each pixel to a predefined target type, based on these values. This allows accurate, reliable and automatic identification of objects of interest based on their physical properties and independent of their shape or orientation.

1. Introduction

Automatic target detection is an important goal, not just for military applications, but in fields such as automatic medical diagnosis, unattended environmental monitoring and intelligent public transport. Despite considerable research, Automatic Target Recognition (ATR) based on object shape has generally failed to produce acceptable performance under a wide range of conditions. This is primarily due to the large differences in the perceived target image when viewed at different ranges and orientations, and when partly obscured. Here we present a complimentary approach, based on the recognition of the material physically comprising the target object, rather than the shape of the object itself. Such a technique can provide a solution in its own right to the detection problem where range/resolution prevents adequate discerning of the target shape, or where the target is of an inherently variable shape. Additionally, it can improve target detection performance dramatically when used in conjunction with traditional ATR techniques. The technique described here also lends itself to real-time implementation, which is important for most unattended applications.

2. Infrared Multispectral Imaging

The use of multispectral data for the enhancement of images or for the automatic extraction of information has been available for some time[1]. Reference to the familiar spectrophotometer reminds us of the power of using spectral information to discover the physical composition of objects. However, most of the work that has been done in the past has concentrated on the visible electromagnetic spectrum (0.2 μ – 0.8μ), or the near infrared (0.8μ ~ 1.5μ), because of the high intensity of radiation available in these bands, and the availability of suitable detectors. In contrast to this, much less energy is available in the thermal infrared (3 μ –12 μ) and detectors are of a lower resolution and more expensive. Despite this, a clear advantage is obtained when working in this spectral region, since much of the radiation received from an object is the result of the Planckian radiation from the object itself, rather than the reflection of luminous external sources such as the sun. As a result, the spectral data contains much more information on the physical properties of the object, less confounded by the source of the illumination. By dividing the spectrum into a number of smaller sub-bands, specifically chosen for the application in question, the information present can be extracted by mathematical analysis.

3. Method

Imaging was carried out using a focal plane array thermal imaging camera (FLIR) sensitive to the 3-5μ band and with a resolution of 256x256 pixels. Spectral discrimination was achieved by introducing a series of bandpass filters into the optical path. Each successive image recorded represented the radiation in one narrow range of wavelengths. For each pixel in the image, it is possible to construct a series of equations describing the radiation received, in terms of the temperature of the object located at that pixel, and a wavelength dependent parameter of the material known as the emissivity[2]

$$I_{\Delta\lambda} = \int_{\Delta\lambda} f(T, \varepsilon_{\Delta\lambda}) d\lambda \qquad (1)$$

where $I_{\Delta\lambda}$ is the intensity of radiation in spectral band $\Delta\lambda$, T is the temperature of the object and $\varepsilon_{\Delta\lambda}$ is the emissivity in the same spectral band. As a result of making measurements in N spectral bands, a system of integral equations is produced with N unknown emissivities and one unknown temperature common to all equations; in other words N equations with $N+1$ unknowns. In order to solve these equations, we note that ε tends to vary smoothly with wavelength in these bands and, since we do not need to derive the precise value of ε for all λ, we can approximate the emissivity by some function whose coefficients are invariant with λ

$$\varepsilon_{\lambda}^{*} = g(\lambda, \vec{p}) \qquad (2)$$

where $\vec{p}$ is a vector of coefficients that fit the observed emissivity function. By transforming the system of equations from $\varepsilon(\lambda)$ (measured) to $\varepsilon^*(\lambda)$ (calculated) a new set of *N*x*N* equations is created, which can be solved. Once values for T and ε^* for each band are obtained, the pixels of the image can be represented as an *N+1* dimensional parametric graph which shows the distribution of image pixels in ε-T space (Fig. 1).

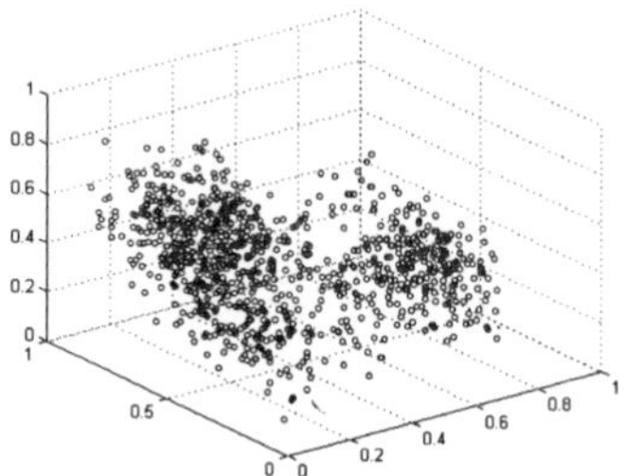

FIG. 1 Pixels from a real infrared image represented in a simplified 3-dimensional ε-T space.

Examination of the data shows that pixels originating from objects of similar composition tend to occupy a particular hypervolume in this space. Association of any point in the hyperspace to a particular object type is carried out by an advanced clustering algorithm, details of which cannot be discussed here due to lack of space. The system is presented with data in which the object of interest has been identified by an expert, and the algorithm performs an iterative learning process. Following this, when presented with subsequent images from different scenes, the algorithm classifies each pixel according to membership or otherwise of each of the target classes. This operation is carried out in real-time even for realistic picture size (256x256) and an adequate number of spectral bands (4-6).

4. Results

As one example of the application of this method, we show here the results of presenting the system with multispectral infrared images of a view of countryside surrounding a vantage point in central Israel. The system was previously trained to recognise (a) buildings of various kinds and (b) trees. The results are shown in Fig. 2. Good sensitivity and selectivity are obtained for both targets.

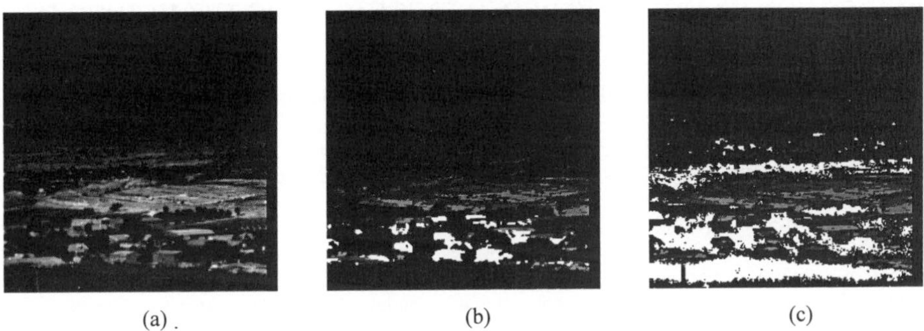

(a) . (b) (c)

FIG. 2 Image (a) shows a scene including hills, buildings and woods. Presenting this data to a system trained to recognise buildings, the results are as shown in (b). Image (c) shows the results where the system has been trained on trees. Pure white represents target areas.

5. Summary

The described approach, based on infrared multispectral analysis provides a fully automated, real-time target detection system which is independent of object shape and orientation. The detection is carried out on the basis of the object's physical properties and thus is unaffected by partial obstruction of the line of sight. A wide range of targets are detected without human intervention; a feature particularly important in practical applications. The system is based on a learning algorithm, which makes in robust and able to adapt itself to changing conditions without loss of performance.

References:

[1] Lewotsky, K. 1994, Hyperspectral imaging: evolution of imaging spectrometry, *Optical Engineering* Nov. 1994

[2] Lloyd, JM. 1975, Thermal radiation theory and atmospheric transmission, chapter 2 in "Thermal Imaging Systems", Plenum Press

RECOGNITION OF MOTION BLURRED TARGETS USING THE METHOD OF MOMENTS

Adrian Stern, Inna Kruchakov, Eitan Yoavi and Norman S. Kopeika

Ben-Gurion University of the Negev
Department of Electrical and Computer Engineering, P. O. Box 653,84105, Beer-Sheva, ISRAEL

Abstract
Image motion causes blur which changes features of objects and therefore complicates the task of automatic recognition. In this work we develop two recognition methods for motion blurred images using a modified moment matching technique. Two cases are discussed. In the first case we assume that the motion function and direction during the exposure are given. In the second case the motion function is not known. The advantage of the suggested methods is that no time-consuming image restoration is required prior to recognition.

1. Introduction

Image motion is often the major cause of image degradation. Images in or of flying aircraft, moving vehicles, objects on conveyor belts, etc. may appear smeared because of motion. The task of recognition of an observed image against images stored in a database becomes more difficult in the presence of image motion degradation because the motion blur changes the image features needed for feature-based recognition techniques.

The classical approach to the recognition of blurred images is by first restoring the blurred image via an inverse filter, Wiener filter (see for example Ref.1) or any other restoration method [2], [3]. Then the restored image is matched against an image database. In this work we present two recognition methods that do not require restoration of the blurred image.

The proposed methods are based on matching of image moments. Moments have been used to distinguish between shapes of different aircraft, character recognition, chromosome recognition, and scene-matching applications [4], [5]. An effective recognition method based on invariant moments appropriate for images distorted by symmetric blur was developed by Flusser, et al. [6]. Their method is not appropriate for motion blurred images since motion blur is not symmetrical.

The first method is developed in Sec. 2. Based on a distinction showing that the motion blur point spread function (PSF) moments are equal to the motion function moments, the central moments of the original image are restored from the observed image central moments and the motion function. The restored moments are compared against the moments of the images in the database to find the best match.

The second recognition method is described in Sec. 3. This method is based on the fact that moments in directions perpendicular to linear motion are invariant to motion blur. It is assumed that the motion is linear (the motion trajectory is on a straight line). This is a reasonable assumption if the exposure period is no more than a few tenths of a millisecond. Unlike the first recognition method, this method does not need the motion function itself, only its direction. The motion direction can be either measured by mechanical sensors or estimated from the blurred image [7]. Practical considerations based on empirical results are suggested. The robustness of the methods is also empirically investigated.

2. Recognition using motion function moments

Our purpose is to recognize an image g, which is a motion-blurred version of an image f_i from an image database $\{f_i\}_{i=1}^{N}$ of N images. In this section we assume that the motion function $\vec{S}(t)=[s_x(t),\ s_y(t)]$ is known. The motion function can be either measured by sensors or estimated from the image [7].

Assuming a noise-free imaging system the following relation between the observed image moments and the original and PSF moments exists [6],[8]:

$$\mu_{p,q}^{(g)}=\sum_{i=0}^{p}\sum_{j=0}^{q}\binom{p}{i}\binom{q}{j}\mu_{i,j}^{(h)}\mu_{p-i,q-j}^{(f_i)} \tag{1}$$

where $\mu_{p.q}^{(g)}$, $\mu_{p.q}^{(fi)}$ and $\mu_{p.q}^{(h)}$ are the 2-D (p+q)th-order central moments of the observed image, original image, and PSF respectively. It can be shown [9] that the PSF moments are equal to the motion function moments $\mu_{p.q}^{(s)}$. Assuming the motion function is linear with the PSF's central moments:

$$\mu_{p.q}^{(h)}=\frac{1}{t_e}\int_{t_0}^{t_0+t_e}\left[s(t)\cos\theta-m_{1,0}^{(h)}\right]^p\left[s(t)\sin\theta-m_{0,1}^{(h)}\right]^q dt\equiv\mu_{p.q}^{(s)} \quad . \tag{2}$$

Eq. (1) together with (2) define a direct relation between the blurred image moments g and the original image moments f_i via the motion function. Based on this, the following algorithm is suggested: 1) calculate $\mu_{p.q}^{(g)}$, $\mu_{p.q}^{(fi)}$ and $\mu_{p.q}^{(s)}$; 2) Estimate the restored central moments of the original image $\hat{\mu}_{pq}^{(fi)}$ by solving (1) for $\mu_{p.q}^{(fi)}$; 3) Compare the restored central moments $\hat{\mu}_{pq}^{(fi)}$ to the central moments of the images in the database $\mu_{pq}^{(fi)}$, i=1..N using the following normalized distance function:

$$\rho(i)=\sum_{(p,q)\in M}\left(\frac{\hat{\mu}_{pq}^{(fi)}\big/\hat{\mu}_{00}^{(fi)}-\mu_{pq}^{(fi)}\big/\mu_{00}^{(fi)}}{\hat{\mu}_{pq}^{(fi)}\big/\hat{\mu}_{00}^{(fi)}}\right)^2 \qquad i=1...N \tag{3}$$

where M is the set of the moments to be used in the comparison. The best match is achieved for the i'th image in the database for which $\rho(i)$ is smallest.

The method was examined by simulating 120 experiments. In each experiment an image from the database shown in Fig. 1 (a) was blurred in a one of ten possible directions and one of three types of motions (constant velocity, parabolic motion and high-frequency vibrations). An example of a database, blurred image, and distance function $\rho(i)$ is shown in Fig. 1. From the simulation we found that the minimal set of central moments $M_1=(\mu_{02}, \mu_{20}, \mu_{03})$ when used in (1) gives satisfactory results.

In order to examine the robustness of the method we added Gaussian white noise to the blurred images. We point out that relation (1) assumes that the observed image is noise-free. Empirically we found that the method fails at SNR of less than approximately 13dB.

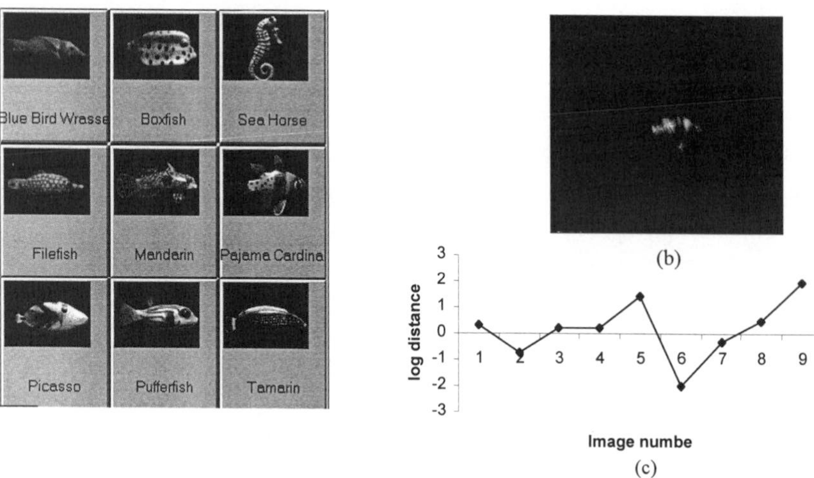

Fig. 1 (a)Image database, (b) the sixth image (Pajama Cardinal fish) blurred by the motion, (c) the distance ρ (on logarithmic scale) between the sixth "unknown" image (b) and the i=1...N images of the database (a).

3. Recognition using moments invariant to motion blur

The second method we introduce is based on the fact that moments normal to motion blur are invariant to the blur. The motion PSF has non-zero values only in the direction of the motion. Therefore moments normal to the motion direction remain unaffected by motion. Based on this distinction we suggest the following procedure to recognize the motion blurred image g in direction θ with respect to the X axis:

1) Define a new coordinate system ($X'Y'$) so that the X' axis is in the motion direction.
2) Calculate the central moments in Y' direction for each image of the database using the following relation between the central moments in X'-Y' system to that in the X-Y coordinate system [8]:

$$\mu'^{(f_i)}_{pq} = \sum_{r=0}^{p}\sum_{s=0}^{q}(-1)^{k-s}\binom{p}{r}\binom{q}{s}(\cos\theta)^{p-r+s} \times (\sin\theta)^{q+r-s}\left(\mu^{(f_i)}_{r+s,p+q-r-s}\right). \tag{4}$$

3) Calculate the central moments $\mu'^{(g)}_{0,q}$ of the observed image (in the Y' direction) .

4) Define a distance $\rho'(i)$ similar to (3):

$$\rho'(i)=\sum_{q=2}^{n}\left(\frac{\mu'^{(g)}_{0q}\Big/\mu'^{(g)}_{00}-\mu'^{(fi)}_{pq}\Big/\mu'^{(fi)}_{00}}{\mu'^{(g)}_{0q}\Big/\mu'^{(g)}_{00}}\right)^2 \qquad i=1...N \tag{5}$$

where n is the highest moment order used in the comparison. The observed image is recognized as the blurred version of the i'th image in the database for which $\rho'(i)$ is smallest.

We examined this method in a similar way to the first method. From the simulations we found that central moments up to order n=5 (μ'_{02}, μ'_{03}, μ'_{04}, μ'_{05}) are sufficient for the comparison using (4). The minimum required SNR threshold using this method was approximately 12dB.

The method was examined also by real (nonsimulated) experiments. Images that were moved by hand (unknown and uncontrolled motion) during the exposure were precisely recognized.

5. Conclusions

Two recognition methods of motion blurred images against image database are proposed. Both methods avoid the need of image restoration prior to recognition. The first method is more computationally efficient but requires measurement of the motion function. The second method does not require knowledge of the motion function and is not sensitive to the type of the motion but is more complex since it requires rotation of images or moments. The methods are demonstrated both by simulated and real experiments.

References:

[1] N. S. Kopeika *A System Engineering Approach to Imaging*, chp. 18, SPIE Optical Engineering Press, Bellingham WA (1998).

[2] A. K Jain. *Fundamentals of digital image processing*. Prentice Hall, New Jersey (1989).

[3] R. L Lagendijk., *Iterative Identification and Restoration of images,* Kluwer Academic Publishers, Boston (1991).

[4] S. Dudani, K. Breeding and R. McGhee, "Aircraft identification by moment invariants", *IEEE Trans. on Computers C-26*, no. **1**(1), pp. 39-45 (1977).

[5] R. Wong and E. Hall, "Scene matching with moment invariants", *Computer Graphics and Image Processing*, **8**, pp. 16-24 (1978).

[6] J. Flusser, T. Suk and S. Saic, "Recognition of blurred images by the method of moments", *IEEE Trans. on Im. Proc.*, **5**(3), pp. 533-538 (1996).

[7] Y. Ytzhaky. I. Mor, A. Lantzman and N. S. Kopeika, "Direct method for restoration of motion-blurred images", *J. Opt. Soc. Am. A*, **15**(6), pp.1512-1519 (1998).

[8] J. Teuber, *Digital Image Processing*, pp. 213-216, Prentice Hall, New York (1993).

[9] A. Stern and N. S. Kopeika," Analytical method to calculate optical transfer function for image motion and vibration using moments", *J. Opt. Soc. Am.* A, **14**(2), pp. 388-396 (1997).

MODELING AND SIMULATION FOR MULTIPLE INFRA-RED BACKGROUNDS

Zeev Zalevsky and Steven Lashansky

IDF, M.P.O Box 02158, Israel

Fax: (972)+3-5694119, Tel: (972)+3-6942723

Abstract

In this paper we present a numerical comparison between procedures for generation of Infra Red (IR) backgrounds. Various statistical models of the power spectrum are assumed for the backgrounds. The images are generated using three different approaches: differences equation, Cholesky factorization and a spectral analytical approach. The differential approach involves expressing the power spectrum as an Infinite Impulse Response (IIR) filter by writing an auto-regressive moving average (ARMA) difference equation involving white noise. Solving the equation yields a stochastic image having the desired power spectrum. The Cholesky factorization approach involves a decomposition of the desired power spectrum into triangular matrix, adding a white noise to it to generate a stochastic image, which fulfills the desired power spectral constrains. The spectral analytic approach assigns the absolute value of the Fourier transform of the image to be the square root of the desired power spectrum. A random phase is added and an inverse Fourier transform is performed.

The outputs of the three approaches are compared with each other and with experimental results. Good correspondence is achieved. The proposed approaches may be used for creation of non-stationary scenes containing various regions having different statistical properties (such as sky, ground etc.). The generated scenes can then be used for testing the performances of IR Warning or IR Search and Track (IRST) systems and of their automatic detection algorithms.

Key words: Statistical background models, IR Warning or IR Search and Track Systems (IRST).

1. Introduction

Detecting a target out of its surrounding background is a major problem in various missile warning (MWS) and infra red search and track (IRST) systems. The detection procedure usually involves a complex algorithm that relies upon both the spatial and the temporal properties that distinguish a target from its background. The detection performance, i.e. detection range and probability of detection are dependent on the applied algorithm as well as the background on which the target appears [1-5].

This paper presents three different methods for simulating various IR background scenarios. Previous works have investigated the statistical properties of IR backgrounds [6-9]. Ben-Yosef demonstrated how basic thermodynamic properties connect the background with a first order statistical Markoff's process [10-12]. Our clutter model assumes a first order statistical Markoff's representation for the background, which is based upon a combination of Gaussian distribution for the irradiance level and a Poisson's statistics for the width of each spatial region. However, the illustrated approach is general and allows the adaptation of any other combination of distributions. Different statistical parameters are used in the model to describe various types of backgrounds. Three approaches are suggested and compared for the background realization: Difference equation, Cholesky factorization and a spectral analytical approach. The difference approach involves expressing the power spectrum as an IIR filter by writing an auto-regressive moving average difference equation involving white noise. Solving the equation yields a stochastic image having the desired power spectrum. The Cholesky factorization approach involves a decomposition of the desired power spectrum into a triangular matrix, adding a white noise in order to generate a stochastic image which fulfills the desired power spectral constraints.

The spectral analytic approach assigns the absolute value of the Fourier transform of the image to be the square root of the desired power spectrum. A random phase is added and an inverse Fourier transform is performed.

Section 2 and 3 deal with the statistical model for the background. The three different approaches for IR background simulation are discussed in section 4. In section 5 some experimental results are presented. Section 6 concludes the paper.

2. General Statistical Model for the Background

We assume that the background noise process is a random set of a 2-D pulses whose amplitude and width obey the Gaussian and Poisson's statistics, respectively:

$$p(z) = \frac{1}{\sqrt{2\pi\sigma^2}} e^{-\frac{(z-\mu)^2}{2\sigma^2}} \qquad (1)$$
$$p(r) = \alpha e^{-\alpha r}$$

where z is the radiance of a point in the x-y plane, μ is its average, σ^2 is the variance, r is the interval length between two adjacent points on the x-y plane, α the reciprocal of the average pulse width and p is the probability density function. Assuming that the random processes x and r are independent of each other and using the formula:

$$E\{z(r+\Delta)z(r)\} = \sum_k E\{z(r+\Delta)z(r)|k\}P_k \qquad (2)$$

where E{} is the ensemble average and k is the examined case of given expectancy. We will distinguish between two cases k=1,2: The two adjacent points whose interval length is Δ belong to the same pulse and the two points do not belong to the same pulse. Thus,

$$R(\Delta) = E\{z(r+\Delta)z(r)\} = \left(\sigma^2+\mu^2\right)P(\Delta)+\mu^2[1-P(\Delta)] = \sigma^2 P(\Delta)+\mu^2 \qquad (3)$$

where $P(\Delta)$ is the probability that two adjacent points on the x-y plane, whose interval length is Δ, belong into the same pulse:

$$P(\Delta) = 1-\int_0^\Delta \alpha e^{-\alpha r} dr = e^{-\alpha\Delta} \qquad (4)$$

resulting in

$$R(\Delta) = \sigma^2 e^{-\alpha\Delta} + \mu^2 \qquad (5)$$

For $\mu=0$, (5) coincides with the expression for the auto correlation of a first order isotropic Markoff process. Note that other distributions p(z) and p(r) may be assumed as well. In those cases, using similar derivation other expression may be obtained for $R(\Delta)$. Using thermodynamic considerations, a physical justification may be found for showing that this statistical model is a good representation of an IR background. Rewriting the expression for the auto correlation of 1-D process yields:

$$R(x) = \sigma^2 e^{-\frac{|x|}{L}}$$

where L is the correlation distance coefficient. The resulted expression for the spectral density is:

$$S(v) \propto \frac{1}{1+L^2(2\pi v)^2} \qquad (6)$$

The 2-D expansion for auto correlation expression is:

$$R(x,y) = \sigma^2 e^{-\sqrt{\frac{x^2}{L_x^2}+\frac{y^2}{L_y^2}}} \qquad (7)$$

which results in the spectral density of

$$S(v_x, v_y) \propto \frac{1}{\left(1+L_x^2(2\pi v_x)^2+L_y^2(2\pi v_y)^2\right)^{3/2}} \qquad (8)$$

A more general spectral distribution may be described by the following expression [13,14]:

$$S(\nu_x,\nu_y)=\frac{2\pi\sigma^2}{B\left(\frac{1}{B^2}+(2\pi\nu_x)^2+(2\pi\nu_y)^2\right)^{\alpha/2}} \tag{9}$$

Note that for α=3 the parameter B is exactly the correlation length.

In order to extract the α parameter from the image one may use a log-log display for the spectral power density:

$$\log[S(\nu_r)]\approx\log\left(\frac{2\pi\sigma^2}{B}\right)-\alpha\log(2\pi)-\alpha\log(\nu_r)$$
$$\nu_r=\sqrt{{\nu_x}^2+{\nu_y}^2} \tag{10}$$

Thus, in a log-log display the power spectrum may be approximated by a linear curve whose slop equals to α. Nevertheless, in practical cases the obtained chart will not be exactly linear. Thus the slope of the curve may be obtained by applying the best fit technique: Given a function y(x), its best linear fit of the from y=ax+b is:

$$a=\frac{\frac{\sum x\sum y}{N}-\sum xy}{\frac{\left(\sum x\right)^2}{N}-\sum x^2}$$
$$b=\frac{\sum y-a\sum x}{N} \tag{11}$$

where N is the number of elements in the vector x and y.

3. Statistical Model for Various Terrain Types

In this paper we will explore three types of terrain: clouds, land and sea. The difference between the various terrain types is obtained by varying the power spectrum and its parameters. The following three tables summarize most of the experimental data available in the literature [13-15]. Note that for sea terrain specified data is missing. From Ref. [15] one may learn only four main insights:

- The image may be approximated by a Gaussian homogeneous random process.
- The spatial correlations of the clutter are small compared with typical dimensions of a tactical target (e.g. a frigate).
- The spatial correlation along the azimuth dimension of the image are much higher then those along the range dimension.
- The spatial properties are highly influenced by the speed of the wind.

No.	Clutter Type	Spectral Range [μm]	Correlation Length [m]	Standard Deviation σ [deg]	Slope of Average Power Spectrum α
1	Cloud (0-10% cover)	8-12	12.8	0.98	1.82
2	Cloud (10-20% cover)	8-12	20.4	0.98	3.04
3	Cloud (20-30% cover)	8-12	25.6	0.98	4.27
4	Cloud (3-40% cover)	8-12	27.6	0.98	4.9
5	Cloud (40-50% cover)	8-12	28	0.98	5.26
6	Cloud (50-60% cover)	8-12	28	0.98	5.3
7	Cloud (60-70% cover)	8-12	27.6	0.98	4.97
8	Cloud (70-80% cover)	8-12	25.6	0.98	4.23
9	Cloud (80-90% cover)	8-12	20.8	0.98	3
10	Cloud (90-100% cover)	8-12	13.2	0.98	1.91

Table 1: Cloudy sky (8-12μm).

No.	Clutter Type	Spectral Range [μm]	Correlation Length [m]	Standard Deviation σ [deg]	Slope of Average Power Spectrum α
11	Cloud (0-10% cover)	3-5	6.4	0.3	1.64
12	Cloud (10-20% cover)	3-5	14.4	0.3	2.3
13	Cloud (20-30% cover)	3-5	22	0.3	3.48
14	Cloud (30-40% cover)	3-5	28	0.3	4.54
15	Cloud (40-50% cover)	3-5	30.4	0.3	4.96
16	Cloud (50-60% cover)	3-5	28.8	0.3	5.07
17	Cloud (60-70% cover)	3-5	27.2	0.3	4.81
18	Cloud (70-80% cover)	3-5	26.4	0.3	3.36
19	Cloud (80-90% cover)	3-5	20.4	0.3	2.42
20	Cloud (90-100% cover)	3-5	8.8	0.3	1.71

Table 2: Cloudy sky (3-5μm).

No.	Clutter Type	Spectral Range [μm]	Correlation Length [m]	Standard Deviation σ [deg]	Slope of Average Power Spectrum α
21	Ground (flat no rocks no vegetation)	3-5	156±65	1.62	3
22	Ground (flat no rocks no vegetation)	8-12	124±60	1.4	3
23	Ground (pebbles, spars and no vegetation)	3-5	51±16	2.39	3
24	Ground (pebbles, spars and no vegetation)	8-12	46±15	3.67	3
25	Ground (rocks, bushes)	3-5	40±22	3.07	3
26	Ground (pebbles, spars and no vegetation)	8-12	29±12	3.19	3
27	Sea	8-12	≈20-30	0.5-1.5	?

Table 3: Land and sea .

For instance, Fig.s 1a and 1b present a sea and a land backgrounds captured with a 8-12μm FLIR. In Figs. 1c and 1d one may see the log-log representation of the obtained power spectrum for the images of Figs. 1a and 1b respectively. Those figures also present the linear best fit curve. The obtained α parameter for the sea image was 3 and for the land 3.5.

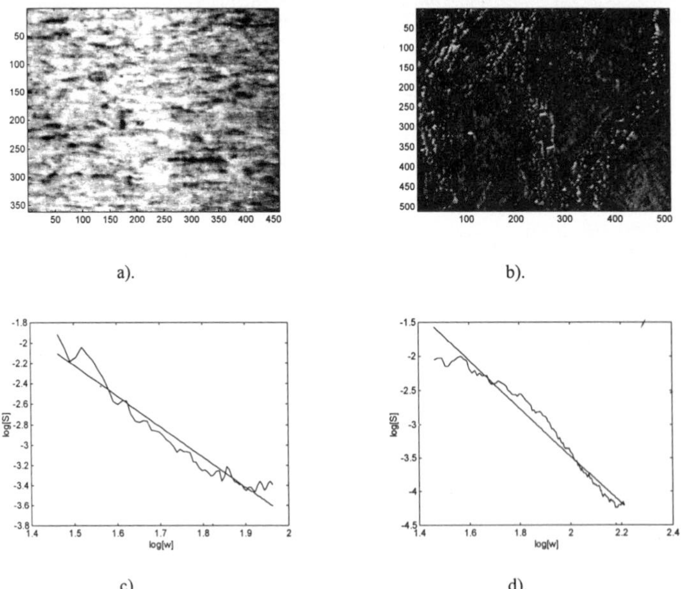

Fig. 1: a). Sea image. b). Land image. c). The log-log display of the power spectrum of 1a. d). The log-log display of the power spectrum of 1b.

4. Background Simulation

4.1 Difference Equation Approach

One of the most basic approaches for the realization of IR backgrounds is obtained by Auto Regressive (AR) first order difference equation. Assuming that $M_{n,m}$ is the n,m pixel with in a IR image having a correlation coefficient of ρ and a standard deviation of σ_x, then:

$$M_{n,m} = \rho M_{n-1,m} + \rho M_{n,m-1} - \rho^2 M_{n-1,m-1} + W_{n,m} \quad \forall n,m \geq 1 \qquad (12)$$

where $W_{n,m}$ is a Gaussian white i.i.d (independent identically distributed) process having a zero mean and a variance of

$$\sigma_w^2 = \sigma_x^2 (1-\rho^2)^2$$

From these equations one may easily obtain the relation required for the 2-D first order Markoff process:

$$R_x(k_1,k_2) = E\{M_{n,m} M_{n+k_1,m+k_2}\} = \sigma_x^2 \rho^{k_1+k_2} \quad \forall\ k_1,k_2 \geq 0 \qquad (13)$$

4.2 Cholesky Factorization Approach

Cholesky factorization allows decomposing a positive definite matrix R into a multiplication of two matrixes:

$$C'_{n,m} \times C_{n,m} = R_{n,m} \qquad (14)$$

where $C_{n,m}$ is an upper triangular matrix. Assuming now that $R_{n,m}$ is the auto-correlation matrix:

$$R_{n,m} = \rho^{|n-m|} \qquad (15)$$

then finding $C_{n,m}$ by the Cholesky technique assists to create a 1-D IR background according to:

$$M_n = \mu_n + C_{n,m} \times z_n \qquad (16)$$

where z_n is a Gaussian random signal having zero average and a standard deviation of σ.

For 2-D IR images the 2-D image is first converted into a long 1-D vector while the various rows of the 2-D image are added one after the other to create the 1-D long vector. An auto-correlation matrix is defined to describe the relations within the long 1-D vector. For instance to create a 2-D N×N image yields:

$$\begin{aligned} R_{n,m} &= \rho^{|k_{n,x}-k_{m,x}|+|k_{n,y}-k_{m,y}|} \quad \forall n,m \leq N \times N \\ k_{n,x} &= n \bmod N \\ k_{n,y} &= \mathrm{Int}(n/N) \\ k_{m,x} &= m \bmod N \\ k_{m,y} &= \mathrm{Int}(m/N) \end{aligned} \qquad (17)$$

where mod is the module operation and Int is the operation of rounding towards the smaller integer. It is evident that one may also fulfill the following correlation function:

$$R_{n,m} = \rho^{\sqrt{(k_{n,x}-k_{m,x})^2+(k_{n,y}-k_{m,y})^2}} \quad \forall n,m \leq N \times N \qquad (18)$$

Similar algorithm is then applied and a long 1-D IR background vector is created. This vector is then transformed into a 2-D image.

4.3 The Spectral Analytic Approach

Another approach for creating the 2-D IR background is by using its power spectrum. A root of the power spectrum is taken. Its inverse Fourier transforming will insure an image having the desired auto-correlation. However, in order to create a stochastic distribution the root of the power spectrum is multiplied by the expression $A(\nu_x, \nu_y)\exp[i\Psi(\nu_x, \nu_y)]$ before applying the inverse Fourier transform. $A(\nu_x, \nu_y)$ is a normal i.i.d. Gaussian distribution variable having zero mean and a standard deviation of 1 and Ψ is uniform distribution distributed within the range of $[-\pi,\pi]$:

$$M_{n,m} = \iint \sqrt{S(\nu_x, \nu_y)} A(\nu_x, \nu_y) \exp\left[i\Psi(\nu_x, \nu_y)\right] \exp\left[2\pi i\left(\nu_x \Delta x\, n + \nu_y \Delta y\, m\right)\right] d\nu_x d\nu_y \quad (19)$$

where Δx and Δy are the spatial sampling distance.

5. Experimental Results

Due to the large variety of data collected by this paper, we will simulate only cases 9,20,22,25,27 from tables 1-3. In order to analyze the statistics of the generated images attention must be paid to the following points:

- The images were smoothed using the Hanning window: The input image was multiplied by cosine square.
- The auto correlation function was extracted by:

  $$R(n,m) = FFT^{-1}\left\{\left|FFT\left[g(n,m)\right]\right|^2\right\}$$

  where g(n,m) is the image and FFT is the discrete fast Fourier transform.
- The power spectrum was computed as $\left|FFT\left[g(n,m)\right]\right|^2$. An angular averaging was applied in order to obtain $S(\nu_r)$.
- A log-log display was applied over $S(\nu_r)$ according to Eq. 10. A best linear fit was obtained using Eq. 11.

A generation and a synthesis of the obtained images was performed for the cases 9,20,22,25,27 of tables 1-3 using each one of the three approaches previously discussed. Figs. 2-12 demonstrate the obtained results. In our simulations we assumed that the gray scale map of the display is 10 gray levels per each temperature degree. We also assumed that each pixels is 1.6 meter.

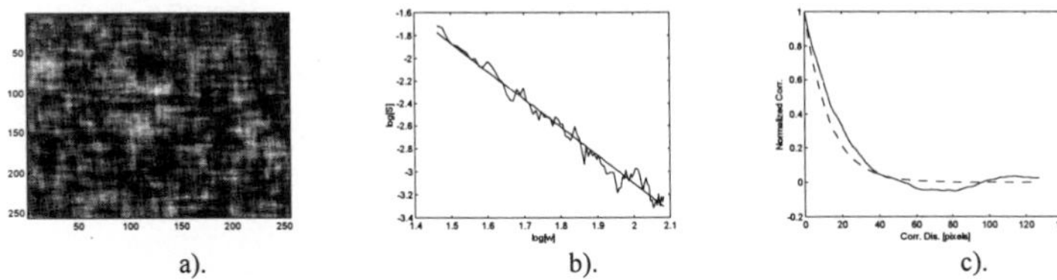

Fig. 2: The difference equation approach with σ=10, Correlation length of 13 pixels and obtained α of 2.47. a). The generated image. b). The log-log display of the its power spectrum and its best fit curve. c). The desired (dashed) and the obtained correlation function (solid).

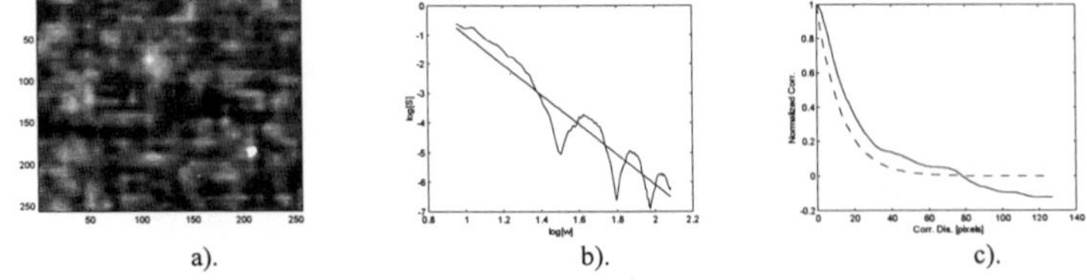

Fig. 3: The Cholesky approach for correlation function of Eq. 17 with σ=10, Correlation length of 13 pixels and obtained α of 5.08. a). The generated image. b). The log-log display of the its power spectrum and its best fit curve. c). The desired (dashed) and the obtained correlation function (solid).

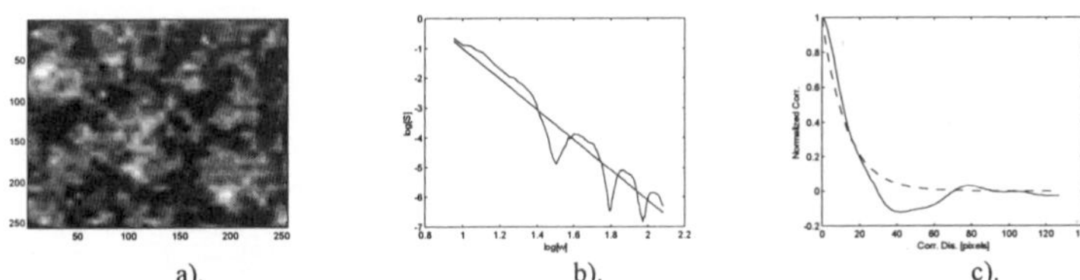

Fig. 4: The Cholesky approach for correlation function of Eq. 18 with σ=10, Correlation length of 13 pixels and obtained α of 5.1. a). The generated image. b). The log-log display of the its power spectrum and its best fit curve. c). The desired (dashed) and the obtained correlation function (solid).

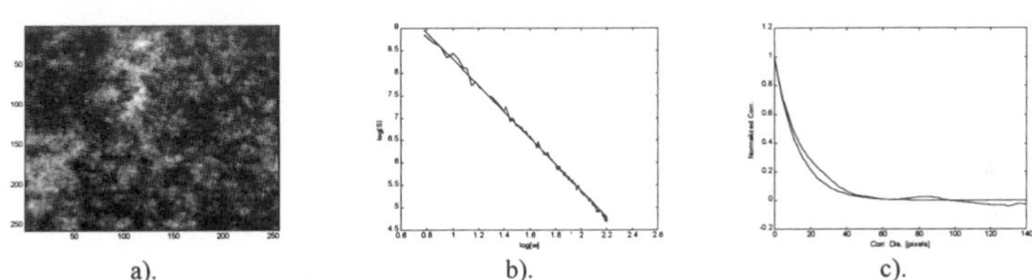

Fig. 5: The spectral analytic approach with σ=10, Correlation length of 13 pixels and α of 3 a). The generated image. b). The log-log display of the its power spectrum and its best fit curve. c). The obtained correlation function.

The best fit for the slope obtained in the log-log plain yields α of 2.92 while the aim of the design was α of 3.

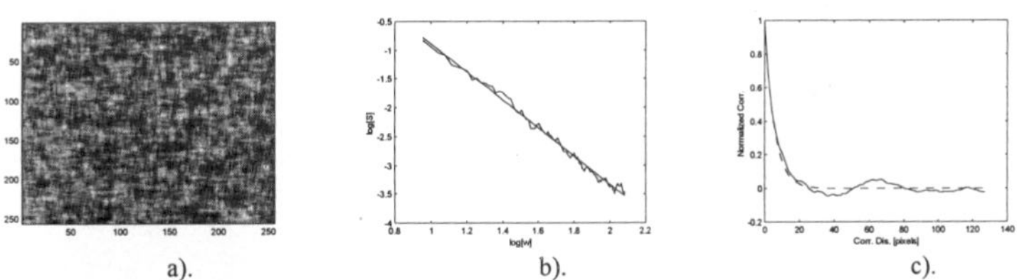

Fig. 6: The difference equation approach with σ=3, Correlation length of 5.5 pixels and obtained α of 2.44. a). The generated image. b). The log-log display of the its power spectrum and its best fit curve. c). The desired (dashed) and the obtained correlation function (solid).

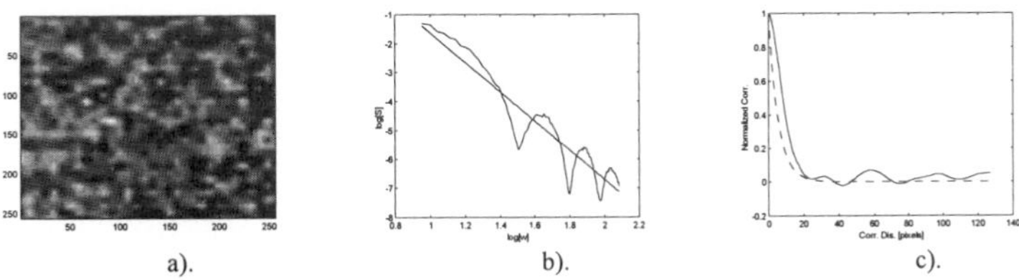

Fig. 7: The Cholesky approach for correlation function of Eq. 17 with σ=3, Correlation length of 5.5 pixels and obtained α of 5.08. a). The generated image. b). The log-log display of the its power spectrum and its best fit curve. c). The desired (dashed) and the obtained correlation function (solid).

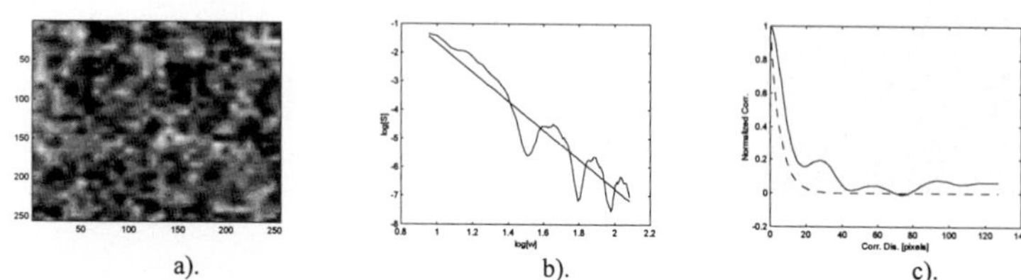

Fig. 8: The Cholesky approach for correlation function of Eq. 18 with σ=3, Correlation length of 5.5 pixels and obtained α of 5.1. a). The generated image. b). The log-log display of the its power spectrum and its best fit curve. c). The desired (dashed) and the obtained correlation function (solid).

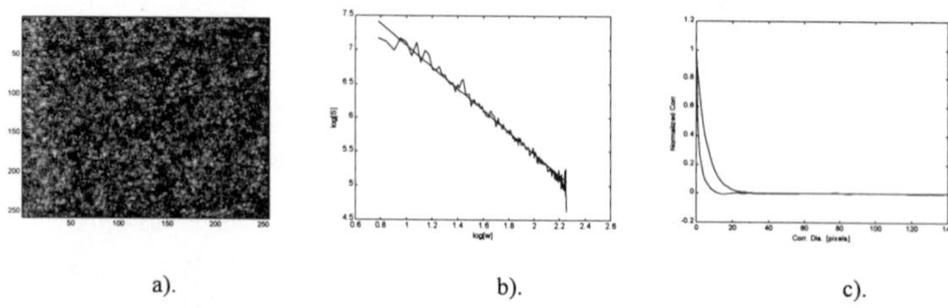

Fig. 9: The spectral analytic approach with σ=3, Correlation length of 5.5 pixels and α of 1.71. a). The generated image. b). The log-log display of the its power spectrum and its best fit curve. c). The obtained correlation function.

The best fit for the slope obtained in the log-log plain yields α of 1.59 while the aim of the design was α of 1.71.

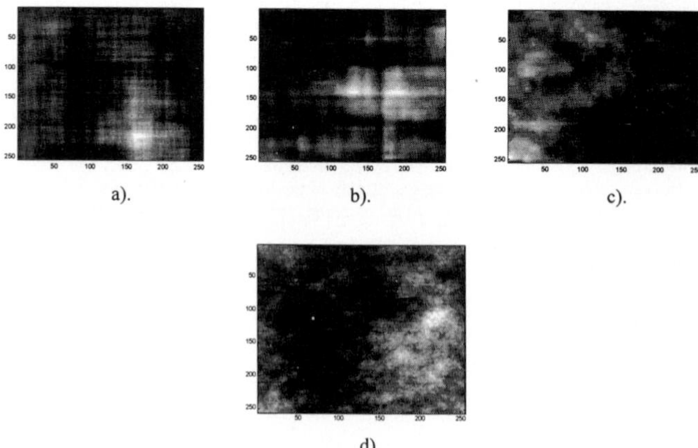

Fig. 10: σ=14, Correlation length of 75 pixels. The generated image using a). Difference equation b). Cholesky technique which implements the correlation function of Eq. 17.c). Cholesky technique which implements the correlation function of Eq. 18 d). the spectral analytic approach for α=3.

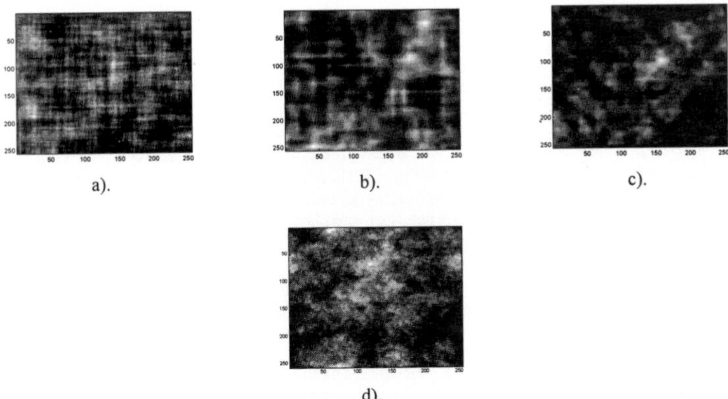

Fig. 11: σ=30.7, Correlation length of 25 pixels. The generated image using a). Difference equation b). Cholesky technique which implements the correlation function of Eq. 17. c). Cholesky technique which implements the correlation function of Eq. 18 d). the spectral analytic approach for α=3.

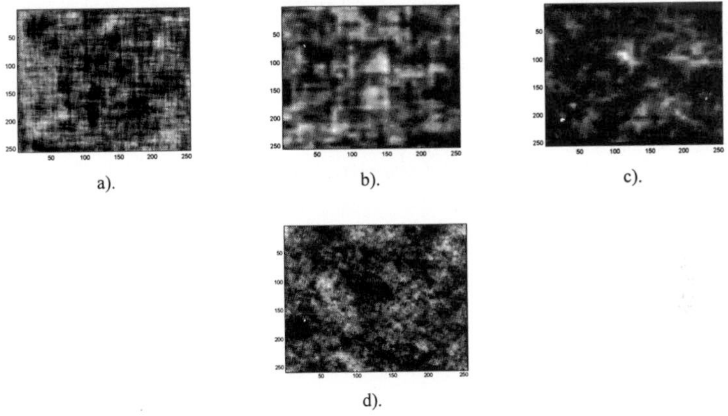

Fig. 12: σ=15, Correlation length of 15.6 pixels. The generated image using a). Difference equation b). Cholesky technique which implements the correlation function of Eq. 17. c). Cholesky technique which implements the correlation function of Eq. 18 d). the spectral analytic approach for α=3.

6. Conclusions

In this paper we have presented three numerical procedures for generation of a mixed Infra Red (IR) backgrounds. The three approaches were the difference equation, the Cholesky factorization and the spectral analytical approach. The outputs of the three approaches were used to generate different auto-correlation distributions corresponding to various scene regions such as cloudy sky, land and sea. The results are compared with each other and with experimental results. Good correspondence is achieved.

References

1. D. B. Reiss, "Spatial signal processing for infrared detection," Proc. SPIE **2235**, 38-51 (1994).
2. J. M. Mooney, J. Silverman and C. E. Caefer, "Point target detection in consecutive frame IR imagery with evolving cloud clutter," Opt. Eng. **34**, 2772-2784 (1995).
3. J. Silverman, J. M. Mooney and C. E. Caefer, "Temporal filters for tracking weak slow point targets in evolving cloud clutter," Infrared Phys. Tech. **37**, 695-710 (1996).
4. M. Burton and C. Benning, "Comparison of imaging infrared detection algorithms," Proceedings of SPIE **302,** Infrared technology for target detection and classification, 26-32 (1981).
5. P. C. Chen and T. Pavlidis, "Segmentation by texture using a co-occurrence matrix and a split and merge algorithm," Computer graphics and image processing **10**, 172-182 (1979).
6. Y. Itakura, S. Tsutsumi and T. Takagi, "Statistical properties of the background noise for the atmospheric windows in the intermediate infrared region," Infrared Physics **14**, 17-29 (1974).
7. J. W. Modestino and R. W. Freis, "Edge detection in noisy images using recursive digital filtering," Computer graphics and image processing **6**, 409-433 (1977).
8. J. M. Modestino, R. W. Fries and A. L. Vickers, "Texture discrimination based upon an assumed stochastic texture model", IEEE Trans. on pattern analysis and machine intelligence **PAMI-3**, 557-580 (1981).
9. N. Ben-Yosef, B. Rahat and G. Feigin, "Simulation of IR images of natural backgrounds," Appl. Opt. **22**, 190-193 (1983).
10. N. Ben-Yosef, K. Wilner, S. Simhony and G. Feigin, "Measurements and analysis of 2-D infrared natural background," Appl. Opt. **24**, 2109-2113 (1985).
11. N. Ben-Yosef, K. Wilner, I. Fuchs, S. Simhony and M. Abitbol, "Natural terrain infrared radiance statistics: Daily variation," Appl. Opt. **24**, 4167-4171 (1985).
12. N. Ben-Yosef, K. Wilner, S. Simhony and M. Abitbol, "Correlation length of natural terrain infrared images: Daily variation," Appl. Opt. **25**, 866-869 (1986).
13. S. Lashansky, N. Ben-Yosef and A. Weitz, "Simulation of ground-based infrared cloudy sky images," Opt. Eng. **32**, 1290-1297 (1993).
14. S. Lashansky, N. Ben-Yosef and A. Weitz, "Spatial analysis of ground-based infrared cloudy sky images," Opt. Eng. **32**, 1272-1280 (1993).
15. I. Wilf and Y. Manor, "Simulation of sea surface images in the infrared," Appl. Opt. **23**, 3174-3180 (1984).

PART 15

MICRO-MECHANICS AND MICRO-OPTICS

Chairperson: *N. Eisenberg, Israel*

EVAPORATING TWO-PHASE FLOW MECHANISM IN MICRO-CHANNELS

Yoav P. Peles, Leonid P. Yarin, Gad Hetsroni
and Zoli Bihari

Department of Mechanical Engineering
Technion-Israel Institute of Technology
Haifa, Israel - 32000
Fax: 972-4-8324533
e-mail: hetsroni@tx.technion.ac.il

Abstract

The basic evaporating two-phase flow mechanism in micro-channels is introduced. This flow is accompanied by a very high heat transfer coefficient and can remove very high heat fluxes at low temperatures, thus, making it attractive for micro device cooling applications. An experimental micro-channels chip was fabricated by means of a photolithography process, in order to study the flow mechanism involved in such flow. The results show that there is periodic behavior in evaporating two-phase flow in micro-channels, with a few harmonics.

Keywords: micro-channel, two-phase flow, unsteady.

1. Introduction

The new development in MEMS technology allows increasing complex micro devices fabrication. Some of the new devices incorporate high heat loads, which should be removed sustaining low temperature. One of the promising methods of cooling is a single or two-phase convective flow through micro-channels. The developments in the semiconductor technology have made possible extraordinary advantages in devices construction and miniaturization. The latter allows for the manufacturing of micro heat exchangers consisting of multiple minute flow conduits with hydraulic diameter ranging from 10 to 1000 μm. These micro heat exchangers are characterized by extremely high surface area per unit volume, low thermal resistance, low mass, volume and inventory of working fluid.

The researches of flow in micro-channels may be divided into two groups related to a single-phase flow or evaporative two-phase flow. Single-phase flow was extensively investigated following Tuckermann and Pease[1]. Two-phase flow is much less understood.

Hsu[2] developed a semi–theoretical model that provides considerable insight into the effects of a non-uniform liquid superheat resulting from transient conduction in liquid during the bubble growth and release process. The model shows that there is a finite range of active cavity sizes on the heating surface, which depends on sub-cooling, pressure, heat flux and physical properties of the flow. For water at atmospheric pressure, temperature difference between the wall and saturation temperatures is 15.6° C, thermal boundary layer thickness of 76 μm near the nucleation site, and heat flux of 30 W/cm^2 the active nucleation sites are those having cavity opening radius of 2.5 – 15 μm. The thermal boundary layer far from the nucleation site was measured by optical means by Yamagata et al.[3] .It was found that the thickness of the thermal boundary layer is of the order of 250 μm. Thus the boundary layers of an adjacent heated micro-channel walls converge in a channel with an hydraulic diameter of the order of 100 μm, causing a higher bulk temperature and therefore a wider range of active nucleate sites. The latter results in higher bubble departure frequency. Furthermore, the bubble departure diameter in channels of regular size is of the order of 1 mm. That shows that the bubble initiation process in micro-channels is strongly controlled by the surrounding walls.

Vapor bubble growth in the inertia-controlled regime, as well as thermally controlled growth has been long studied in many researches. Forster and Zuber[4], Plessent and Zwick[5] and others analyzed the mechanism of momentum exchange during bubble growth. Mikic and Rohsenow[6] studied heat transfer controlling bubble growth.

Peng et al.[7] investigated the flow boiling through micro-channels with a cross-section of 0.6×0.7 mm. They observed that no partial nucleate boiling existed and that the velocity and liquid sub-cooling have no obvious effect on the flow nucleate boiling. Peng and his co-workers[8-10] conducted additional experimental investigations on flow boiling in micro channels with rectangular cross-sections ranging from $0.1 \times 0.3\text{mm}$ to $0.6 \times 0.7\text{mm}$. Peng et al.[11] suggested a dimensionless parameter for nucleate boiling in micro-channels, which agree well with the experimental data

Bowers and Mudawar[12] performed an experimental study of pressure drop and critical heat flux (CHF) in mini-channels with circular cross-section (d = 2.54mm and d = 510μm) using R-113. A CHF correlation proposed by Katto[13] was presented as $q_{CHF} / Gh_f = 0.16We^{-0.19}(L/D)^{-0.54}$ (q_{CHF} is the critical heat flux, G the mass velocity, L the channel length, D the channel hydraulic diameter, h_f the latent heat of evaporation, We the Weber number defined as $We=\sigma/(\rho_l u^2 L)$, were σ the surface tension [N/m], ρ_l the liquid density [kg/m^3], u the inlet velocity [m/s]). As the channel size decreases the capillary forces acting in the longitudinal direction become significant. Therefore, it would be expected that the hydraulic diameter would be a better choice for the Weber number than the channel length. The experiment yields CHF values about 200 W/cm^2. However, the pressure dorp for the mini-channels were less than 0.01 bar compared to 0.23 bar for the micro channels.

Peles et al.[14] developed a one-dimensional model for a two-phase laminar flow in a heated capillary excited by liquid evaporation from the interface. This model takes into account the multistage character of the process, as well as the effect of capillary, friction and gravity forces on flow development. Khrustalev and Faghri[15] considered a two-dimensional model of steady state evaporation of liquid-vapor meniscus in a heated capillary slot.

The objective of this work is to investigate experimentally the evaporation two-phase flow mechanism in parallel triangular micro-channels. The experimental model was fabricated by means of a photolithography process, and an unsteady flow was observed.

2. Experimental Apparatus.

The experimental set used to study the flow in a heated micro-channel is shown in figure 1. Previous to the experiment the water had passed through a $1/2$ μm cartridge filter, and was used to fill up the reservoir. Water was circulated through the flow loop by a peristaltic pump having a frequency of 6 Hz. As the fluid leaves the reservoir, it passes through a silicon rubber tube, enters a flow regulator and then flows into the test module and leaves through the second outlet silicon rubber tube. A special Teflon cartridge was manufactured to hold the module in place and to move it in the x-y plane. The temperature and pressure were measured at both the inlet and exit of the test section in the tubing upstream and downstream of the test module using a 0.3mm copper Constantine thermocouple and silicon pressure gauges, respectively. The thermocouples and pressure gauge output voltages were recorded by a multiscan 1200 data acquisition module, which in turn was connected to a PC via a data acquisition card. The input voltage was then translated to degrees Celsius and Kilo Pascal's, by a calibration process prior to the test runs.

A DC power supply with an operating range of 0 to 60 volts was used to supply power to the chip aluminum resistor. The Teflon module cartridge was placed underneath a high speed CCD camera mounted on a microscope, which delivered the boiling process occurring in the micro-channels to the PC and a video recorder. At the same time an IR camera measured the resistor temperature on the other side of the chip. Special software was used to analyze the IR results and to compute the average and maximum temperature on the chip.

The test module is shown in figure 2. It consists of three distinct parts. The first and most important is a 525 μm micro-channel silicon substrate bonded to the 500 μm micro-channel cover Pyrex, and the last is the PEEK flow diverger and collector. The micro-channel silicon substrate was fabricated by photolithography process on one side and consists of 16 mm long triangular micro-channels (figure 3) having hydraulic diameter ranging from 50 μm to 200 μm . On the other side an $1\times1\ cm^2$ aluminum resistor was evaporated (figure 4), forming three distinct channel regions: inlet adiabatic, heat flux dissipation and outlet adiabatic. The channels were sealed by a 7400 Pyrex substrate to form the chip test module (figure 5).

While the PEEK collector/diverger was fabricated by plain molding process, the micro-channels substrate fabrication is quite complicated and is achieved by a multistage process. The following main stages were used in the process: i) Double side oxidation of a 525 μm <100>-Silicon substrate to $1000^\circ A$, ii) Single side $1200^\circ A$ Silicon nitride deposition, iii) Silicon nitride channels template opening by RIE (Reactive Ion Etching), iv) Channels template oxidation BOE (Buffer oxide etching), v) Silicon etching by TMAH (Tetramethyl Ammonium Hydroxide), vi) $4000^\circ A$ double side oxidation, vii) Silicon nitride etching by RIE, viii). $1000^\circ A$ oxidation layer etching, ix) Silicon/Pyrex anodic bonding, x) Aluminum thin film evaporation, xi) Aluminum template etching xii) Wafer sawing.

Once the micro-channels were fabricated the flow collector and diverger were bonded to the substrate by epoxy glue, and the resistor was wired by silver-based epoxy electric conductor and painted black by 1 μm thin paint for the IR camera.

The experiments were conducted on a heat exchanger described above, having 17 triangular micro-channels with hydraulic diameter of 157 μm and mass flow rates of 5.7 ml/min. The applied heat flux ranged from 0 W/cm^2 up to 35 W/cm^2. The characteristic value of the dimensionless groups at which the observations were carried out are: $Eu\sim10^6,\ Pe_2\sim7\cdot10^2, \vartheta_2\sim0.3,\ Ja\sim1.8$, were Eu is the Euler number defined as $Eu=P_{20}/(\rho_{20}u^2_{20})$, Pe_2 the liquid Peclet number $Pe_2=u_{20}L/\alpha_2$, ϑ_2 the dimensionless heat flux $\vartheta_2=qL/(\rho_{20}u_{20}Cp_{20}T_{20})$, and Ja the Jacob number $Ja=h_{lv}/(Cp_2T_{20})$, q the wall heat flux, L the channel length, ρ_{20} the liquid density, u_{20} the inlet liquid velocity, Cp_{20} the inlet liquid heat capacity, T_{20} the inlet liquid temperature, α_2 the inlet liquid thermal diffusivity and P_{20} the inlet liquid pressure, h_{lv} the latent heat of evaporation. That allows us to study the flow in a heated micro-channel outside the domain of the steady flows, as predicated by Peles et al.[16] for $\vartheta_2<Ja$.

3. Results and Discussion

The development of the two-phase evaporation flow in a heated capillary at different Peclet number is illustrated figure 6a,b. It shows that different mechanisms of two-phase flow formation may occur depending on the value of Pe_2. At small Pe_2 the fine bubbles formation (on the micro-channel wall) plays a dominant role. Grow of these bubble leads to a blockage of the micro-channel, to a sharp change of the hydraulic resistance and ultimately to an unstable gas/liquid flow with a fast void explosion inward the channel. This effect is negligible at large Peclet number. In this case liquid evaporation is accompanied by formation of the two-phase flow with distinct vapor/liquid evaporation front dividing the domains of the liquid and the vapor. Depending on the values of the governing parameters such flow may be steady or unsteady. A specific flow at Pe_2 $\approx$500 and $\vartheta_2\simeq0.3$ (the condition correspond the boundary of the non-steady flow domains) are shown in figure 7. A number of successive photos of the flow are presented. They show that the position of the evaporation front changes during the observation period. Thus at the given values of parameters the flow in a heated micro-channel is non-steady. The latter agree with the estimations of the boundary of steady state given by Peles et al.[16].

In the initial stage of the vapor/liquid front the penetration velocity into the channel is very high. As the front penetrates into the channel its velocity reduces until a halting point. The penetration velocity is on the order of 1 m/s, which corresponds to 10 ms penetration time. The time needed for the liquid resident in the channel to evaporate, assuming all heat is absorbed for evaporation, is approximately 1s. It follows that the front penetration process is accommodated by liquid flow toward the inlet. The evacuation time of the front from the channel toward the outlet is much greater than the front penetration time. The total resident time of the front in the channel is lower than 0.2 [s]. Thus, the process frequency resembles the pump frequency. An insight into the inlet pressure reveals that the pump frequency dominates the process frequency. Figure 8 shows the inlet pressure as a function of time for flow in 13 micro-channels with hydraulic diameter of 207 μm , mass flux of 5.76 g/min and heat load of 35 w/cm^2. It can be observed that the amplitude of the inlet pressure fluctuation reaches 50%. Observation of the frequency plane (figure 9) reveals that the main frequency is 6 [Hz] and there are two minor frequencies of 20[z] and 40 [Hz]. The lower frequency of 6 [Hz] is the pump frequency and the 20 [Hz] is the halting front frequency. Thus it can be concluded that altering the pump frequency can control the process frequency.

The hydraulic diameter of the micro-channel is smaller than the bubble departure diameter, leading to a considerably distinct evaporation mechanism in the two-phase flow, as compared to larger sized channels. Two distinct phase domains, one for the liquid and another for the vapor, were observed, with a very short (of the order of the hydraulic diameter) section of two-phase mixture between them. This implies that the outlet vapor mass quality for a steady flow can take only values of zero (a single-phase liquid flow) or larger than unity (superheated vapor). The energy observed for the flow with zero vapor outlet mass quality is much lower than the energy observed for vapor mass quality larger than unity. Thus an energy gap is formed between those energy levels, for which steady evaporation two-phase flow exists does not exist. If the applied heat flux is sufficient to initiate evaporation at quality lower than unity an unsteady flow is expected with an outlet phase flow fluctuation corresponding to some time-average mass quality lower than one, as was showed to be the case in the present experiment.

Conclusion

In the present paper an experimental investigation was conducted in order to investigate evaporation two-phase flow in triangular micro-channels. A different flow mechanism, compared to channels of conventional size, was observed, namely a formation of two-phase flow with a distinct vapor/liquid front dividing the domains of the liquid and the vapor. This causes an unsteady flow with a few harmonics, which causes a fluctuation in the surrounding temperature. The pump frequency dominates the process frequency, and thus the temperature fluctuation frequency and amplitude can be controlled by the frequency of the pump.

Two-phase flow in micro-channels with an outlet quality lower than unity is inherently unstable, since the evaporation flow has two distinct flow domains. The outlet flow can exhibit only single phase either liquid or vapor. A quality lower than one means alternating outlet phase with an time-average quality lower than one.

Acknowledgement

This work was supported by the Ministry of Science and the Arts (State of Israel). The Israel Council supports L.P. Yarin for Higher Education.

Reference

1. D.B., Tukermann, R.F.W. Pease, "High performance heat sink for VLSI", *IEEE Electron Device Letters*, Vol. EDI-2, No. 5, pp. 126-129, 1981.
2. Y.Y., Hsu, "On the size Range of Active Nucleating Cavities on a Heating Surface", *Journal of Heat Transfer*, 184, 1. 207-213, 1962,.
3. Hironai, F., Yamagata, K., Mishikawa, and H., Matsouka, "Nucleate boiling of Water on the Horizontal Heating surface", Memoirs of the Faculty of Engineering, Kyushu University, Japan, Vol. 15, p. 98, 1955.
4. H.K., Forester, and N., Zuber, "dynamic of vapor bubbles and boiling heat transfer", *AIChE J.*, Vol.1, pp. 531-535, 1955.
5. M.S. Plesset and S.A., Zwick, "The growth of vapor bubbles in superheated liquids", *J. Appl. Phys.*, vol. 25, pp. 493-500, 1954.
6. B.B., Mikic, and W.M., Rohsenow, "Bubble growth rates in non-uniform temperature filed", *Prog. Heat and Mass Transfer*. Vol. 2, pp. 283-292, 1969.
7. X.F. Peng, and B.X., Wang, "Forced convection and flow boiling heat transfer for liquid flowing through micro-channels", *Int. J. Heat Mass Transfer*, 36(14), 3421-3427, 1993.
8. X.F. Peng, and B.X., Wang, "Cooling characteristics with micro-channeled structures", *Journal of enhanced Heat Transfer*, 1(4), 315-326, 1994a.
9. X.F., Peng, B.X., Wang, G.P. Peterson, and H.B.,Ma, "Experimental investigation of heat transfer in flat plates with rectangular micro-channels", *Int. J. Heat Mass Transfer*, 37(1), 127-137, 1994b.
10. X.F., Peng, G.P. Peterson, and B.X., Wang, "Flow boiling of binary mixtures in Micro-channels plates", *Int. J. Heat Mass Transfer*, 39, 1257-1263, 1996.
11. X.F., Peng, H.Y. Hu, and B.X., Wang, "Boiling Nucleation During Liquid flow in Micro-channels", *Int. J. Heat Mass Transfer*, 41, No. 1, 101-106, 1998.
12. M.B., Bowers, I., Mudawar, "High flux boiling in low flow rate, low pressure drop mini-channel and micro-channel heat sink", *Int. J. Heat Mass Transfer*, 37(2), 321-332, 1994.
13. Y., Katto, "a generalized correlation for critical heat flux for the forced convection boiling in vertical uniformly heated round tubes", *Int. J. Heat Mass Transfer*, 21, 1527-1542, 1978.
14. Y.P., Peles, L.P., Yarin, G., Hetsroni, "Heat Transfer of Two-phase Flow in a Heated Capillary", 11th International Heat Transfer Conference, Kyongju, Korea, 1998.
15. D., Khrustalev, A. Faghri., "Fluid flow effect in evaporation from liquid-vapor Meniscus", *J. Heat Transfer*. Vol. 118, pp. 725-730, 1995.
16. Y.P., Peles, L.P., Yarin, G., Hetsroni , "Steady and unsteady flow in a heated capillary with distinct vapor/liquid interface", submitted to *J. of multi-phase flow*, 1999.

Figure captions

Fig. 1. Experimental set up.

Fig. 2. The test module.

Fig. 3. The triangular micro-channels and unpainted resistor.

Fig. 4. The $1 \times 1\ \mathrm{cm}^2$ aluminum resistor.

Fig. 5. The micro-channel silicon substrate bonded to the 500 $\mu\mathrm{m}$ micro-channel cover Pyrex.

Fig. 6. Flow types in a heated capillary.
One) bubble formation
Two) liquid/vapor front

Fig. 7. Unsteady flow at heated capillary at $\vartheta_2 = 0.3$ $\mathrm{Ja} = 1.83$ $\mathrm{Pe}_2 = 598$.

Fig. 8. The inlet pressure as a function of time for flow in 13 micro-channels with hydraulic diameter of 207 $\mu\mathrm{m}$, mass flux of 5.76 g/min and heat load of 35 w/cm^2.

Fig. 9. The frequency plane.

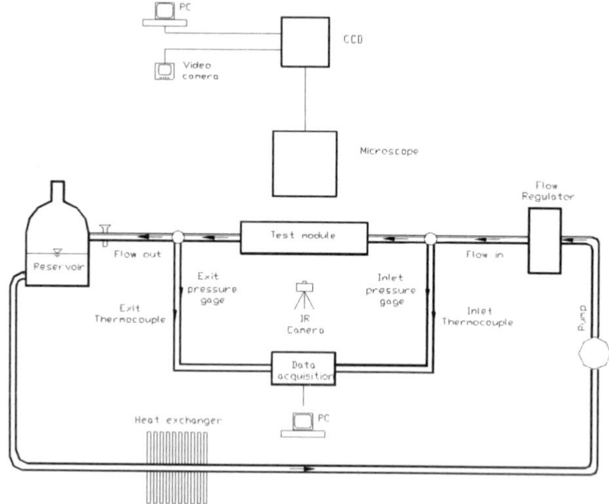

Fig. 1. Experimental set up.

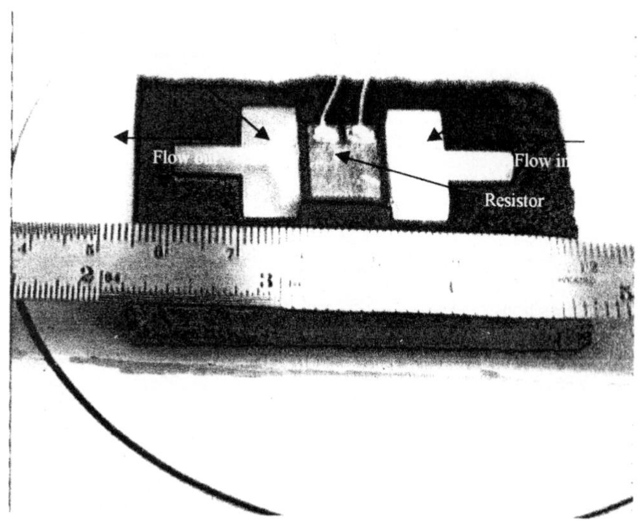

Fig. 2. The test module.

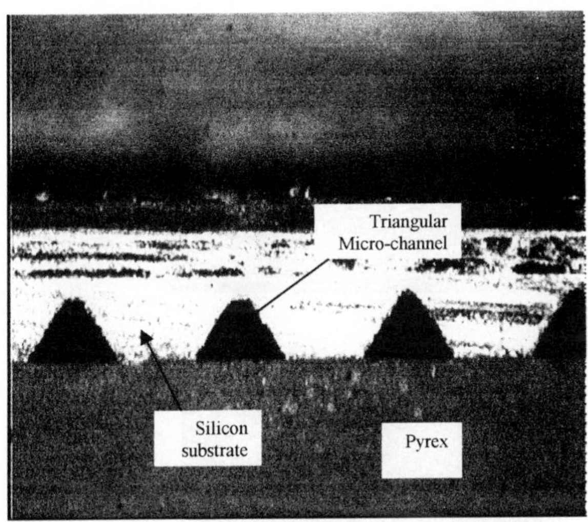

Fig.3. The triangular micro-channels and unpainted resistor.

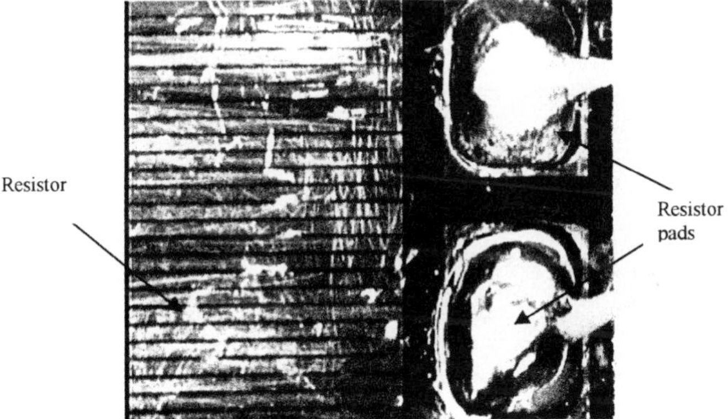

Fig. 4. The 1×1 cm^2 aluminum resistor.

The bubble initiation

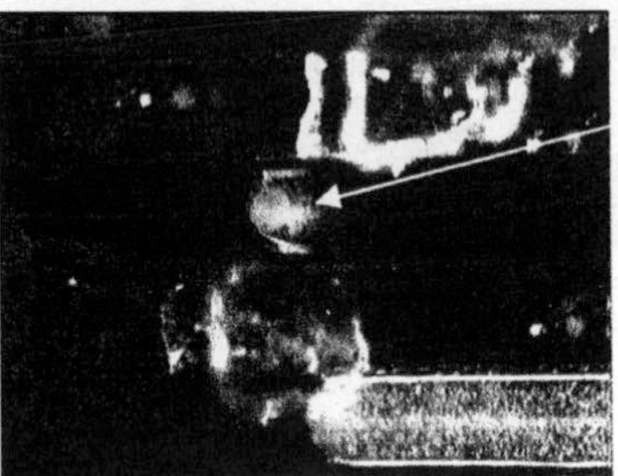

The bubble blocks the channel

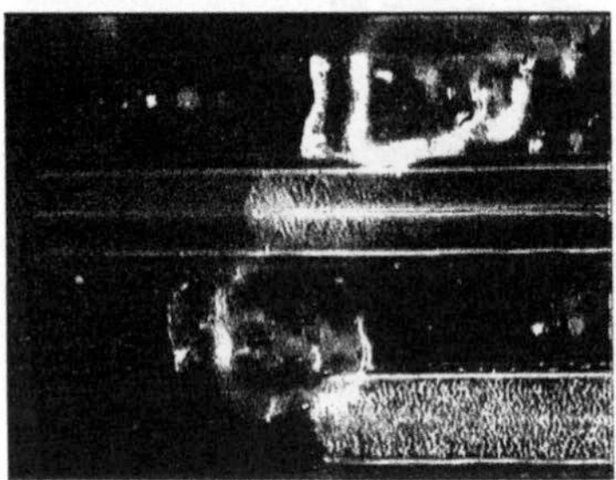

The fast void explosion

Fig. 6a. Flow types in a heated capillary-bubble formation.

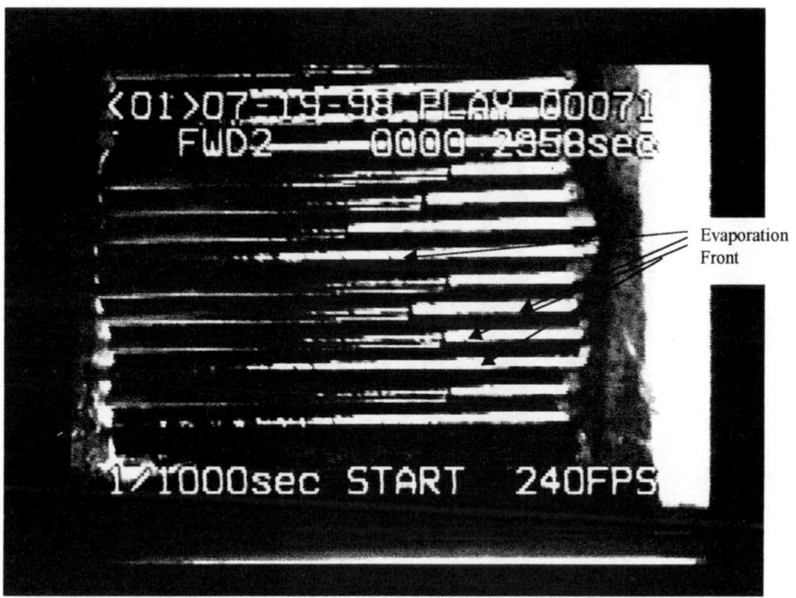

Fig. 6b. Flow types in a heated capillary- liquid/vapor front.

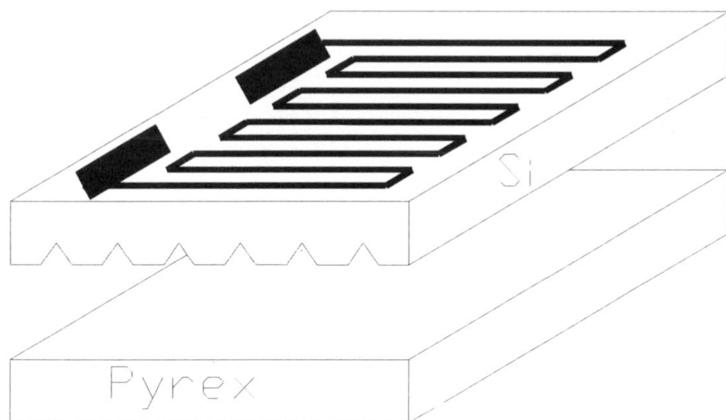

;. 5. The micro-channel silicon substrate bonded to the 500 μm micro-channel cover Pyrex.

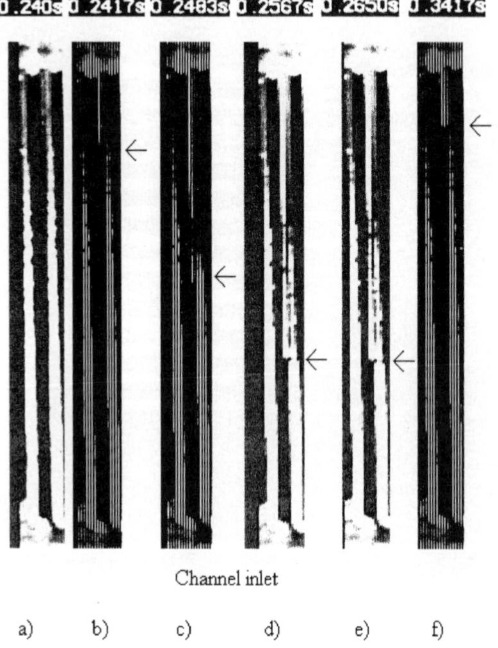

Fig. 7. Unsteady flow at heated capillary at $\vartheta_2 = 0.3$ $\mathrm{Ja} = 1.83$ $\mathrm{Pe}_2 = 598$.

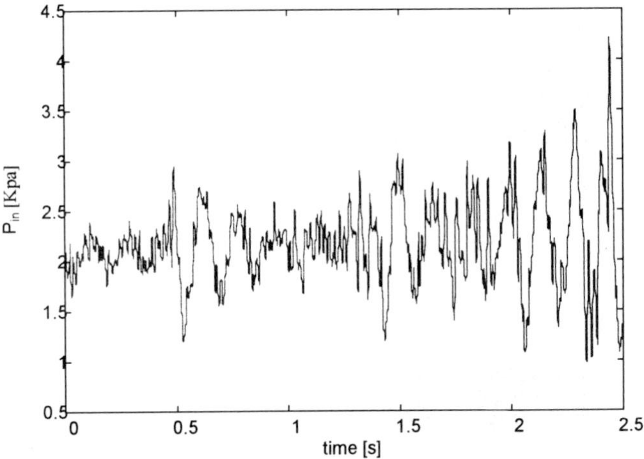

Fig. 8. The inlet pressure as a function of time for flow in 13 micro-channels with hydraulic diameter of 207 μm , mass flux of 5.76 g/min and heat load of 35 w/cm^2.

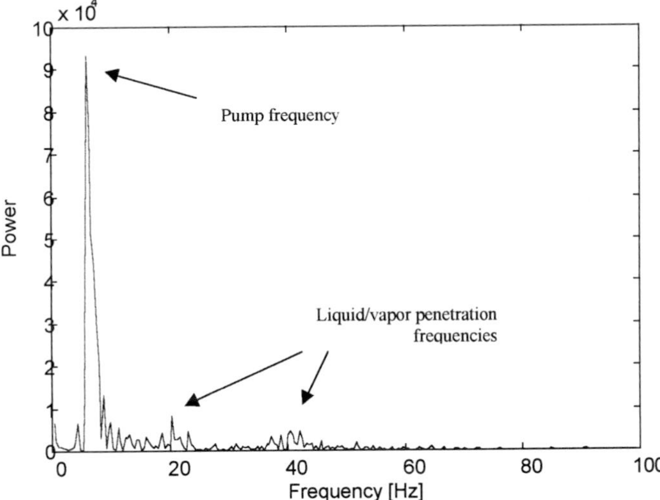

Fig. 9. The frequency plane.

HIGH POWER LASER DIODE FAST AXIS COLLIMATION USING MICROLENSES

Ophir Eyal and Jonathan Gelberg

ELOP Electro-optics Industries
P.O.Box 1165, Rehovot 76111, Israel
Phone: 972-8-9386912, Fax: 972-8-9386045
e-mail: elop11@netvision.net.il

Abstract

Radiation from bars and stacks of laser diodes diverges at large angles of approximately 40° in the fast axis. The use of a single microlens enables the reduction of the fast axis divergence to the order of 1°. Efficient light collection, high transmission and low optical aberrations are required for the collimation so as to preserve good beam quality and high brightness, characteristic of the fast axis. A range of different microlenses, such as graded indexed cylindrical and aspheric lenses, are currently available for collimation of laser diode radiation. This paper will report some results obtained from the initial work, which was carried out in Elop to characterize and compare microlenses of different types and sizes using criteria such as beam divergence, beam profile and sensitivity to adjustment.

1. Introduction

Microlenses are used for the collimation of laser diodes as a first step in optical systems such as range finders[1], high power illuminators [2], fiber coupled laser diodes [3] and in side pumped laser modules [4] and end pumped modules [5].

The divergence and energy distribution obtained by using a microlens is influenced by several factors. Concerning the microlens, there are three contributing factors that are convoluted with each other to achieve the resultant output: the aberrations, diffraction and the gaussian nature of the source. In addition, laser diode manufacturing induces positioning errors of the emitters, which in turn affects the collimation. The major positioning errors are related to array planarity (whereby not every emitter is precisely located at the focal plane of the lens) and bar curvature (also known as the smile effect). Both planarity and smile will increase the beam divergence. Therefore, it is very important to specify tight tolerances on the laser diode packaging and also to find collimating lenses that are less sensitive to packaging errors.

There are several types of microlenses, which can be divided into two main groups of aspheric lenses and gradient index lenses. The graded-index, GRIN, lenses are drawn from gradient index preform between 60 microns and 3 millimeters in diameter. These lenses have the shape of a fiber or a rod and, when illuminated from the side, act as perfect cylindrical lenses with a NA of 0.5. Aspheric lenses are drawn from a preform or manufactured by ultraprecision grinding and are characterised usually by a NA of 0.8.

2. Experimental setup

A schematic of the experimental test setup is shown in Fig.1. A CW diode bar (OPTO POWER CORPORATION model OPC-A010) was used. The divergence of the laser was determined by the knife-edge test. In the fast axis, 86% of the energy of the gaussian beam was contained within 38° of the divergence.

Two CCD cameras, the top and the side camera, equiped with magnifier lenses, were used to align the microlens with respect to the laser diode. A third CCD camera (COHU 4810) and a beam profiler [6] (Spiricon LBA 300 PC) were used to view and analyze the far field profile. The distance, z, between the laser diode and the screen was 166 cm. This distance is longer than the Rayleigh range, Z_R, for most of the lenses (usually the far field is defined at a distance of 5 times the Rayleigh range or at the focal plan of a lens) and we can assume that the observed radiation is representative of the far field. This set up gives real time displays of beam profiles, thereby shortening the lens alignment procedure significantly. Since measurements with a CCD camera may suffer from different errors such as variation in the baseline offset or zero signal level, the accuracy of measurement was further verified with the well known knife edge method. The error of the measurement depended on the specific lens that was measured, but was less than 3 mrad in each case. The test setup has been used to compare the performances of several microlenses. The lens translator included 5 degrees of freedom. The lens could be translated along the optical axis, the perpendicular y-axis and also along the x-axis (which is actually redundant in this application, as the lens is, in principle, consistent in optical characteristics along the x-axis). The lens can also be orientated with two angular degrees of freedom. It can be rotated about the optical z-axis and the y-axis. Rotation about the x-axis, that is, the center of the lens, is not possible with this set up, however this would have no effect with regards to the GRIN lenses as they have circular symmetry about the center of the lens. As for its effect on the aspheric lens, improper orientation of this degree of freedom would result in the translation of the entire energy distribution and not in its dispersion. The lens translator was adjusted so as to obtain the maximum on axis intensity on the screen. This involves operating the laser and observing the focused beam during the alignment process.

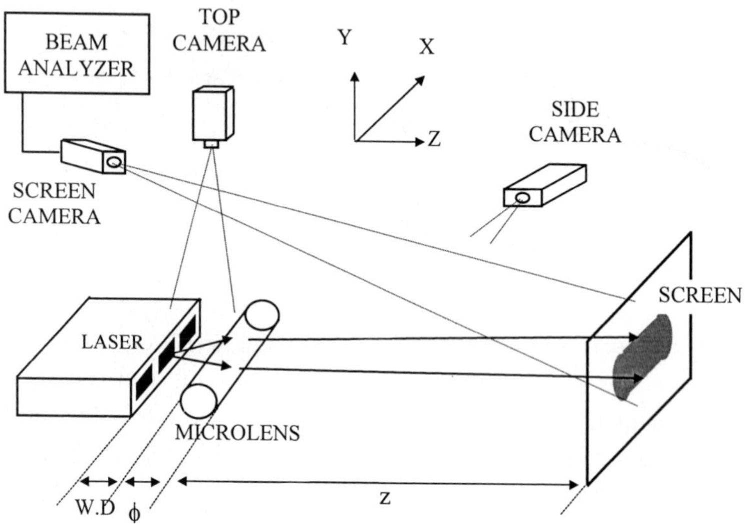

Figure 1: schematic of the experimental test setup

3. Results

Fig. 2 shows profiles obtained with Doric's GRIN lens [7] of 250 microns in diameter. A divergence (90 % of the energy) of 12 mrad was obtained. A 5 microns displacement in the positive z axis direction reduces the beam quality and the divergence measured 25 mrad. This is shown in Fig.3. Fig.4 shows profiles obtained with a Limo aspheric lens, the divergence is 13 mrad. This lens has a larger focal length and is therefore less sensitive to changes in alignment. The profile obtained after a 30 microns displacement in the positive z axis direction is shown in Fig.5. The divergence is 33 mrad . The graph in Fig. 6 summarizes the divergences (FWHM) results that were obtained with several microlenses and includes sensitivities to adjustment. We should mention that quite different results are obtained when the divergence is defined as the angle that contains 90 % of the energy. This is due to the energy in the side lobes of the beam.

For the GRIN lenses of Doric, lens diameter = f/0.65. For the Limo lenses the dimensions are: 1.5 x 1.5 x 10 mm^3 ,NA=0.85 and the index of refraction, n =1.82. BlueSky aspheric lenses have center thickness of 407 microns, width of 380 microns, NA= 0.7 and the index of refraction is, n=1.61.

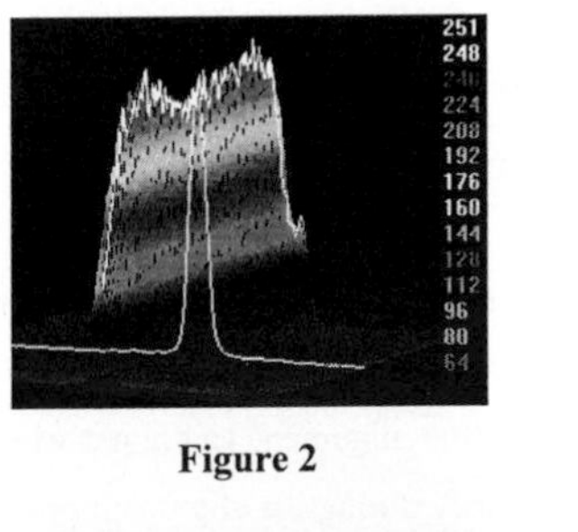

Figure 2

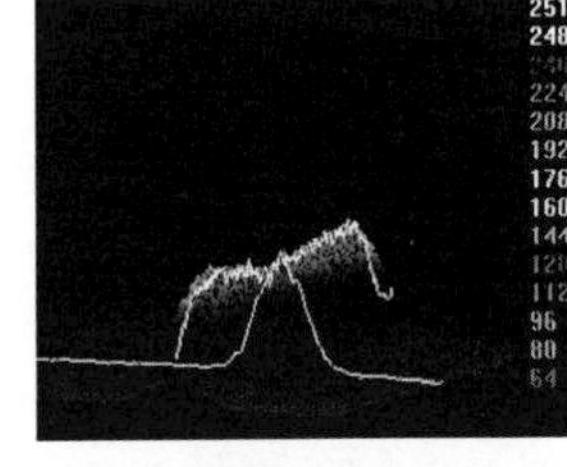

Figure 3

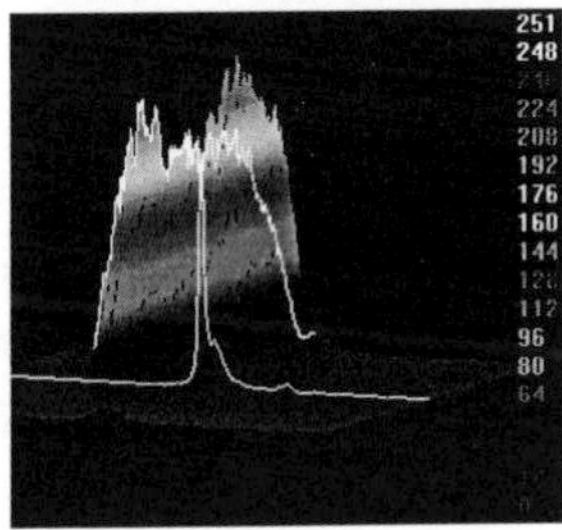

Figure 4

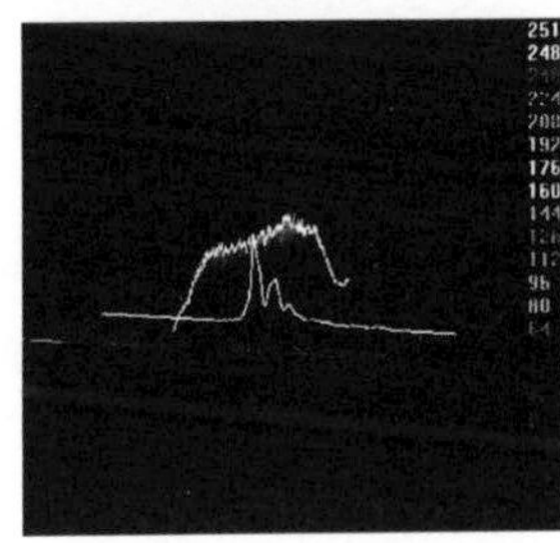

Figure 5

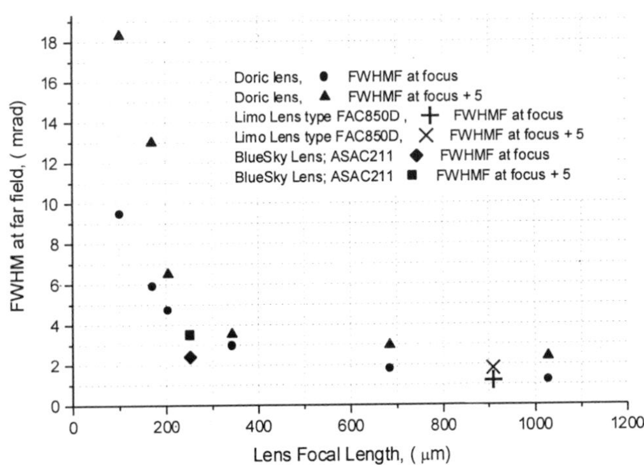

Figure 6

4. Summary

This work presents initial work that was carried out in Elop to characterize and compare microlenses of different types and sizes using criteria such as beam divergence, beam profile and sensitivity to adjustment.

The use of a single microlens enables the reduction of the fast axis divergence to the order of 1°. However, the optical system requires critical alignment and positioning to micron tolerances, especially when thin and short focal length microlenses are used. An active alignment technique which involves operating the laser and observing the focused beam simultaneously during the alignment process must be used to align the microlens to the laser diode when short focal length microlenses are used and when low divergence is required. Passive alignment technique can be used when long focal length microlenses are used and when only moderate divergence is required, as for this case, the system is less sensitive and, as such, the iterative procedure of alignment is not required to the same degree. In order to make active alignment practical and cost effective, automation is necessary. Such a system needs high precision motion controllers, machine vision and computer automations for performing alignments of the microlens to the laser diode.

GRIN lenses are cost effective have cylindrical symmetry and are available in very small diameters and can therefore be used even for collimation of laser diode stacks with very small pitch. GRIN lenses give good results, although there are slight differences in the performances of lenses of the same type. This is probably due to differences in the preforms and the drawing procedure. Aspheric lenses, such as the Limo lenses, are more expensive, but each lens comes with a data sheet which describes the performance of the lens as measured by the manufacturer. For better than 1° fast axis collimation two or more microlenses should be used in a telescope configuration.

A computer simulation, which is able to simulate the optical behaviour of microlenses of different types, has now been written. Such a tool will be extremely useful in the design of future systems incorporating microlenses and even in the design of microlenses themselves.

This work was done as part of the "LESHED" consortium supported by the Israeli Ministry of Trade and Industry. The authors would like to thank Mark Shinmerman and David Chuk for their technical help in this work.

5. References

1. A. Lee, " High powered laser diode transmitter with integrated microlens for fast axis collimation, " SPIE Vol. 3289, p. 33, 1998.

2. Andre Parent, " Laser diode array collimation for scene illumination purposes, " SPIE Vol. 3267, p.84, 1998.

3. H. G. Treusch, " Fiber coupling technique for high power diode laser arrays, " SPIE Vol. 3267, p. 98, 1998.

4. B. Labranche, " Side pumped eyesafe laser, "OSA Proceedings on advanced solid state lasers, Vol. 20, p. 151, 1994.

5. G. Feugnet," High efficiency TEM_{00} Nd: YVO_4 laser longitudinally pumped by a high power array, " Opt. Lett. Vol.20, No.2, p. 157-159, 1995.

6. Method and apparatus for improving dimensional measurements made with video cameras, U.S. pat. no. 5,440,338.

7. Luneburg lens with a graded index core and homogeneous cladding, U.S. pat. no. 5,638,214.

PART 16

SPECIAL SESSION ON START-UPS

Chairperson: *Y. Amitay, Israel*

Annals of the Israel Physical Society, v. 14

ELECTRO-OPTICS and MICROELECTRONICS
Eds: Raphael LAVI and Ehud AZULAY

NEW POLYMER/ THERMOPLASTIC GLASSES COMPOSITIONS AND THEIR PROPERTIES

Edward Bormashenko[1], Semion Sutovsky[2], Roman Pogreb[1] and A.Voronel[3]

[1]*The College of Judea and Samaria, The Research Institute, Ariel*
[2]*Polytris LTD*
[3]*Tel-Aviv University*

Abstract

Physical properties of composite material based on low-density polyethylene and thermoplastic chalcogenide glasses were investigated. Compositions, which contain up to 30% of chalcogenide glass dispersed in a polymer matrix, were obtained by single- and twin-screw extrusion processes. It was proved that the controlled structure of a composite can be derived. Highly oriented structures involving chalcogenide glass fibers immersed in the polymer matrix were obtained. Perfect spherical particles of glass of 1 μm diameter were dispersed in the polymer matrix with a high homogeneity as well. Optical properties of new composite were studied.

Introduction

Chalcogenide glasses were under intensive study recently because of their unique optical properties: high transparency in the middle and far IR-bands, extremely high refraction index, non-linear optical effects, were revealed [1]. At the same time chalcogenide glasses suffer from essential disadvantages which limit their feasibility. Chalcogenides are toxic, expensive, fragile materials. The authors developed a new non-toxic, inexpensive composite that allows production of flexible films of 20-100μm thickness [2]. The developed process permits variation the structure of the obtained composite: from highly oriented structures up to spherical particles of 1μm diameter dispersed in a polymer matrix.

Experiment and discussion

We used chalcogenide glass $Se_{57}I_{20}As_{18}Te_{3}Sb_{2}$ and polyethylene as basic components of our composite material. Properties of chalcogenide glass were studied thoroughly at the first stage of our investigation. The IR-spectrum of glass was investigated by FTIR spectrometer (Nicolet, model 5PC). The rise of transparency of glass as wavelength increases can be recognized from the spectral data. At the 10.6 μm the sample of 2 mm thickness permits the passage of 60% of IR radiation. The most abundant chalcogenide glasses (As_2S_3, etc produced by Amorphous Materials Inc.) are more transparent in the middle and far IR-bands, on the other hand they have relatively high softening temperatures and do not permit obtaining the controlled homogeneity. The chalcogenide glass under discussion does not demonstrate essential dispersion in the middle and far IR (3-15 μm) band and has the refraction index 2.4.

We studied the thermal capacity, conductivity and mechanical properties (loss modulus, storage modulus) and temperature dependencies. It was established that in the temperature range of 35-40 °C physical properties of chalcogenide glass change dramatically and as a result it was concluded that the glassification point lies in cited above temperature range. We established that the viscosity of cited above chalcogenide glass is close to the viscosity of polyethylene in the 150-250 °C temperature range. This fact gave hope that the effective mixing of polymer glass and polymer is possible.

We tried different methods of mixing the glass under discussion with the low, middle and high-density polyethylenes. Dispersive mixing by twin-screw co-rotating extruder (APV laboratory one) and distributive mixing by single screw extruder equipped by mixing section (Randcastle microtruder) were tried and it was concluded that an excellent homogeneity can be obtained. Single screw extrusion of composite allowed obtaining the very narrow size distribution of glass particles dispersed in the polymer matrix. The shape of particles is strictly spherical. 99% of particles are smaller than 2 μm in diameter. Particles dimensions distribution can be effectively controlled by parameters of extrusion: temperature regime, pressure, cooling rate. Compositions that contain up to 30% of chalcogenide glass distributed uniformly in the middle density polyethylene matrix were obtained; it was shown that these compositions could be processed into flexible films of 20-100 μm thickness, fiber, lens etc.

The Randcasle microtruder extruder (screw diameter 12.5 mm, L/D ratio – 25) equipped with cast film die was used for the development of the process of manufacture of oriented films based on our composite. The extrusion process attended by the drawing of the film brings into existence the highly oriented structures. A "fiberlike" structure was formed under the drawing process. "Fibers" of chalcogenide glasses of 1-10 μm diameter and up to 5 cm length were dispersed in polymer matrix in strictly ordered way. Diameter of "fibers" and distance between them can be controlled by parameters of the extrusion process.

We studied the optical properties of the developed composite. It was established that obtained films are opaque in the visible light and transparency of films increases dramatically as wavelength increases and reaches 86% (the thickness of the sample 40 μm) in the far IR (at the 25 μm wavelength). The films that have oriented structure show distinct polarization properties in the far IR-band. Films of 100-μm thickness demonstrated 80% of polarization extent at the 10.6 μm wavelength. At the same time the transparency of cited above oriented films is not satisfactory yet and we concentrate our efforts on the improvement of this parameter.

Conclusions

Our work demonstrated that new composites developed by our group based on polymer and chalcogenide glasses offer promising physical properties. Composites under investigation allow to obtain flexible films that have controlled structure: from spherical particles of chalcogenide glass of 1-2 μm dispersed in polymer matrix uniformly up to highly oriented structures. Oriented films demonstrate pronounced polarization properties in the far IR-band. Such oriented films can be used as IR polarizers, waveguides, and immersion materials as well. Effective flexible optical filters transparent in IR and opaque in the UV and visible bands of the spectrum could be based on such composites.

Acknowledgments

The authors wish to thank the Israeli Ministry of Absorption for their generous support of this work and Professor Abraham Katzir for his inestimable contribution in our experimental activity.

References

[1] Materials Science and Technology, Volume 9, Glasses and Amorphous Materials, edited by J. Zarzycki, chapter 7, chalcogenide glasses.

[2] Patent of Israel No 128381 4, Feb., 1999

AUTHOR INDEX

Annals of the Israel Physical Society, v. 14

ELECTRO-OPTICS and MICROELECTRONICS
Eds: Raphael LAVI and Ehud AZULAY